Emil Zolotoyabko

Light and X-Ray Optics

Also of interest

Close-Range Photogrammetry and 3D Imaging
4th Edition
Luhmann, Robson, Kyle, Boehm, 2023
ISBN 978-3-11-102935-1, e-ISBN (PDF) 978-3-11-102967-2,
e-ISBN (EPUB) 978-3-11-102986-3

Weak Light Detection in Functional Imaging.
Vol. 1: Theoretical Fundaments of Digital SiPM Technologies and PET
Vol. 2: Applications of Digital SiPM technologies and PET
D'Ascenzo, Xie, 2023 / 2024 (Vol. 2)
ISBN 978-3-11-060396-5, e-ISBN (PDF) 978-3-11-060577-8,
e-ISBN (EPUB) 978-3-11-060415-3

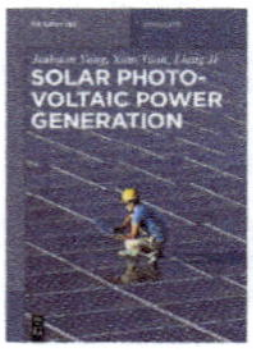

Solar Photovoltaic Power Generation
Yang, Yuan, Ji, in Cooperation with Publishing House of Electronics Industry, 2020
ISBN 978-3-11-053138-1, e-ISBN (PDF) 978-3-11-052483-3,
e-ISBN (EPUB) 978-3-11-052542-7

Quantum Electrodynamics of Photosynthesis.
Mathematical Description of Light, Life and Matter
Braun, 2020
ISBN 978-3-11-062692-6, e-ISBN (PDF) 978-3-11-062994-1,
e-ISBN (EPUB) 978-3-11-062700-8

Multiphoton Microscopy and Fluorescence Lifetime Imaging.
Applications in Biology and Medicine
König (Ed.), 2018
ISBN 978-3-11-043898-7, e-ISBN (PDF) 978-3-11-042998-5,
e-ISBN (EPUB) 978-3-11-043007-3

Emil Zolotoyabko

Light and X-Ray Optics

Refraction, Reflection, Diffraction, Optical Devices, Microscopic Imaging

DE GRUYTER

Author
Emil Zolotoyabko
Professor Emeritus
Department of Materials Science and Engineering
Technion-Israel Institute of Technology
Haifa 32000, Israel
zloto@technion.ac.il

ISBN 978-3-11-113969-2
e-ISBN (PDF) 978-3-11-114010-0
e-ISBN (EPUB) 978-3-11-114089-6

Library of Congress Control Number: 2023931744

Bibliographic information published by the Deutsche Nationalbibliothek
The Deutsche Nationalbibliothek lists this publication in the Deutsche Nationalbibliografie; detailed bibliographic data are available on the Internet at http://dnb.dnb.de.

Cover image: Petrovich9/iStock/Getty Images Plus
Typesetting: Integra Software Services Pvt. Ltd.
Printing and binding: CPI books GmbH, Leck

www.degruyter.com

To the wonderful physicians, Prof. Leonid Eidelman, Prof. Petachia Reissman, Prof. Ofer Nativ, and Dr. Olga Valkovsky, who saved my life three times over the last 25 years and, therefore, facilitated the book publishing

Preface

Optics is one of the oldest branches of science. Its history goes back to ancient times when people began to use magnifying glasses and bowed metallic mirrors for viewing small objects and focusing light. Optics' development, however, started accelerating only in the seventeenth century owing to the invention of the microscope by **Galileo Galilei** and several Dutch researchers and craftsmen. Among the latter, most frequently mentioned is **Anton van Leeuwenhoek**, who, around 1668, achieved a record 300 times magnification. A bit earlier, in 1660, **Robert Hooke** used a handcrafted microscope for successful visualization of a microorganism. In addition, he was the first to visualize cork cells and hence discovered the cellular form of life. Also in the seventeenth century, the law of refraction was established by **Willebrord Snellius**, while **Isaac Newton** discovered the color spectrum of white light, utilizing its refraction by a glass prism.

Over a span of more than a hundred years, progress in this field advanced in the framework of so-called geometrical optics, that is, a sort of particle (corpuscle) theory, in which light propagation is controlled by the principles borrowed from classical mechanics of particle movement. Only at the beginning of the nineteenth century, **Thomas Young** and **Augustin-Jean Fresnel** performed crucial interference and diffraction experiments, supported by theoretical developments, which showed that light has certain wave properties. Further progress continued in the framework of wave theory, reaching its magnificent quintessence in the middle of the nineteenth century when the fundamental **Maxwell** theory of electromagnetism was created. According to this theory, light (and its propagation) is a partial case of a high-frequency electromagnetic field.

Changing the optics paradigm was a rather slow and painful process. In one of his lectures, **James Clerk Maxwell** noted that in the beginning scientists spoke of the corpuscle theory of light, which over time was replaced by the wave theory of light; though people who believed in the former theory did not accept wave theory, they simply died with time.

Progress in science, however, sometimes takes unforeseen directions. At the beginning of the twentieth century, owing to the brilliant ideas of **Planck** and **de Broglie**, **Einstein** and **Bohr**, **Schrödinger** and **Heisenberg**, as well as many others, which led to the development of quantum mechanics, the paradigm again changed, to what today is known as wave–particle duality. This means that light can reveal wave properties or behave as a ray beam or ensemble of particles (photons) depending on specific measurement conditions. Furthermore, it was understood that photons, together with electrons, are indeed the quanta participating in electromagnetic interactions, the quanta that not only have energy and momentum but also spin. Later, these revolutionary ideas sparked the development of quantum optics, including manipulation with single photons, quantum entanglement, and quantum computing.

https://doi.org/10.1515/9783111140100-202

In 1895, X-rays were discovered by **Wilhelm Conrad Röntgen**. In 1905, **Charles Barkla** demonstrated X-ray polarization. In 1912, X-ray diffraction in crystals was observed in seminal experiment proposed by **Max von Laue** and performed by **Walter Friedrich** and his assistant **Paul Knipping**. This fundamental discovery gave rise to numerous X-ray diffraction techniques and imaging methods. All this allowed optics to expand over a wide spectrum of electromagnetic waves with thousands of times shorter wavelengths than those of visible light.

During the twentieth century, conventional light optics also progressed rapidly. Completely new fields were developed, such as electro-optics, nonlinear optics, the confocal microscopy with the ability to image 3D objects, and phase-manipulating methods such as different variants of phase-contrast microscopy and holography. Invention of near-field microscopy and fluorescent microscopy allowed scientists to shatter one of the cornerstones of nineteenth-century optics, that is, the diffraction limit in the resolving power of optical instruments. All this led to tremendous progress in nanotechnology and life sciences.

Manipulation of materials with the aid of nanotechnology, especially the invention of metamaterials, facilitated the establishment of stunning fields of photonics and plasmonics.

The last several decades have been marked by the development of X-ray focusing methods that are crucial to the advancements of X-ray microscopy, nanotomography, and ptychographic coherent diffractive imaging at synchrotron beamlines. This was not an easy task because the refractive index for X-rays is less than one, that is, in the X-ray wavelength range, materials are less optically dense than air.

It is important to admit the tremendous impact on optics wrought by the inventions of lasers and powerful synchrotron radiation sources, which facilitated the achievement of an enormous increase in the intensity and coherence of light and X-ray beams. I believe that in the forthcoming years, we will witness further amazing discoveries and technological breakthroughs.

In this book, I discuss in detail the common principles and essential distinctions between the optics of visible light and X-ray optics. Apart from fundamental issues, the emphasis is on modern experimental techniques utilizing interference, diffraction, refraction, reflection, and imaging schemes. I endeavor to touch upon many of the abovementioned subfields, bearing in mind the rather wide-ranging knowledge in optics that students and young researchers, especially those trained as materials scientists and engineers, require.

Since today's materials science and engineering (MSE) is an important part of scientific activities in many different disciplines, I believe that this book may also be helpful for undergraduate and graduate students studying in the departments of physics, chemistry, biology, chemical engineering, biomedical engineering, mechanical engineering, airspace engineering, in addition to MSE departments.

Haifa 2023 Emil Zolotoyabko

Contents

Preface —— VII

Introduction —— 1

Chapter 1
Foundations of geometrical optics —— 3
1.1 Electromagnetic waves in a homogeneous medium: the refractive index —— **3**
1.2 Inhomogeneous medium: eikonal equation and mirage formation —— **6**

Chapter 2
Fermat's principle: light reflection and refraction —— 10
2.1 Changing the light trajectory at the boundary between two media —— **11**
2.2 Total internal reflection of visible light —— **16**
2.3 Total external reflection of X-rays —— **18**

Chapter 3
Fermat's principle: focusing of visible light and X-rays —— 21
3.1 General approach —— **21**
3.2 Spherical lenses —— **25**
3.2.1 Focusing X-rays by a void —— **26**
3.3 Focusing by reflection —— **27**
3.3.1 Focusing and defocusing by a parabolic mirror —— **28**
3.3.2 Kirkpatrick–Baez mirrors for focusing the synchrotron radiation —— **29**
3.4 Thin lenses —— **31**
3.5 Limitations of Fermat's principle —— **35**
3.6 Conventional light microscope —— **37**
3.6.1 Köhler illumination and Abbe theory of image formation —— **38**

Chapter 4
Refractive index in anisotropic crystals —— 42

Chapter 5
Polarization, birefringence, and related phenomena —— 49
5.1 The Brewster angle for visible light and X-rays —— **49**
5.2 Birefringence in anisotropic crystals —— **51**
5.3 Producing polarized light —— **56**
5.4 Circular polarization and rotation of the polarization plane —— **62**

5.5 Induced birefringence — 66
5.6 Optical activity — 67
5.6.1 Faraday effect — 69
5.6.2 Magnetic circular dichroism and circular polarization of X-rays — 69
5.7 Polarization microscopy — 70
Appendix 5.A Fresnel equations — 74
Appendix 5.B Light ellipsometry and X-ray reflectivity — 77

Chapter 6
Strong frequency effects in light optics — 83
6.1 Skin effect in metals — 86
6.2 Light reflection from a metal — 87
6.3 Plasma frequency — 89
6.4 Negative materials and metamaterials — 91

Chapter 7
Interference phenomena — 97
7.1 Newton's rings — 99
7.2 Interference in thin films — 101
7.3 Structural colors — 104
7.4 Lippmann's color photography — 106
7.5 Gabor holography — 109
7.6 Cherenkov radiation — 112

Chapter 8
Light and X-ray interferometers — 114
8.1 Michelson interferometer — 114
8.2 Fabry–Pêrot interferometer — 116
8.2.1 Stimulated light emission — 120
8.3 Mach–Zehnder interferometer — 125
8.4 Bonse–Hart X-ray interferometer — 129

Chapter 9
Phase-contrast microscopy — 133
9.1 Zernike microscope — 133
9.2 Nomarski microscope — 135
9.3 Modulation-contrast microscopy — 136
9.4 Phase-contrast X-ray imaging — 138

Chapter 10
Fraunhofer diffraction — 141
10.1 Diffraction gratings — **141**
10.2 Kinematic diffraction of X-rays in the Bragg scattering geometry — **143**
10.3 Producing parallel X-ray beams with a Göbel mirror — **148**
10.4 Fraunhofer diffraction of light by a circular hole — **151**
10.5 Rayleigh criterion and diffraction limit of optical instruments — **155**

Chapter 11
Beyond diffraction limit — 159
11.1 Confocal microscopy — **159**
11.2 Near-field microscopy — **166**
11.3 New ideas with negative materials — **170**

Chapter 12
Fresnel diffraction — 172
12.1 X-ray scattering amplitude from an individual crystallographic plane: the Fresnel zone approach — **172**
12.2 Fresnel zone construction in transmission geometry — **176**
12.3 Light and X-ray focusing by Fresnel zone plates and Bragg–Fresnel lenses — **181**
12.4 X-ray microscopy — **183**
12.5 The Talbot effect — **190**
12.5.1 Employing diffraction gratings for phase-contrast X-ray imaging — **194**

Chapter 13
Optics of dynamical diffraction — 196
13.1 Wave propagation in periodic media from symmetry point of view — **196**
13.2 The concept of reciprocal lattice and quasi-momentum conservation law — **199**
13.3 Diffraction condition and Ewald sphere — **201**

Chapter 14
Dynamical diffraction of quantum beams: basic principles — 205
14.1 The two-beam approximation — **207**
14.2 Novel interference phenomena: Pendellösung and Borrmann effects — **213**
14.3 Diffraction profile in the Laue scattering geometry: extinction length and thickness fringes — **218**
14.4 Diffraction profile in the Bragg scattering geometry: total reflection near the Bragg angle — **223**

Chapter 15
Specific features of dynamical X-ray diffraction — 229
15.1 Dynamical X-ray diffraction in two-beam approximation — **234**
15.1.1 Taking account of X-ray polarization — **235**
15.1.2 The four-branch isoenergetic dispersion surface for X-ray quanta — **238**
15.1.3 X-ray extinction length — **240**
15.1.4 Isoenergetic dispersion surface for asymmetric reflections — **241**
15.2 The phase-shift plates for producing circularly polarized X-rays — **245**
15.3 X-ray beam compression using highly asymmetric reflections — **248**
15.4 The use of multicrystal diffractometers for phase-contrast X-ray imaging — **249**

Chapter 16
Optical phenomena in photonic structures — 255
16.1 Key ideas — **255**
16.2 Light localization in photonic structures — **257**
16.2.1 Helmholtz and paraxial wave equations — **258**
16.3 Slowing light — **260**
16.4 Photonic devices — **264**

List of scientists — 271

Index — 281

Introduction

What do visible light and X-rays have in common? Both entities are electromagnetic waves (though greatly differing in frequencies and wavelengths), which are transversely polarized and obey **Maxwell** equations. Consequently, if the sizes of characteristic structural features in an object are much larger than the radiation wavelength λ, the wave propagation can be described in the framework of geometrical optics, governed by **Fermat**'s principle and the spatial distribution of the refractive index. This is the basis of the design principles of most classical optical devices, from focusing lenses to conventional microscopes. In contrast, when these features are comparable in size to the radiation wavelength λ, diffraction phenomena (**Fraunhofer** or **Fresnel** diffraction) play a leading role, imposing a fundamental limitation on the resolving power of optical devices (about $\lambda/2$). In turn, diffraction is itself employed in many applications such as diffraction gratings used for spectroscopy and imaging, focusing **Fresnel** zone plates, and X-ray interferometers, providing superb angular resolution for phase-contrast X-ray imaging. Conventional X-ray diffraction offers unsurpassed precision in measuring crystal lattice parameters, which is the basis of structural analysis in materials science, chemistry, and biology.

What are the differences between visible light and X-rays? First, X-ray wavelengths are about 5,000 times shorter than those of visible light. For this reason, X-ray microscopy, in principle, can provide much better spatial resolution than light microscopy. Furthermore, the interactions of visible light and X-rays with materials are quite different due to great disparities in wave frequencies (about 6×10^{14} vs. 3×10^{18} Hz, respectively). Because of this, the refractive index n for visible light in most materials is around 1.5, while for X-rays, it is slightly less than 1. The closeness of n to 1 explains severe problems in fabricating X-ray optical devices based on X-ray refraction. The fact, however, that in the X-ray domain $n < 1$ makes a lot of sense, being responsible for the total external reflection of X-rays instead of the total internal reflection of visible light widely used for producing polarized light (Chapter 5). The total external reflection of X-rays also has several practical applications, for example, for X-ray focusing (Chapter 3), which are discussed in this book.

Another important issue is the enormous difference in the energy of light and X-ray quanta, which can reach four orders of magnitude (and more) in favor of the latter. For this reason, lasers deliver vastly higher photon flux than laboratory X-ray instruments. Development of innovative synchrotron X-ray sources and, especially, free-electron lasers reduced this gap considerably, but at the expense of requiring huge amount of electrical power on a tens of MW scale.

Note that in some situations, the commonalities and distinctions meet, as for example, in the case of interferometers. Both light and X-ray interferometers are in use (Chapter 8), but in the former, the light reflection is utilized to split and merge the light beams, while in the latter it is done with the aid of X-ray diffraction.

https://doi.org/10.1515/9783111140100-001

These commonalities and essential distinctions are thoroughly discussed in the present book. Besides fundamental issues, the special focus is on the application of refraction, reflection, interference, and diffraction phenomena to modern microscopy and imaging techniques.

The book is composed of 16 chapters. The first three are devoted to the description of reflection, refraction, and focusing phenomena in the framework of geometrical optics. In Chapter 4, we introduce the tensor of the refractive index, essential for light propagation in anisotropic media (crystals). In Chapter 5, light polarization in crystals of different symmetry is comprehensively discussed, including birefringence and other means of producing polarized light and its use in polarization microscopy. Chapter 6 focuses on frequency effects in light optics, for example, the skin effect in metals and light reflection from metallic surfaces. Here, we also introduce negative materials and metamaterials. In Chapter 7, we discuss different interference phenomena such as classical **Newton**'s rings, structural colors in nature, **Lippmann** photography, and **Gabor** holography. In Chapter 8, the design of light and X-ray interferometers is analyzed together with their applications, as for example an application of the **Mach–Zehnder** interferometer to the quantum entanglement problem. In addition, the concept of stimulated light emission is introduced in relation to the **Fabry–Pêrot** interferometer. Chapter 9 is devoted to phase-contrast techniques, from classical **Zernike** microscopy to advanced phase-contrast X-ray imaging. In Chapter 10, we consider different aspects of **Fraunhofer** diffraction of light and X-rays and derive the classical diffraction limit of the resolving power of optical instruments. In Chapter 11, we describe contemporary techniques that allow us to break the diffraction limit, such as confocal and near-field microscopies, as well as new ideas of using negative materials for this purpose. Chapter 12 is devoted to **Fresnel** diffraction with the focus on **Fresnel** zone construction and practical application of **Fresnel** zone plates, especially for X-ray microscopy. The **Talbot** effect is also introduced here, and its usage for the phase-contrast X-ray imaging is discussed. In Chapters 13–15, we analyze the optics of dynamical X-ray diffraction, introducing the concepts of reciprocal lattice and the quasi-momentum conservation law (Chapter 13). In addition, new interference phenomena, such as the Pendellösung and **Borrmann** effects, are discussed, and the fundamental quantities of extinction length and wave vector gap in reciprocal space are presented (Chapter 14). In Chapter 15, a four-branch isoenergetic surface for two polarizations of X-ray quanta in a crystal is analyzed, which helps us to quantitatively understand the total reflection of X-rays at **Bragg** angles, as a result of forbidden quantum states in the reciprocal space. These ideas have now been expanded to photonic structures, in which forbidden states for light propagation exist in the energy (frequency) domain (Chapter 16). In the last Chapter 16, we also focus on light slowing down and its complete localization, as well as describe the functioning of certain photonic devices.

Chapter 1
Foundations of geometrical optics

1.1 Electromagnetic waves in a homogeneous medium: the refractive index

The refractive index n of the medium, in which light propagates, is of crucial importance to the vast variety of optical phenomena. Further, we calculate the refractive index using a general approach based on **Maxwell**'s equations:

$$\operatorname{div}\boldsymbol{D} = \rho_{\mathrm{f}} \tag{1.1}$$

$$\operatorname{div}\boldsymbol{B} = 0 \tag{1.2}$$

$$\operatorname{rot}\boldsymbol{E} = -\frac{\partial \boldsymbol{B}}{\partial t} \tag{1.3}$$

$$\operatorname{rot}\boldsymbol{H} = \boldsymbol{J} + \frac{\partial \boldsymbol{D}}{\partial t} \tag{1.4}$$

Here ρ_{f} is the density of free (noncompensated) charges, which for nonintentionally charged materials equals zero:

$$\rho_f = 0 \tag{1.5}$$

Differential operators in eqs. (1.1.) and (1.2) are called divergence ($\nabla \cdot$), whereas those in eqs. (1.3) and (1.4) are rotor or curl ($\nabla \times$). The displacement field $\boldsymbol{D}$ and magnetic induction $\boldsymbol{B}$ are connected, correspondingly, to the electric field $\boldsymbol{E}$ and magnetic field $\boldsymbol{H}$ as follows:

$$\boldsymbol{D} = \varepsilon \boldsymbol{E} = \varepsilon_{\mathrm{m}} \varepsilon_0 \boldsymbol{E} \tag{1.6}$$

$$\boldsymbol{B} = \mu \boldsymbol{H} = \mu_{\mathrm{m}} \mu_0 \boldsymbol{H} \tag{1.7}$$

where ε_0 and μ_0 are, respectively, the dielectric permittivity and magnetic permeability of a vacuum, while ε_{m} and μ_{m} are, respectively, the dielectric and magnetic constants of a medium. Note that in anisotropic media, parameters ε_{m} and μ_{m} are, in fact, tensors of second rank. We will discuss this issue in more detail in Chapter 4, in relation to light polarization in anisotropic crystals. Here we consider isotropic materials such as glass, which is often used for fabricating conventional optical components. In this case, parameters ε_{m} and μ_{m} are scalars, though frequency dependent.

To proceed, we recall the linear relationship between the vectors of current density $\boldsymbol{J}$ and electric field $\boldsymbol{E}$ (**Ohm's** law), which are interconnected via the specific electrical conductivity σ:

https://doi.org/10.1515/9783111140100-002

$$\boldsymbol{J} = \sigma \boldsymbol{E} \tag{1.8}$$

Together with eqs. (1.5)–(1.8), this yields:

$$\operatorname{div} \boldsymbol{D} = \operatorname{div} \boldsymbol{E} = 0 \tag{1.9}$$

$$\operatorname{div} \boldsymbol{B} = \operatorname{div} \boldsymbol{H} = 0 \tag{1.10}$$

$$\operatorname{rot} \boldsymbol{E} = -\mu_0 \mu_{\mathrm{m}} \frac{\partial \boldsymbol{H}}{\partial t} \tag{1.11}$$

$$\operatorname{rot} \boldsymbol{H} = \boldsymbol{J} + \frac{\partial \boldsymbol{D}}{\partial t} = \sigma \boldsymbol{E} + \varepsilon_{\mathrm{m}} \varepsilon_0 \frac{\partial \boldsymbol{E}}{\partial t} \tag{1.12}$$

Applying a rot (curl) vector operator ($\nabla \times$) to both sides of eq. (1.11) and using eq. (1.12), one obtains:

$$\operatorname{rot}(\operatorname{rot} \boldsymbol{E}) = -\mu_0 \mu_{\mathrm{m}} \left[\sigma \frac{\partial \boldsymbol{E}}{\partial t} + \varepsilon_{\mathrm{m}} \varepsilon_0 \frac{\partial^2 \boldsymbol{E}}{\partial t^2} \right] \tag{1.13}$$

Since $\operatorname{rot}(\operatorname{rot} \boldsymbol{E}) = \operatorname{grad}(\operatorname{div} \boldsymbol{E}) - (\operatorname{grad} \cdot \operatorname{grad})\, \boldsymbol{E}$, then with the aid of eq. (1.9) we find:

$$\operatorname{rot}(\operatorname{rot} \boldsymbol{E}) = -(\operatorname{grad} \cdot \operatorname{grad})\, \boldsymbol{E} = -\nabla^2 \boldsymbol{E} \tag{1.14}$$

where $\nabla^2 = \frac{\partial^2}{\partial x^2} + \frac{\partial^2}{\partial y^2} + \frac{\partial^2}{\partial z^2}$ is the **Laplace** operator. Substituting eq. (1.14) into eq. (1.13) yields

$$\nabla^2 \boldsymbol{E} = \mu_0 \mu_{\mathrm{m}} \left[\sigma \frac{\partial \boldsymbol{E}}{\partial t} + \varepsilon_{\mathrm{m}} \varepsilon_0 \frac{\partial^2 \boldsymbol{E}}{\partial t^2} \right] \tag{1.15}$$

In a vacuum, $\mu_{\mathrm{m}} = \varepsilon_{\mathrm{m}} = 1$, whereas the conductivity $\sigma = 0$. Therefore, eq. (1.15) converts to the standard wave equation

$$\nabla^2 \boldsymbol{E} - \mu_0 \varepsilon_0 \frac{\partial^2 \boldsymbol{E}}{\partial t^2} = 0 \tag{1.16}$$

with the light velocity in a vacuum equal to

$$c = \frac{1}{\sqrt{\mu_0 \varepsilon_0}} \tag{1.17}$$

In fact, substituting the plane wave

$$\boldsymbol{E} = \boldsymbol{E}_0 \exp[i(\boldsymbol{kr} - \omega t)] \tag{1.18}$$

into eq. (1.16) yields the dispersion law, that is, the relationship between the light frequency ω and wave vector $\boldsymbol{k}$

$$\omega^2 = \frac{1}{\mu_0 \varepsilon_0} \boldsymbol{k}_0^2 \tag{1.19}$$

where $\boldsymbol{k}_0$ is the wave vector in vacuum. One can rewrite the dispersion law (1.19) in its more familiar linear form

$$\omega = \frac{1}{\sqrt{\mu_0 \varepsilon_0}} |\boldsymbol{k}_0| = c|\boldsymbol{k}_0| = ck_0 \tag{1.20}$$

justifying definition (1.17).

In the case of a homogeneous medium, differing from a vacuum (i.e., with $\mu_m \neq 1$, $\varepsilon_m \neq 1$, and generally nonzero conductivity $\sigma > 0$), a similar procedure leads to the modified wave equation:

$$c^2(\nabla^2 \boldsymbol{E}) = \mu_m \left(\varepsilon_m \frac{\partial^2 \boldsymbol{E}}{\partial t^2} + \frac{\sigma}{\varepsilon_0} \frac{\partial \boldsymbol{E}}{\partial t} \right) \tag{1.21}$$

Substituting the plane wave (1.18) into eq. (1.21) yields

$$c^2 \boldsymbol{k}^2 = \mu_m \left(\varepsilon_m \omega^2 + i \frac{\sigma \omega}{\varepsilon_0} \right) \tag{1.22}$$

and

$$|\boldsymbol{k}| = k = \frac{\omega}{c} \left(\mu_m \varepsilon_m + i \mu_m \frac{\sigma}{\varepsilon_0 \omega} \right)^{1/2} \tag{1.23}$$

Expression (1.23) is, in fact, a new dispersion law for an electromagnetic wave in a medium. This dispersion law (1.23) is very different from that of an electromagnetic wave in a vacuum (1.20):

$$k_0 = \frac{\omega}{c} \tag{1.24}$$

Using eqs. (1.23) and (1.24), one can calculate the refractive index $n = (k/k_0)$, which determines the key optical phenomena for electromagnetic waves interacting with a material

$$n = \frac{k}{k_0} = \left(\mu_m \varepsilon_m + i \frac{\mu_m \sigma}{\varepsilon_0 \omega} \right)^{1/2} \tag{1.25}$$

Note that the refractive index comprises the real and imaginary parts, being responsible, respectively, for light refraction and attenuation. We elaborate further on this issue in Chapter 6.

In dielectrics (e.g., in a glass), the conductivity $\sigma = 0$; correspondingly, the wave equation (1.21) transforms into

$$\frac{\partial^2 \boldsymbol{E}}{\partial t^2} = V_\mathrm{p}{}^2 \nabla^2 \boldsymbol{E} \tag{1.26}$$

with the phase velocity of light in a medium equal to

$$V_\mathrm{p} = \frac{c}{\sqrt{\mu_\mathrm{m} \varepsilon_\mathrm{m}}} = \frac{c}{n_\mathrm{d}} \tag{1.27}$$

In other words, the refractive index

$$n_\mathrm{d} = \sqrt{\mu_\mathrm{m} \varepsilon_\mathrm{m}} \tag{1.28}$$

shows in what proportion the phase velocity of light propagation in a medium (V_p) is decreased with respect to that in a vacuum c.

For nonmagnetic materials ($\mu_\mathrm{m} = 1$), eqs. (1.25) and (1.28) are transformed into

$$n = \frac{k}{k_0} = \left(\varepsilon_\mathrm{m} + i \frac{\sigma}{\varepsilon_0 \omega} \right)^{1/2} \tag{1.29}$$

and further to the well-known expression for nonmagnetic dielectrics ($\sigma = 0$):

$$n_\mathrm{d} = \sqrt{\varepsilon_\mathrm{m}} \tag{1.30}$$

In most materials used for visible light optics, n_d is around 1.3 – 1.8. In contrast, the refractive index for X-rays is a bit less than 1, that is, $n = 1 - \delta$, with $\delta \approx 10^{-5} - 10^{-6}$. Despite the smallness of δ, the fact that $n < 1$ leads to some interesting optical phenomena that are absent in the optics of visible light. We discuss the related consequences in more detail in the following chapters (e.g., in Chapters 2 and 3).

1.2 Inhomogeneous medium: eikonal equation and mirage formation

As we will learn later in this book, the wave aspects in light propagation through material objects are not important (or play a little role) when the light wavelength $\lambda = (2\pi/|\boldsymbol{k}|)$ is much smaller than the characteristic sizes of these objects. In this limit ($\lambda \to 0$), which is called **geometrical optics**, the basic phenomena in light propagation, such as reflection and refraction, are completely determined by the refractive index n only.

In this context, let us consider the more general case of light propagation through an inhomogeneous medium in which the refractive index is the coordinate-dependent function, $n(x, y, z)$. In this case, the solution of the wave equation (1.26) is not more the plane wave. For simplicity, we use the wave equation for the scalar function $f(x, y, z)$:

$$\frac{\partial^2 f}{\partial t^2} = V_p{}^2(x,y,z)\left(\frac{\partial^2 f}{\partial x^2} + \frac{\partial^2 f}{\partial y^2} + \frac{\partial^2 f}{\partial z^2}\right) \tag{1.31}$$

and considering the elastic scattering processes ($\omega = \text{const}$) only, we seek its solution in the form

$$f = Ae^{-i\omega t}e^{ik_0\psi} \tag{1.32}$$

where both parameters A and ψ are coordinate-dependent, while $k_0 = (2\pi/\lambda)$ again is the wave vector in a vacuum. To further proceed, we assume that both A and ψ slowly change with distance; therefore, their spatial derivatives are small, for example, $(\partial A/\partial x) \ll Ak_0$ and $(\partial^2 A/\partial x^2) \ll Ak_0^2$. Now, substituting eq. (1.32) into eq. (1.31) and keeping the largest term (quadratic one) over k_0 (i.e., in the limit, $\lambda \to 0$; $k_0 \to \infty$) yield

$$\omega^2 = V_p{}^2 k_0{}^2\left[\left(\frac{\partial\psi}{\partial x}\right)^2 + \left(\frac{\partial\psi}{\partial y}\right)^2 + \left(\frac{\partial\psi}{\partial z}\right)^2\right] = V_p{}^2 k_0{}^2(\text{grad}\psi)^2 \tag{1.33}$$

Recalling that

$$V_p{}^2 = \frac{c^2}{n^2} \tag{1.34}$$

one finally obtains the so-called eikonal equation for the eikonal function ψ:

$$(\text{grad}\psi)^2 = n^2(x,y,z) \tag{1.35}$$

In proximity to a given spatial point $\boldsymbol{r}(x,y,z) = 0$, the eikonal function ψ can be expanded into a series:

$$\psi = \psi_0 + \boldsymbol{r}\,\text{grad}\psi \tag{1.36}$$

Correspondingly, the coordinate-dependent phase part $e^{ik_0\psi}$of our wave function (1.32) gets the following form:

$$\exp[ik_0(\psi_0 + \boldsymbol{r}\,\text{grad}\psi)] \tag{1.37}$$

Furthermore, in small spatial regions across its trajectory, each wave can be considered as a plane wave

$$\exp[i(\boldsymbol{kr} + \alpha)] \tag{1.38}$$

Comparing eqs. (1.37) and (1.38) yields

$$\boldsymbol{k} = k_0\,\text{grad}\psi \tag{1.39}$$

It follows from eq. (1.39) that the eikonal function ψ represents a wave surface of constant phase, whereas the direction of light rays is given by gradψ, that is, locally being normal to the surface of constant phase. Combining eqs. (1.35) and (1.39), one finds

$$\left(\frac{\boldsymbol{k}}{k_0}\right)^2 = (\mathrm{grad}\psi)^2 = n^2 \tag{1.40}$$

or

$$\frac{\boldsymbol{k}}{k_0} = \hat{\boldsymbol{e}} \cdot n(x, y, z) \tag{1.41}$$

where $\hat{\boldsymbol{e}}$ is a unit vector along the normal to the local surface of the constant wave phase. If $n = \mathrm{const}$, that is, $\mathrm{grad}\psi$ is the same across the space domain, light will propagate along a straight line with the same wave vector $\boldsymbol{k}$. This agrees with the particle theory of light propagation, since, in mechanics, the particle momentum (or wave vector) does not change in a homogeneous medium. This conclusion follows from the momentum conservation law as the consequence of the homogeneity of space.

In nonhomogeneous media, the light trajectory will bend toward the regions with a higher refractive index. In fact, let us consider light propagation along the x-direction in some hypothetic medium, in which the refractive index $n(y)$ increases in the y-direction (see Figure 1.1). Using eq. (1.40), one obtains

$$k^2 = k_x{}^2 + k_y{}^2 = k_0^2 n^2(y) = k_0^2[n_0 + \Delta n(y)]^2 \tag{1.42}$$

Assuming that the k_x projection is only weakly changed, that is, $k_x \simeq k_0 n_0$ yields

$$k_y{}^2 \simeq 2k_0^2 n_0{}^2 \left[\frac{\Delta n(y)}{n_0}\right] \tag{1.43}$$

It denotes that the k_y projection is growing with $\Delta n(y)$, which means that the trajectory bending is indeed governed by the gradient of refractive index n.

Figure 1.1: Illustration of the trajectory bending toward $\mathrm{grad}(n)$.

This result explains several famous optical phenomena observable in nature, for example, the Sun that is visible some period after sunset. It occurs because the upper air layers in the atmosphere are less dense (having a lower refractive index) than the denser bottom layers (having a higher refractive index), which compels the light trajectories to bend appropriately (see Figure 1.2).

Another example of this kind of phenomena is a mirage. Here we deal with opposite situation when the air layer adjacent to the hot surface is less dense than the upper layers. This may be the layer located close to the asphalt road in very hot day or close to hot sand in a desert. In these cases, the light rays will be bent in opposite

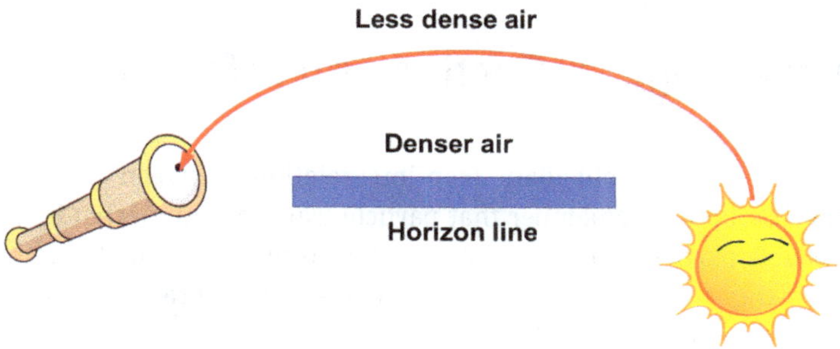

Figure 1.2: Seeing the Sun after sunset.

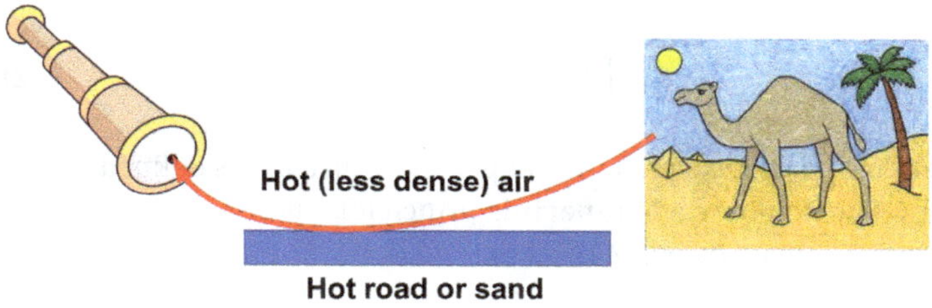

Figure 1.3: Illustration of a mirage formation.

way, as compared to the previous example, and the observer will see the images of objects situated far away from him/her (see Figure 1.3).

In charged particle physics, trajectory bending by alternating gradients of magnetic fields (which can be described in terms of alternating refractive indices) is used for beam focusing in some kinds of particle accelerators.

Chapter 2
Fermat's principle: light reflection and refraction

Eikonal approximation allows us to elucidate deep interrelations between classical mechanics and geometrical optics. Remember that particle trajectory in mechanics is determined by the variable particle momentum $\boldsymbol{P}$, which is locally directed along the tangent to the trajectory line. In turn, momentum $\boldsymbol{P}$ is defined as the gradient of fundamental quantity S, called action:

$$\boldsymbol{P} = \mathrm{grad} S \tag{2.1}$$

Correspondingly,

$$S = \int \boldsymbol{P}\, d\boldsymbol{l} \tag{2.2}$$

where $d\boldsymbol{l}$ is an element of the particle path. Actual particle trajectory is defined by the least action principle (**Hamilton** or **de Maupertuis** principle), that is, minimizing the variation of the functional S given by eq. (2.2):

$$\delta S = \delta \int \boldsymbol{P}\, d\boldsymbol{l} = 0 \tag{2.3}$$

In optics, $(\boldsymbol{k}/k_0) = \mathrm{grad}\psi$ (see eq. (1.39)) and, therefore, eikonal function (or phase function) ψ plays the role of action S (compare with eq. (2.1), while wave vector $\boldsymbol{k}$ is used instead of momentum $\boldsymbol{P}$ (compare eqs. (2.1) and (1.39)). In other words

$$\psi = \int \frac{\boldsymbol{k}}{k_0} d\boldsymbol{l} \tag{2.4}$$

Respectively, we can formulate the least action principle in optics similar to that expressed by eq. (2.3):

$$\delta\psi = \delta \int \frac{\boldsymbol{k}}{k_0} d\boldsymbol{l} = 0 \tag{2.5}$$

Recalling eq. (1.41) yields the so-called **Fermat**'s principle

$$\delta\psi = \delta \int n(x, y, z)\, \hat{\boldsymbol{e}} \cdot d\boldsymbol{l} = 0 \tag{2.6}$$

Fermat's principle determines the light trajectory in optically inhomogeneous media having a refractive index of $n = n(x, y, z)$. The true trajectory provides zero-variation $\delta\psi$ for virtual optical paths S_0 around it:

https://doi.org/10.1515/9783111140100-003

$$\delta\psi = \delta S_0 = 0$$
$$S_0 = \int n(x,y,z)dl \tag{2.7}$$

where dl is the length of the light trajectory element projected in the direction of the local wave vector $\boldsymbol{k}$. Let us illustrate **Fermat**'s principle by some working examples.

In a homogeneous medium, $n(x,y,z) = n =$ const, and the optical path is expressed as follows:

$$S_0 = n\int dl \tag{2.8}$$

Therefore, zero-variation $\delta\psi = \delta S_0 = 0$ corresponds to the trajectory that provides a minimum length between two points A and B, that is, a straight line (see Figure 2.1). This result is sometimes used as a definition of geometrical optics.

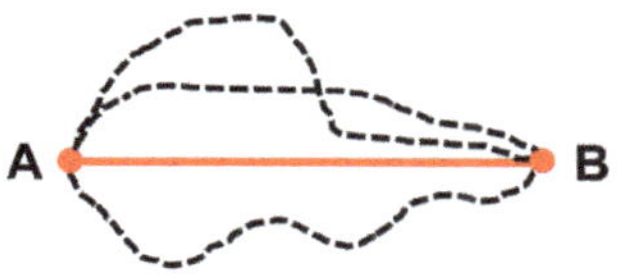

Figure 2.1: In a homogeneous medium (n = const), electromagnetic waves (including visible light and X-rays) propagate along a straight line, AB, which provides the shortest path between points A and B.

2.1 Changing the light trajectory at the boundary between two media

Staying within the same approach, we can consider light reflection from a flat interface between two homogeneous media: let us say air ($n = 1$) and a material with refractive index $n \neq 1$ (see Figure 2.2). Setting the light source in point A and the detector in point B, both in the material with refractive index n, we can find the actual light trajectory via the interface point O. Designating the heights of points A and B, that is, the segments AC and BD as h_1 and h_2, respectively, and the entrance and exit angles (counted from the interface) as Θ_1 and Θ_2, respectively, we find the optical path (2.8):

$$S_0 = AO + OB = n\left(\frac{h_1}{\sin\Theta_1} + \frac{h_2}{\sin\Theta_2}\right) \tag{2.9}$$

Application of **Fermat**'s principle yields

$$h_1\frac{\cos\Theta_1}{\sin^2\Theta_1}\Delta\Theta_1 + h_2\frac{\cos\Theta_2}{\sin^2\Theta_2}\Delta\Theta_2 = 0 \tag{2.10}$$

We find the second equation using the fact that $CO + OD = \left(\frac{h_1}{\tan\Theta_1}\right) + \left(\frac{h_2}{\tan\Theta_2}\right) = \text{const}.$

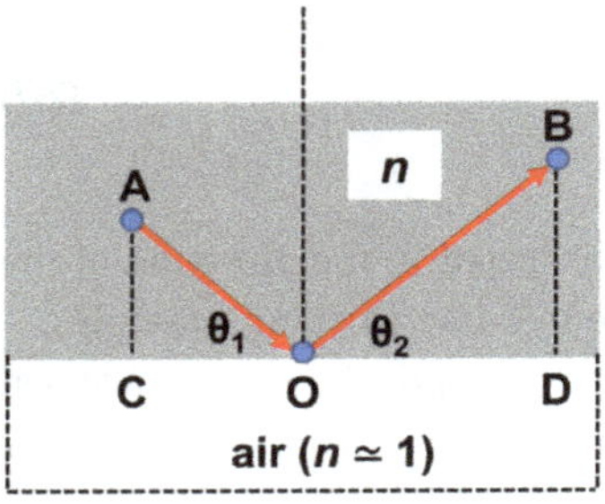

Figure 2.2: Light reflecting from a flat interface between two homogeneous media.

Differentiating the latter relationship, one obtains:

$$h_1 \frac{\Delta \Theta_1}{\sin^2\Theta_1} + h_2 \frac{\Delta \Theta_2}{\sin^2\Theta_2} = 0 \tag{2.11}$$

Combining eqs. (2.10) and (2.11) leads to the well-known result

$$\Theta_1 = \Theta_2 \tag{2.12}$$

that is, the equality of the entrance and exit angles (specular reflection). It is worth emphasizing that this result straightforwardly follows from the momentum or wave vector conservation law along the interface:

$$k\cos\Theta_1 = k\cos\Theta_2 \tag{2.13}$$

As another example, let us consider the reflection of a parallel light beam by a spherical mirror with radius R. Already in ancient times, people knew that a spherical mirror converts a parallel beam to a rather small spot with high light intensity (Figure 2.3). Using **Fermat**'s principle and applying it to the planar cross section of the mirror passing through its center of curvature and parallel to the incident rays, we easily calculate the focal length F of such a simple optical device to be nearly a half of the mirror radius of curvature, that is, $F = R/2$. In fact, viewing Figure 2.3 and bearing in mind that all the rays are propagating in the air ($n = 1$), we can say that to satisfy eq. (2.7), the two ray trajectories, *ABF* and *FOF*, should be equal, that is,

$$F - y + \sqrt{x^2 + (F - y)^2} = 2F \tag{2.14}$$

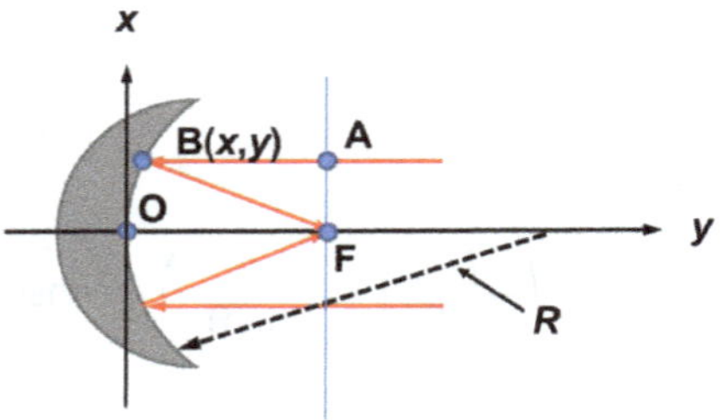

Figure 2.3: Focus of a spherical mirror.

or

$$\sqrt{x^2 + (F - y)^2} = F + y \tag{2.15}$$

Note that point $B(x,y)$ is taken as an arbitrary point on the mirror surface, whereas point $O(0,0)$ is the origin of our coordinate system. Squaring both sides of eq. (2.15) one obtains:

$$x^2 = 4Fy \tag{2.16}$$

Recalling the equation of the circle with the pole at point (0, 0) (see Figure 2.3)

$$x^2 + (y - R)^2 = R^2 \tag{2.17}$$

and using eqs. (2.16) and (2.17) yields

$$4F = -y + 2R \tag{2.18}$$

At small $y \ll R$, that is, near the pole (0,0), we indeed find

$$F = \frac{R}{2} \tag{2.19}$$

Taking the same approach, we can quantitatively describe the light refraction at the boundary between two homogeneous media with different refractive indices n_1 and n_2 (see Figure 2.4). In fact, light propagates along straight lines in both media, with the respective segments of their trajectories being AO and OB. **Fermat**'s principle again provides the actual trajectory with a minimum optical path between two points A and B, that is, the appropriate position of point O at the interface between two media. Designating the heights $AC = h_1$, $BD = h_2$, and the angles between the rays and the normal to the interface as α and β (see Figure 2.4), we find the respective optical path $S_o = n_1\, AO + n_2\, OB$:

$$S_o = n_1 \frac{h_1}{\cos\alpha} + n_2 \frac{h_2}{\cos\beta} \tag{2.20}$$

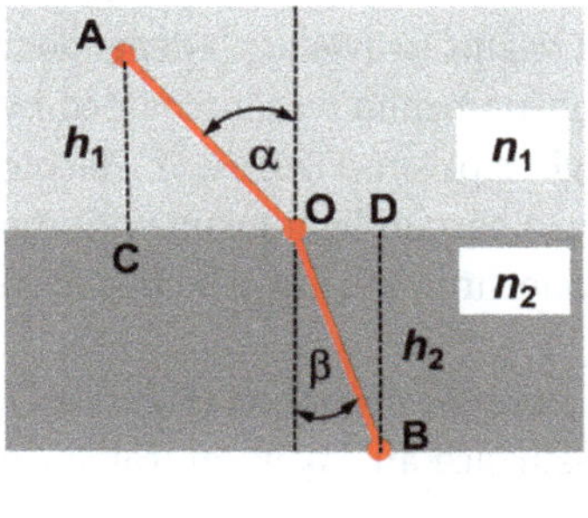

Figure 2.4: Illustration of **Snell**'s law.

and its variation

$$\delta S_0 = n_1 h_1 \frac{\sin\alpha}{\cos^2\alpha}\Delta\alpha + n_2 h_2 \frac{\sin\beta}{\cos^2\beta}\Delta\beta = 0 \tag{2.21}$$

Noting that $CO + OD = h_1 \tan\alpha + h_2 \tan\beta = \text{const}$ and differentiating the latter relationship over angles α and β yields

$$h_1 \frac{\Delta\alpha}{\cos^2\alpha} + h_2 \frac{\Delta\beta}{\cos^2\beta} = 0 \tag{2.22}$$

Combining eqs. (2.21) and (2.22), we finally obtain the well-known **Snell**'s law:

$$n_1 \sin\alpha = n_2 \sin\beta \tag{2.23}$$

This result can also be derived from the conservation law of the wave vector component along the interface. Considering the magnitudes of wave vectors k_1 and k_2 as real numbers (no absorption) yields

$$k_1 \sin\alpha = k_2 \sin\beta \tag{2.24}$$

or

$$k_0 n_1 \sin\alpha = k_0 n_2 \sin\beta \tag{2.25}$$

If an incident light is coming from the air ($n_1 \approx 1$) and enters the medium with refractive index $n_2 = n > 1$, then

$$\sin\beta = \frac{\sin\alpha}{n} \tag{2.26}$$

and $\beta < \alpha$. In the opposite case, when light is coming from an optically denser medium, for example, water with $n = 1.33 = 4/3$, into air, we have $\sin\alpha = (\sin\beta/n)$ and $\beta > \alpha$. Everyone has experienced this phenomenon when looked at a coin on the bottom of a pool. Due to the light refraction at the water/air boundary, the coin is seen as if lying at a shallower depth H^* than the real depth of the pool H. Using Figure 2.5 and assuming small angles α, $\beta \ll 1$, we find $\frac{H^*}{H} = \frac{\tan\alpha}{\tan\beta} \approx \frac{\sin\alpha}{\sin\beta} = \frac{1}{n} = \frac{3}{4}$.

Furthermore, because of the refraction phenomenon, we can resolve the colored spectrum of visible light. In fact, since $n = \sqrt{\varepsilon_m}$ (see eq. (1.30)) is frequency-dependent, the white light will split into spatially separated colored beams (wavelength-dependent light dispersion) that refract at different angles β. This phenomenon was discovered by **Isaac Newton** when investigating light refraction by a glass prism (Figure 2.6). Placing a moving slit beyond the refracted light, one can pick up a certain light beam with the desired wavelength. This is one possible way of light monochromatizing, which is used in different optical devices.

One more example of the discussed phenomenon is rainbow formation (Figure 2.7). In this case, the sunlight obeys refraction in the near-surface air layer saturated by water droplets formed just after the rain.

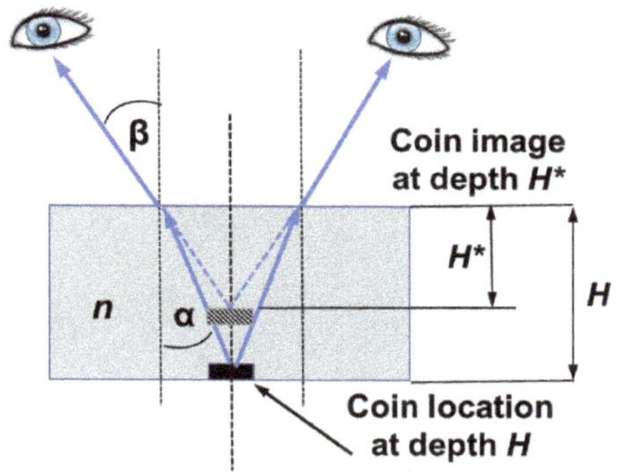

Figure 2.5: Erroneous perception of the visible position of a coin lying on the bottom of a pool.

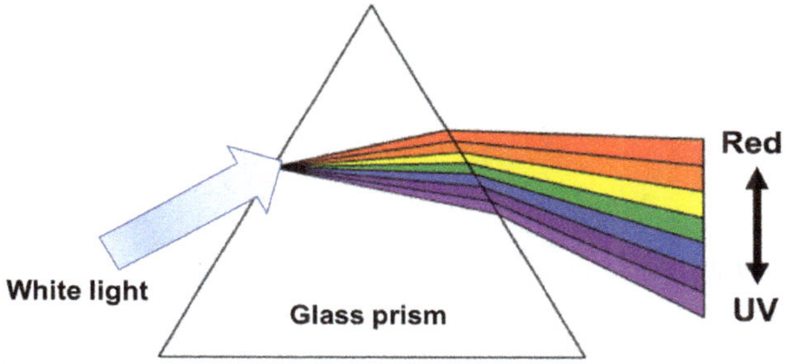

Figure 2.6: Wavelength-dependent light dispersion revealed through the refraction phenomenon.

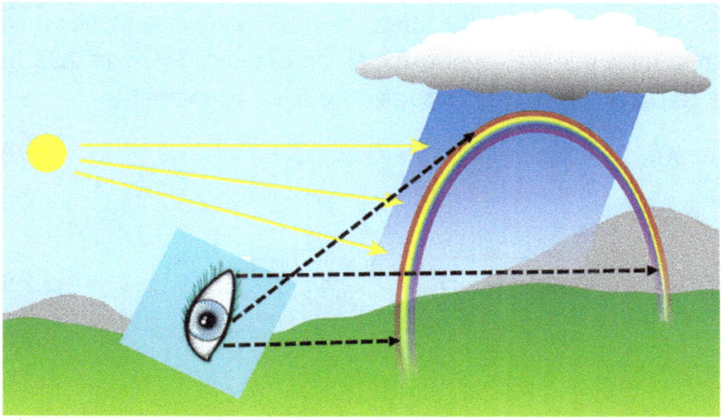

Figure 2.7: Illustration of the formation of a rainbow.

2.2 Total internal reflection of visible light

An interesting situation arises when light is entering an optically less dense medium (n_2) from a denser medium (n_1), that is, when $n_1 > n_2$. Using eq. (2.23), we find $\sin\beta > 1$ at the incident angles α larger than some critical value α_c defined by the condition $\sin\beta = 1$, that is,

$$\sin\alpha_c = \frac{n_2}{n_1} \tag{2.27}$$

For visible light, the critical angle is $\alpha_c \approx 49°$ for entering air from water while $\alpha_c \approx 42°$ when going from ordinary glass into air. At this critical angle (and at larger angles), the refracted light beam propagates along the interface without significant penetration into the second medium (see Figure 2.8). In other words, under total internal reflection, refracted light is concentrated within a narrow layer in the proximity of the interface.

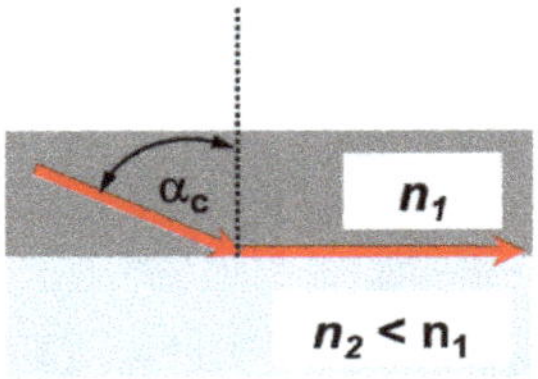

Figure 2.8: Total internal reflection of light at the interface between an optically denser material (n_1) and a less dense material ($n_2 < n_1$).

What does it really mean that the incoming light does not enter the second medium? To what extent does this happen? To answer this important question, let us follow more carefully the fate of the refracted wave in this case. For this purpose, we choose the coordinate system in which the x-axis is parallel to the interface in the scattering plane while the z-axis is perpendicular to the interface toward the second material (Figure 2.9). In this coordinate system, the refracted wave (with wave vector $\boldsymbol{k}_2$) is expressed as

$$\exp(i\boldsymbol{k}_2\boldsymbol{r}) = \exp[ik_2(x\sin\beta + z\cos\beta)] \tag{2.28}$$

Since $\beta \approx 90°$, $\sin\beta \approx 1$ and we have

$$\exp(i\boldsymbol{k}_2\boldsymbol{r}) \approx \exp\left(ik_2x + ik_2z\sqrt{1-\sin^2\beta}\right) = e^{-k_2z\sqrt{\sin^2\beta-1}}\exp(ik_2x) \tag{2.29}$$

Using **Snell**'s law (2.23) and eq. (2.27) yields

$$\exp(i\boldsymbol{k}_2\boldsymbol{r}) \approx e^{-k_2z\sqrt{\sin^2\beta-1}}\exp(ik_2x) = \exp(ik_2x)\exp\left[-k_2z\sqrt{\frac{\sin^2\alpha-\sin^2\alpha_c}{\sin^2\alpha_c}}\right]$$

$$= \exp(ik_2x)\exp\left[-\frac{2\pi}{\lambda}n_2z\sqrt{\frac{\sin^2\alpha-\sin^2\alpha_c}{\sin^2\alpha_c}}\right] \tag{2.30}$$

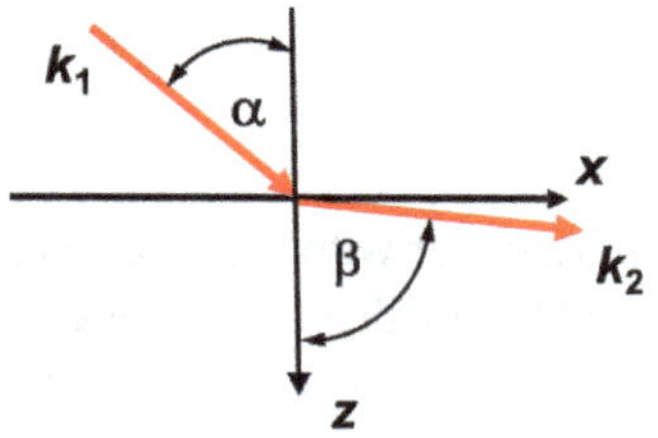

Figure 2.9: The coordinate system used for analyzing evanescent wave propagation. Wave vectors $\boldsymbol{k}_1$ and $\boldsymbol{k}_2$ are related to the incident and refracted waves, respectively.

We see that at $\alpha > \alpha_c$, the refracted wave indeed propagates along the interface (term $\exp(ik_2x)$), but with an amplitude strongly attenuated in the z-direction. Its e-times decay takes place at the characteristic penetration depth:

$$L_c = \frac{\lambda}{2\pi n_2\sqrt{\frac{\sin^2\alpha - \sin^2\alpha_c}{\sin^2\alpha_c}}} = \frac{\lambda}{2\pi n_2} \cdot \frac{\sin\alpha_c}{\sqrt{\sin^2\alpha - \sin^2\alpha_c}} \tag{2.31}$$

with $\sin\alpha_c = (n_2/n_1)$, given by eq. (2.27). For rather large angles $\alpha > \alpha_c$, the magnitude of L_c may be less than the light wavelength λ. Waves propagating near material surfaces or interfaces and rapidly attenuating in the material's depth are called evanescent waves. We will meet with this entity several times across this book.

Concentration of light energy in thin material layers under total reflection is used in different applications. As an example, one can mention the waveguides for optical communication systems. A change in the refractive index in the near-surface layer of a nonlinear crystal (such as $LiNbO_3$) is achieved, for example, by ion implantation or metal diffusion, which prevents light penetration into the crystal bulk. Advanced methods of waveguide fabrication are described in Section 16.4.

Another field is light propagation in optical fibers, which are composed of an optically denser core covered by a less dense cladding layer. Rays entering close to the fiber axis ($\alpha \approx 90°$) are confined within the fiber and are unable to escape (Figure 2.10).

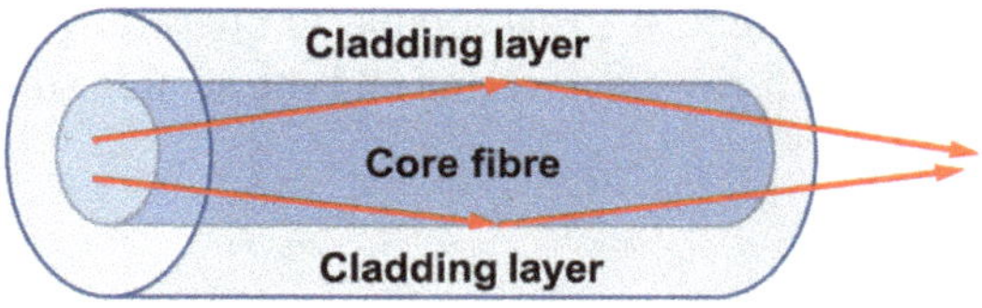

Figure 2.10: Total internal reflection at the interface between the core and cladding layer utilized for light propagation in optical fibers.

If the refractive index of the fiber core is n_1 and that of the cladding layer is $n_2 < n_1$, then for a characteristic depth (2.31), we find (also using eq. (2.27))

$$L_c = \frac{\lambda}{2\pi\sqrt{n_1^2 \sin^2\alpha - n_2^2}} \tag{2.32}$$

which may be substantially smaller than the wavelength λ. Together with the discovery of the ultra-low intensity losses for some wavelengths in highly purified fused silica by **Charles Kao** (2009 Nobel Laureate in Physics), these findings have revolutionized the entire field of communication technology.

2.3 Total external reflection of X-rays

In contrast to visible light ($380 < \lambda < 740$ nm), X-rays are short-wavelength electromagnetic waves with $0.1 \le \lambda \le 10$ Å. Remember that 1 Å equals 0.1 nm. Because of specific features of the dielectric polarizability in this wavelength (energy) region (see Chapter 15), the real part of refractive index for X-rays is less than 1

$$n = 1 - \delta \tag{2.33}$$

with $\delta \approx 10^{-5} - 10^{-6}$. Due to the smallness of δ, the direct effect of refraction on the X-ray trajectories is very weak (i.e., $\alpha \approx \beta$ in **Snell**'s law). The principal fact that $n < 1$, however, leads to very interesting phenomena in X-ray optics. The first one is the so-called total external reflection instead of total internal reflection in the optics of visible light for which $n > 1$. In most practical cases, X-rays enter materials from the air side. Correspondingly, **Snell**'s law (2.23) transforms into

$$\sin\alpha = (1 - \delta)\sin\beta \tag{2.34}$$

Setting $\beta = 90°$, we find the critical angle for total reflection to be defined by the condition

$$\sin\alpha_c = 1 - \delta \tag{2.35}$$

This means that for incident angles $\alpha_c < \alpha < 90°$, X-rays are unable to enter materials at all (even in case of zero absorption).

When describing the total external reflection, it is acceptable to operate with small angles $\gamma = 90° - \alpha$ between the incident X-ray wave vector and the sample surface (see Figure 2.11).

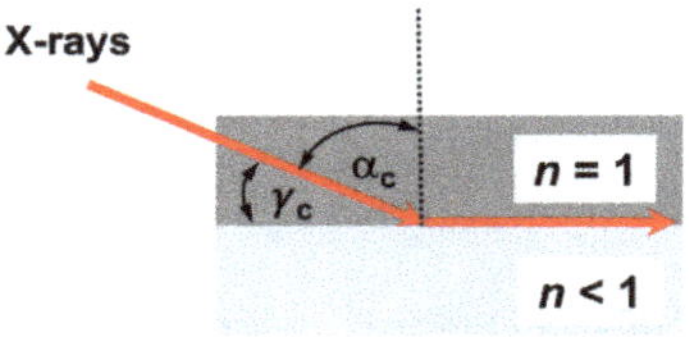

Figure 2.11: Illustration of the total external reflection of X-rays (red arrows) at the interface between air ($n = 1$) and a material ($n < 1$).

Therefore

$$\cos\gamma_c = 1-\delta;\ 1-\frac{\gamma_c 2}{2} \approx 1-\delta \tag{2.36}$$

and, finally

$$\gamma_c = \sqrt{2\delta} \tag{2.37}$$

Typical critical angles are rather small, for example, for $\delta \approx 10^{-5}$, $\gamma_c \approx 4.5$ mrad or 0.25°. In fact, the γ_c value depends on the electron density of a material and the X-ray energy (or wavelength) in use (see Chapter 15). Specifically, the critical angle increases with atomic number Z and decreases with the X-ray energy. Roughly, the product of the critical angle (mrad) and the X-ray photon energy (keV) is a material's constant (nearly 30 keV·mrad for glass and 80 keV·mrad for gold).

As we already mentioned, if the entrance angle with respect to the sample surface is less than the critical angle γ_c, X-rays do not penetrate the sample depth. They are concentrated within a very narrow layer beneath the sample's surface. Expressing again the refracted wave as $\exp(i\boldsymbol{k}_2\boldsymbol{r}) = \exp[ik_2(x\sin\beta + z\cos\beta)]$ and setting $\sin\beta \approx 1$, one obtains

$$\exp(i\boldsymbol{k}_2\boldsymbol{r}) = \exp\left(ik_2x + ik_2z\sqrt{1-\sin^2\beta}\right) = e^{-k_2z\sqrt{\sin^2\beta-1}}\exp(ik_2x) \tag{2.38}$$

Using **Snell**'s law yields

$$\exp(i\boldsymbol{k}_2\boldsymbol{r}) = e^{-k_2z\sqrt{\sin^2\beta-1}}\exp(ik_2x) = e^{-k_2z\sqrt{\frac{\sin^2\alpha-n^2}{n^2}}}\exp(ik_2x) \tag{2.39}$$

For total reflection, $\sin^2\alpha \approx 1$ and $n = 1-\delta$. Hence,

$$\exp(i\boldsymbol{k}_2\boldsymbol{r}) = e^{-k_2z\sqrt{2\delta}}\exp(ik_2x) = e^{-\frac{2\pi}{\lambda}z\sqrt{2\delta}}\exp(ik_2x) \tag{2.40}$$

Therefore, under total external reflection, the evanescent X-ray waves are confined within a very tiny characteristic depth beneath the interface

$$L_c = \frac{\lambda}{2\pi\sqrt{2\delta}} \approx 5\,\text{nm} \tag{2.41}$$

Practically, for glancing angles $(\gamma/\gamma_c) \le 1$, all X-rays are pushed out from the surface and a very strong specular reflection is observed. As described in Chapter 3, total external reflection is used in some X-ray focusing devices.

Concentration of X-ray flux within a narrow layer of thickness L_c beneath the sample surface is very high, and this fact allows us to probe the atomic structure of materials and overall electron density in the ultra-thin near-surface layers by, respectively, the grazing incidence diffraction and X-ray reflectivity techniques. The latter is described in more detail in Appendix 5.B.

Total external reflection is also employed to produce very narrow X-ray beams utilizing capillaries. Using a glass capillary with the proper internal diameter, most of the incoming X-rays will meet the reflecting walls at angles less than the critical angle (see Figure 2.12). In this way, the size of X-ray beams can be reduced to about 10 μm while maintaining the high X-ray flux.

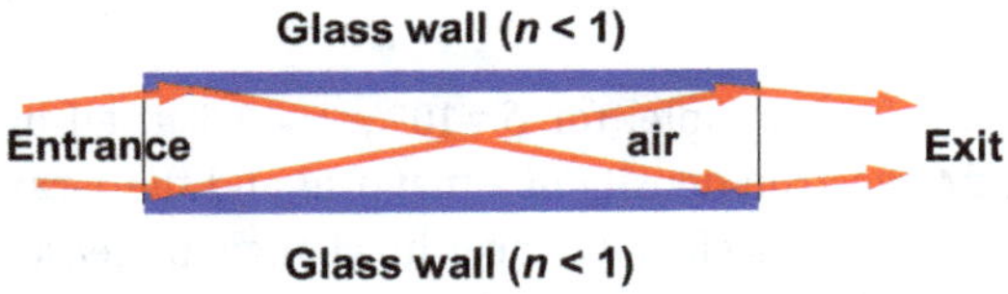

Figure 2.12: Producing narrow X-ray beams by capillary optics utilizing the total external reflection at the interface between air and glass walls.

Chapter 3
Fermat's principle: focusing of visible light and X-rays

3.1 General approach

Light focusing and defocusing by curved interface shapes between different materials is widely used in lots of optical devices. The focusing effect in geometrical optics can also be described with the aid of **Fermat**'s principle. In this framework, focusing means that many of the light trajectories (all trajectories in the ideal case), issuing from point source S and ending at focal point K, have the same optical path, that is, $\delta S_0 = 0$ (see eq. (2.7)). Let us apply this approach to the situation when point S is in air and point K is in a medium with refractive index n (see Figure 3.1). Using **Fermat**'s principle, we can find the analytical function $f(x,y,z)$ describing the boundary between these two media, which will be the shape of an ideal focusing element.

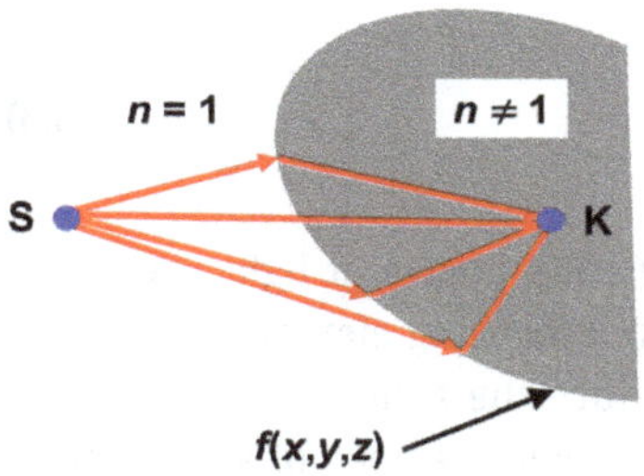

Figure 3.1: Application of the **Fermat**'s principle to the focusing problem.

For the sake of simplicity, let us consider the two-dimensional problem, that is, the light focusing in the (x,y) plane (see Figure 3.2). According to **Fermat**'s principle, in this case, focusing means that two optical paths, SOK and SMK, are equal. Taking the origin of the coordinate system in point O and designating the segments $SO = a$ and $OK = b$, one obtains

$$\sqrt{(x+a)^2 + y^2} + n\sqrt{(b-x)^2 + y^2} = a + nb \tag{3.1}$$

We see that, generally, the boundary curve is of the fourth order with respect to the coordinates (x,y). If, however, the angular deviations from the central ray (SOK) are small, that is, $x, y \ll a, b$, the solution becomes much simpler. Rewriting eq. (3.1) as

$$a\sqrt{1 + \frac{2x}{a} + \left(\frac{x}{a}\right)^2 + \left(\frac{y}{a}\right)^2} + nb\sqrt{1 - \frac{2x}{b} + \left(\frac{x}{b}\right)^2 + \left(\frac{y}{b}\right)^2} = a + nb \tag{3.2}$$

and expanding the square roots in series up to the second-order terms yield

https://doi.org/10.1515/9783111140100-004

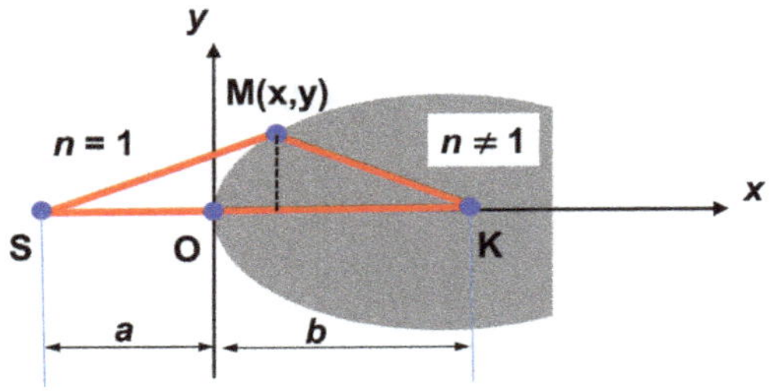

Figure 3.2: Deriving the shape of an ultimate focusing surface.

$$a\left(1+\frac{x}{a}+\frac{x^2}{2a^2}+\frac{y^2}{2a^2}-\frac{1}{8}\cdot\frac{4x^2}{a^2}\right)+nb\left(1-\frac{x}{b}+\frac{x^2}{2b^2}+\frac{y^2}{2b^2}-\frac{1}{8}\cdot\frac{4x^2}{b^2}\right)=a+nb$$

or

$$x+\frac{y^2}{2a}-nx+n\frac{y^2}{2b}=0$$

and finally

$$x=\frac{y^2}{2(n-1)}\left(\frac{1}{a}+\frac{n}{b}\right) \tag{3.3}$$

This is a parabolic equation. It means that if the deviation angles of light rays from the optical axis *SK* are small, the border curve of such a focusing element is a parabola. It is much easier, however, to mechanically produce the spherical surface of a material, having the same curvature, $1/R$, as the curvature in the apex of a parabola:

$$\frac{1}{R}=\left(\frac{d^2x}{dy^2}\right)\frac{1}{\left[1+\left(\frac{dx}{dy}\right)^2\right]^{3/2}}\bigg|_{y=0}=\frac{1}{(n-1)}\left(\frac{1}{a}+\frac{n}{b}\right) \tag{3.4}$$

or

$$\frac{1}{a}+\frac{n}{b}=\frac{n-1}{R} \tag{3.5}$$

Using eq. (3.5), we can say that for a parallel incident beam ($a=\infty$), the incoming light will be focused on the focal point *K*, located at distance $F=b$ from point *O* (see Figure 3.3):

$$F=b=\frac{nR}{n-1} \tag{3.6}$$

Certainly, the usage of spherical shapes instead of parabolic ones in the optical systems is a source of the so-called spherical aberrations. In other words, the refracted

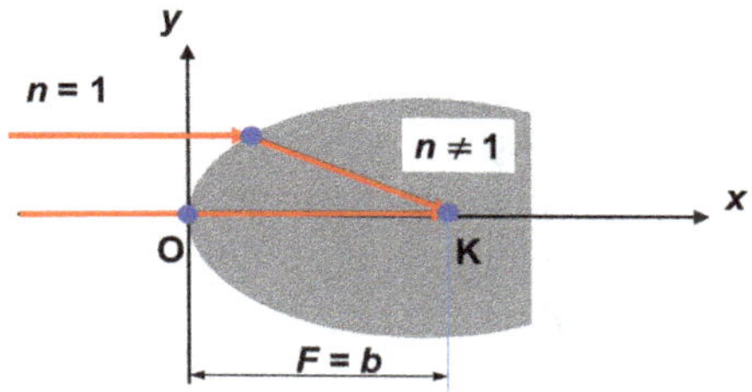

Figure 3.3: Focusing of a parallel light beam.

rays will cross the optical axis at different points depending on the height H of the incoming parallel beam (see Figure 3.4).

Another type of aberration in the optical systems is chromatic aberration, which is associated with light dispersion, that is, with the fact that the refractive index n in eq. (3.6) is frequency (or wavelength) dependent. For this reason, the focal spot becomes blurred when using nonmonochromatized light.

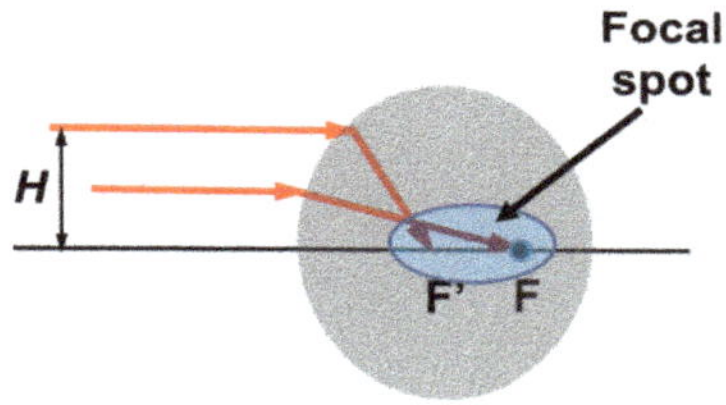

Figure 3.4: Illustration of the spherical aberration formation.

For three-dimensional surfaces, there is an additional kind of aberration called astigmatism. In fact, in this case, the surface element near a given spatial point is characterized by two radii of curvature R_1 and R_2 (see Figure 3.5), instead of a single radius R for border curves in the two-dimensional case. For example, for a cylindrical surface, these radii are $R_1 = R$ (i.e., that one for a circular cross section) and $R_2 = \infty$. Correspondingly, such a surface will provide a linear focus (parallel to the cylinder axis) instead of a point focus in the planar situation depicted in Figure 3.2. In the general

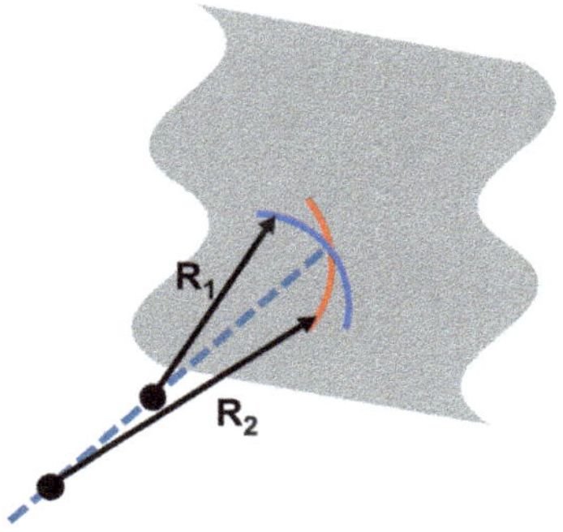

Figure 3.5: Illustration of the astigmatism formation.

case, we will have one surface in space (locus) for all points designating the ends of local vectors $\boldsymbol{R}_1$ and the second surface – for $\boldsymbol{R}_2$. These surfaces are called caustics when serve for light concentration. In the unique case of a spherical surface, both caustics converge into a single focal point defined by the sphere radius R (see eq. (3.6)).

Further analyzing eq. (3.6), we find that for $n = 2$, the focal point $F = 2R$ will touch the back surface of the sphere with radius R (Figure 3.6). By covering the back surface with reflecting silver paste or silver reflective tape, one can organize a 180° reflection of the incoming light. This effect is exploited in road retroreflectors (cat's eyes), helping drivers travel in darkness. If $1 < n < 2$, the focal point is moving away from the backward spherical surface ($F/R > 2$; see Figure 3.7) and, hence, such spherical lenses can be used for focusing purposes.

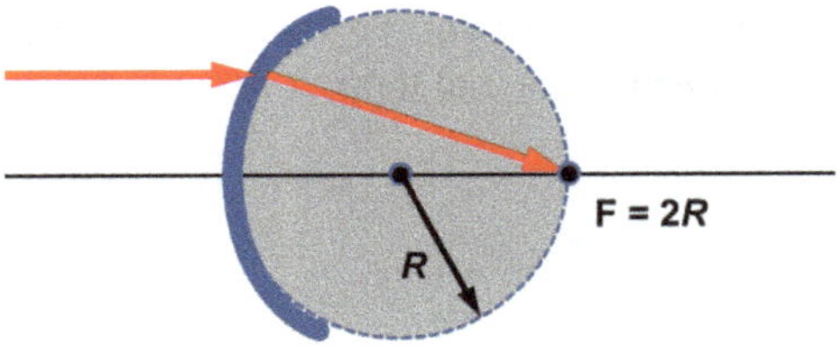

Figure 3.6: Light focusing on spherically bent surface for $n = 2$.

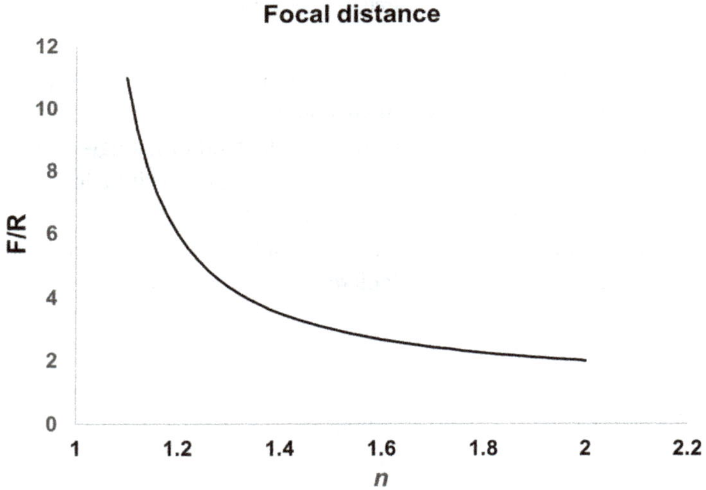

Figure 3.7: Increasing the focal distance by reducing refractive index from $n = 2$ to $n \simeq 1$.

3.2 Spherical lenses

Let us prove that a spherical lens has a focal length two times shorter than that for the single spherical surface given by eq. (3.6). In fact, using Figure 3.8 and eqs. (3.5) and (3.6), we find that

$$AD = \frac{nR}{n-1}$$
$$BD = AD - 2R = R\frac{2-n}{n-1} \tag{3.7}$$

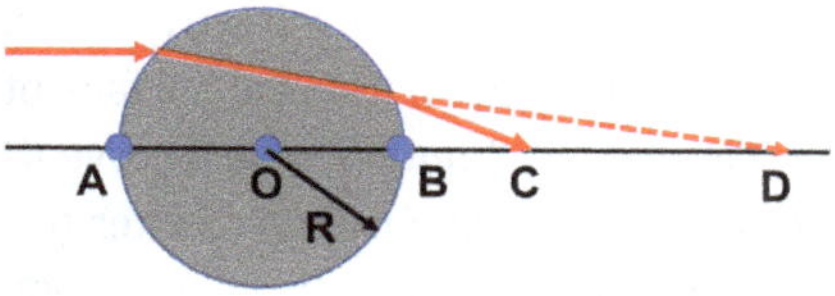

Figure 3.8: Calculating the focal distance of a spherical lens.

Furthermore, for the second refraction event, that is, transforming virtual point D into real focus C, we apply again eq. (3.5)

$$\frac{n}{BD} - \frac{1}{BC} = -\frac{n-1}{R} \tag{3.8}$$

Note that the second term in the left part of eq. (3.8) has a negative sign because the "image" point C is at the same side of the lens as the "source" D. The negative sign of the right-hand term in eq. (3.8) originates in the negative curvature of the backward surface of the lens, as compared to its frontal surface. In the result, one obtains

$$BC = R\frac{(2-n)}{2(n-1)} \tag{3.9}$$

The last step is to recalculate the focal length BC with respect to the lens center O. Doing this, we find that the actual focal length $F = OC$,

$$F = OC = BC + OB = \frac{Rn}{2(n-1)} \tag{3.10}$$

is indeed two times shorter than that for a single spherical surface (see eq. (3.6)). This result fits the more general statement, which can be formulated as follows: the optical power P_o of an optical system is the sum of the optical powers of its elements P_{oj}:

$$P_o = \sum_j P_{oj} \tag{3.11}$$

In turn, the power of an individual optical element is inverse to its focal length

$$P_{oj} = \frac{1}{F_j} \tag{3.12}$$

In other words, the more powerful the optical element is (i.e., the stronger it bends the trajectory of an incoming light), the shorter its focal length.

Amazing examples of the usage of spherical lenses, as components in photosensory organs, are found in living organisms. In marine brittle stars, for example, the "eyesight" system includes an array of single-crystalline calcite microlenses, each having radius $R \approx$ 10 μm (see Figure 3.9). As was shown by **Joanna Aizenberg** group from Bell Laboratories/Lucent Technologies and Harvard University, these microlenses focus the low-intensity light coming from the upper water layers. Using eq. (3.10) and refractive index of calcite ($n = 1.66$), the estimated focal distance (counted from the backward surface of the lens) is about 5 μm. This coincides with the location of nerve bundles, which serve as primary photoreceptors. Enhancing weak light intensity in this way facilitates better orientation of brittle stars in underwater darkness and, hence, faster escape from predators.

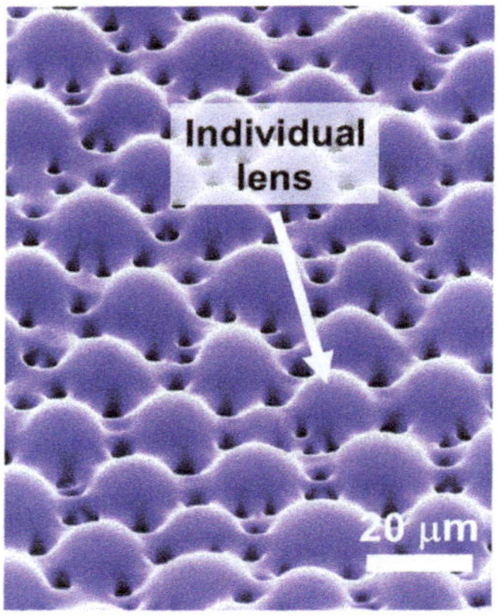

Figure 3.9: An array of single-crystalline calcite microlenses in brittle stars. Courtesy of **Joanna Aizenberg** (Harvard University).

3.2.1 Focusing X-rays by a void

A fascinating application of spherical lenses can be found in the field of X-ray optics. Recalling that air ($n = 1$) is an optically denser medium than, for example, a metal ($n = 1 - \delta$, see eq. (2.33)), we immediately understand that a hole drilled into a metal will work as a converging lens. Using eqs. (2.33) and (3.10), one obtains, for the focal length F, the following expression:

$$F = \frac{nR}{2(1-n)} = \frac{R}{2\delta} \tag{3.13}$$

For a single hole with radius $R =$ 300 µm drilled in a material with $\delta = 5 \times 10^{-6}$, the focal distance is $F = 30$ m, which is too long for any practical application. If, however, several holes N_h are drilled, the overall focal length F_a can be drastically reduced (see Figure 3.10). If the distance d_h between holes is much smaller than the hole diameter ($d_h \ll 2R$), the optical powers of individual lenses will be summated (see eqs. (3.11) and (3.12)); therefore, the actual focal length of the entire assembly will be reduced to

$$F_a = \frac{R}{2N_h\delta} \tag{3.14}$$

Taking $N_h = 30$, one obtains $F = 1$ m, which is quite reasonable for synchrotron beamlines.

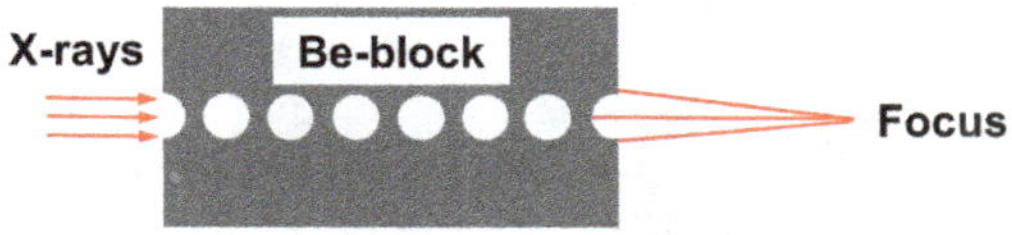

Figure 3.10: X-ray focusing by compact lenses.

Nowadays, such compact lenses are produced within beryllium blocks to enable a substantial reduction of X-ray absorption. The holes have parabolic profiles, which help to eliminate spherical aberrations of such optical elements. This design provides a fine focal spot, only a few microns in size. Compact lenses, however, suffer from rather strong chromatic aberrations since parameter δ is proportional to λ^2. In fact, as we will see in Chapter 15

$$\delta = \frac{r_0\rho}{2\pi}\lambda^2 \tag{3.15}$$

where r_0 is the fundamental constant known as the classical radius of electron, while ρ is the electron density, that is, the number of electrons per unit volume. As follows from eqs. (3.14) and (3.15), the focal distance is inversely proportional to the X-ray wavelength in square ($1/\lambda^2$). Due to this fact, compact lenses can only be used for monochromatized X-rays with fixed wavelength λ.

3.3 Focusing by reflection

Chromatic aberration can be avoided, for example, when all light trajectories are in air ($n = 1$). This can be achieved by utilizing curved reflecting surfaces. We encountered this situation earlier when describing light focusing by a spherical mirror. Let us now consider the more general case of reflecting surfaces having parabolic or elliptic shapes.

3.3.1 Focusing and defocusing by a parabolic mirror

Light focusing by a parabolic surface is illustrated in Figure 3.11.

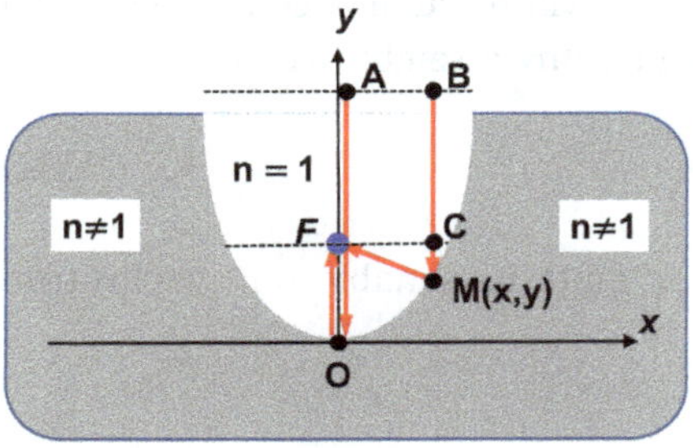

Figure 3.11: Focusing of parallel X-rays (to point *F*) by a parabolic surface.

Applying **Fermat**'s principle $\delta S_0 = 0$ and requiring equal optical passes for all ray trajectories *BMF* (or *CMF*), touching an arbitrary point *M*(*x*,*y*) on the parabola (as was done for the spherical mirror in Section 2.1), yield

$$FOF = 2F = CMF = F - y + \sqrt{(F-y)^2 + x^2} \tag{3.16}$$

and

$$y = \frac{x^2}{4F} \tag{3.17}$$

Recalling the parabolic equation

$$y = px^2 \tag{3.18}$$

we obtain the focal length as

$$F = \frac{1}{4p} \tag{3.19}$$

Since the curvature in the pole of the parabola is

$$\frac{1}{R} = \left(\frac{d^2y}{dx^2}\right)\frac{1}{\left[1+\left(\frac{dy}{dx}\right)^2\right]^{3/2}}\Big|_{x=0} = 2p \tag{3.20}$$

then

$$F = \frac{R}{2} \tag{3.21}$$

as for a spherical mirror close to its optical axis (see Section 2.1). Using parabolic mirrors, one can redirect parallel beams of light to point focus or, oppositely, to convert spherically symmetric distribution of light rays issuing from a point source into a parallel beam. In the field of X-rays, the latter variant (sometimes called **Göbel** mirror) is widely used to provide nearly parallel X-ray beams for high-resolution X-ray diffraction (see Figure 3.12). In more detail, the functioning of **Göbel** mirror is discussed in Section 10.3.

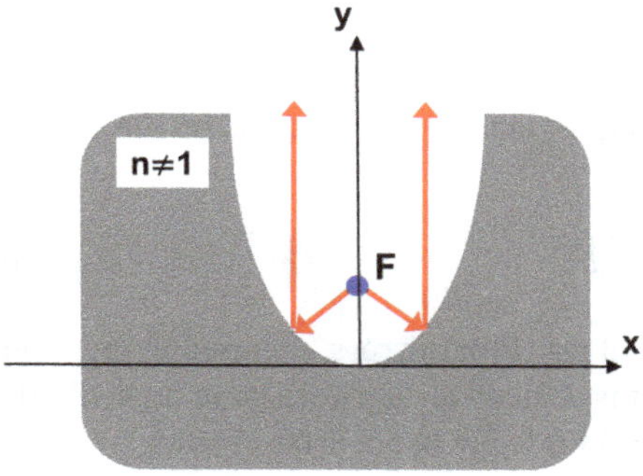

Figure 3.12: Converting X-rays from a point source *F* to a parallel beam using a parabolic mirror.

3.3.2 Kirkpatrick–Baez mirrors for focusing the synchrotron radiation

The second option is to redirect the rays issuing from point source F_1 to focal point F_2 by using the reflecting surface of the elliptic shape (see Figure 3.13).

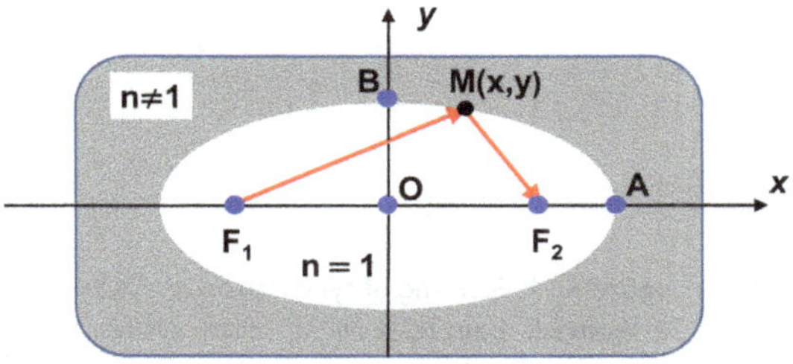

Figure 3.13: Focusing of X-rays by means of elliptic surface: all X-rays issuing from a point source F_1 are automatically collected at the focal point F_2.

In this case, **Fermat**'s principle ($\delta S_0 = 0$) is automatically fulfilled for all ray trajectories $S_0 = F_1MF_2$ touching an arbitrary point $M(x, y)$ on the elliptic surface. This statement follows from the ellipse definition, as the set of points for which the total length of two segments connecting a given point $M(x, y)$, with two ellipse focuses, F_1 and F_2, is constant:

$$S_o = F_1M + MF_2 = \text{const} \tag{3.22}$$

Let us write the ellipse equation as

$$\frac{x^2}{a^2} + \frac{y^2}{b^2} = 1 \tag{3.23}$$

with $OA = a$ and $OB = b$ (see Figure 3.13), and designate the focal coordinates as $(-c, 0)$ for F_1 and $(c, 0)$ for F_2. Applying eq. (3.22) for extreme ellipse points, A and B, yields

$$2\sqrt{c^2 + b^2} = 2a \tag{3.24}$$

and, finally, the x-coordinates of the focal points via the half-lengths of the ellipse axes:

$$c = \pm\sqrt{a^2 - b^2} \tag{3.25}$$

Large-size elliptic mirrors, called **Kirkpatrick–Baez** (K-B) mirrors, are used to focus on X-rays at synchrotron beamlines. Practically, it is difficult to fabricate large area elliptic surfaces in 3D. It is much easier to produce a cylindrically bent surface having an elliptic cross section. Within such 2D projection, all X-rays propagate in air from point F_1 to point F_2, experiencing reflection from a mirror at points $M(x, y)$ (see Figure 3.13). For an ideal elliptic shape, there are no spherical or chromatic aberrations. To achieve a sharp focal spot in space, we need two elliptic mirrors, subsequently acting in horizontal and vertical scattering planes. This configuration is often called a K-B focusing mirror system (see Figure 3.14).

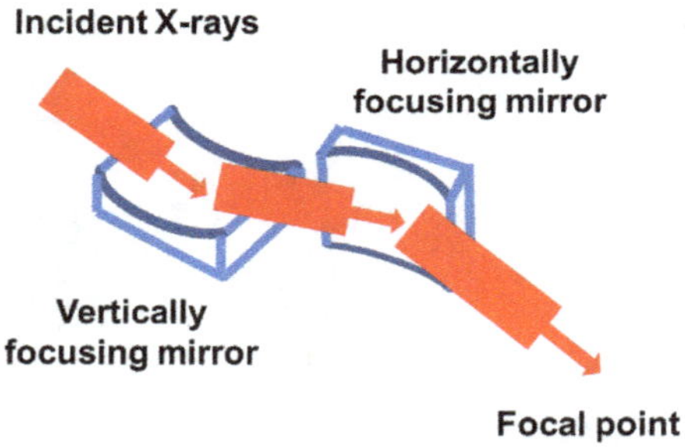

Figure 3.14: Focusing of synchrotron X-rays by a **Kirkpatrick–Baez** (K-B) elliptic mirror system.

To supply high photon flux, the incident X-ray angles for both mirrors should be below the critical angle γ_c given by eq. (2.37). For a fixed wavelength λ, one can increase the critical angle by using materials with high electron density (high Z-materials) such as gold. The gold surface should be carefully polished towards diminishing the surface roughness down to the sub-nm level. The K-B mirrors allow focusing X-rays to a 50–100 nm spot size.

The elliptic shapes of capillary shells (see Figure 3.15) are employed in the capillary concentrators of X-rays (described in Section 2.3) with the objective of diminishing the focal spot size. To substantially increase the exit X-ray flux, polycapillary optical elements are used (see Figure 3.15).

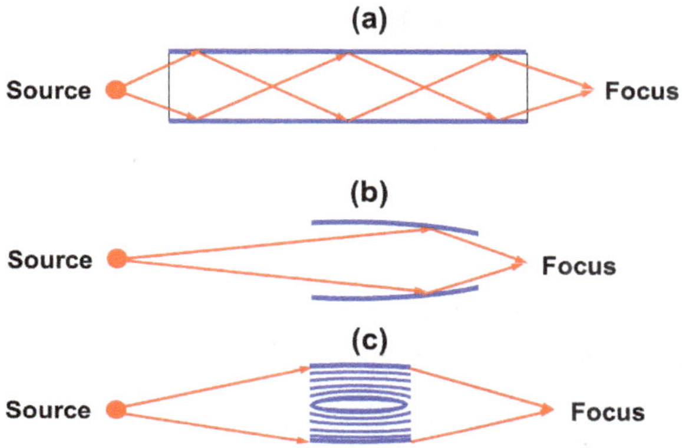

Figure 3.15: Focusing of X-rays using capillary optics: (a) cylindrical; (b) elliptical; and (c) polycapillary.

3.4 Thin lenses

Let us now consider the functioning of so-called thin lenses, which are essential parts of many optical devices. In thin lenses, a material with the refractive index n is confined between two spherical surfaces having radii of curvature R_1 and R_2, and the distance d between them being much smaller than R_1, R_2 (see Figure 3.16).

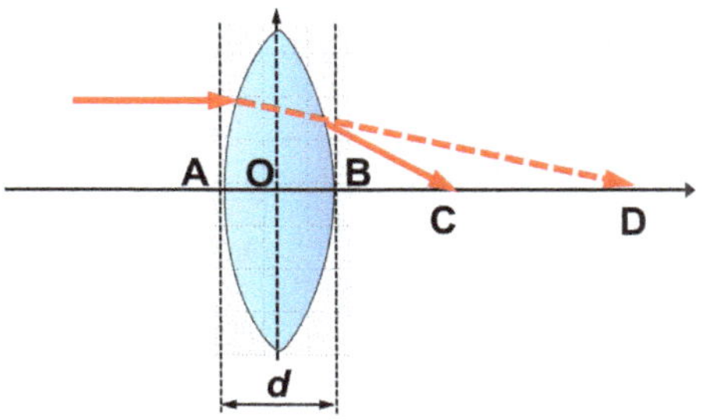

Figure 3.16: Illustration of the functioning of a thin converging lens.

Repeating calculations for spherical lenses presented in Section 3.2 and bearing in mind the two different curvatures of the front and back surfaces as well as $d \approx 0$, we find the focal length F:

$$AD \approx OD \approx BD = \frac{nR_1}{n-1}$$

$$\frac{n}{BD} - \frac{1}{BC} = -\frac{n-1}{R_2}$$

$$\frac{1}{F} = \frac{1}{OC} \approx \frac{1}{BC} = (n-1)\left(\frac{1}{R_1} + \frac{1}{R_2}\right) \tag{3.26}$$

This equation effectively illustrates the principle of summating optical powers (see eqs. (3.11) and (3.12)) when describing light refraction in complex systems. Correspondingly,

$$F = \frac{R_1R_2}{(n-1)(R_1+R_2)} \tag{3.27}$$

Note that when R_1 and R_2 are both negative, that is, both spherical surfaces have a concave shape (see Figure 3.17), the focal length is also negative. This implies that such a lens will diverge light, in contrast to converging convex lenses shown in Figure 3.16. Combining converging and diverging lenses expands the abilities of optical systems, for example, correcting and diminishing different types of lens aberrations. If both curvatures are equal, that is, $R_1 = R_2 = R$, then the focal length is

$$F = \frac{R}{2(n-1)} \tag{3.28}$$

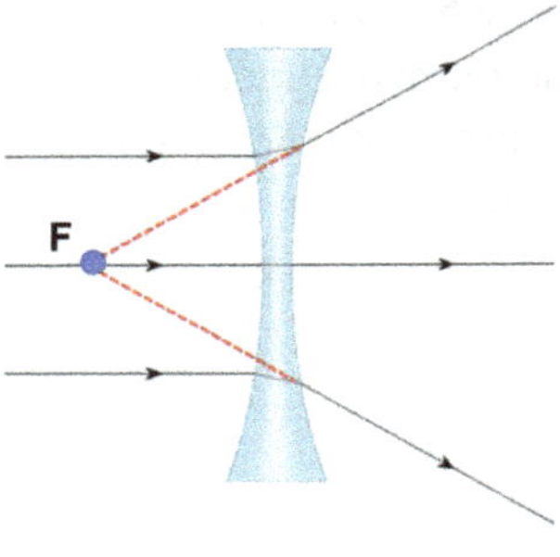

Figure 3.17: Action of a concave (diverging) lens.

Lens functioning is illustrated in Figure 3.18. Let us consider an object with height Y, located on the left-hand side of the lens. We know that light propagating parallel to the optical axis, when being refracted by the lens, crosses the right-hand focal point F. At the same time, light propagating via the left-hand focal point F will change its trajectory to be parallel to the optical axis on the right side of the lens. These two refracted rays meet at some point after lens, thus producing an image with height Y'. Moreover, for thin lenses, the rays passing through the lens center do not experience refraction. To further proceed, we introduce the object/lens and image/lens distances as a and b, respectively, and the object/image distances to the nearest focal points as X and X', respectively.

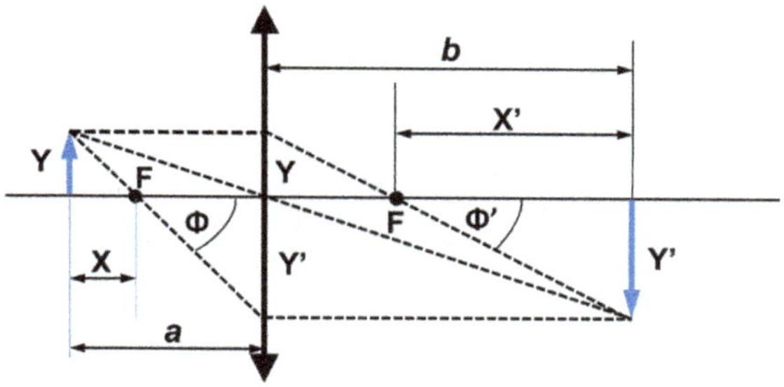

Figure 3.18: Derivation of the **Newton** formula.

In other words,

$$X = a - F;\ X' = b - F \tag{3.29}$$

Considering similar triangles in Figure 3.18, we find

$$\frac{Y}{X} = \frac{Y'}{F} \tag{3.30}$$

$$\frac{Y'}{X'} = \frac{Y}{F} \tag{3.31}$$

and finally

$$XX' = F^2 \tag{3.32}$$

This equation is called the **Newton** formula. Applying definitions (3.29), one can rewrite eq. (3.32) in the more accustomed form

$$\frac{1}{a} + \frac{1}{b} = \frac{1}{F} \tag{3.33}$$

Using the **Newton** formula (3.32), it is easy to understand the basic principles of lens functioning. If $X > 0$, that is, when an object is located left of the left-hand focal point, then also $X' > 0$, and we have the situation depicted in Figure 3.18. We see that in this case, the magnified image ($Y' > Y$) is real but inverted with respect to the object. If $X < 0$, that is, when an object is located right of the left-hand focal point, also $X' < 0$. In this case, the magnified image ($Y' > Y$) is virtual and upright (see Figure 3.19).

In both cases, linear magnification, also called transverse or lateral magnification, is defined by the ratio $M_t = \left(Y'/Y\right)$. Applying eqs. (3.30) and (3.31) yields

$$M_t = \frac{Y'}{Y} = \frac{F}{X} = \frac{X'}{F} \tag{3.34}$$

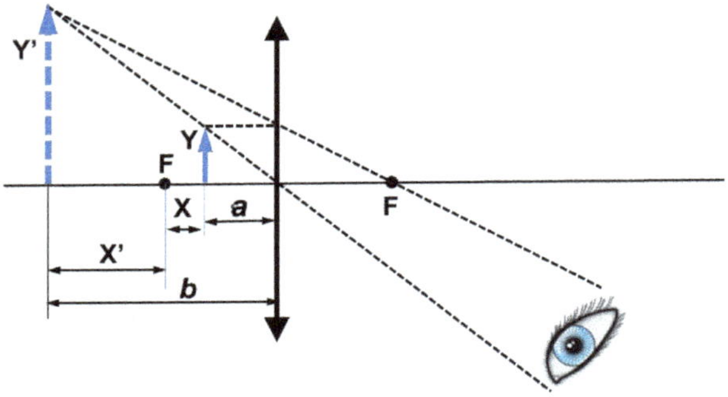

Figure 3.19: Formation of a virtual image for $X < 0$.

In other words, the closer the object is to the focal point (smaller X-value), the greater is the magnification magnitude. At constant X, higher magnification is achieved for a larger focal distance F. This approach is used, for example, in long-focus photographic cameras. Note that when magnification is changed with the distance y from the optical axis, optical devices encounter specific aberrations called distortions. Distorted images of the square in this case are shown in Figure 3.20. If the linear magnification is growing with y, the distortion has a pincushion-like (or pillow-like) shape. In the opposite case, when magnification is diminishing with y, the distortion has a barrel-like shape.

Figure 3.20: Distortion-type aberrations in optical systems.

Real objects have finite dimensions not only vertically but also along the optical axis. Correspondingly, different points of X will be transformed into image points with different values of X'. To characterize this fact, the longitudinal magnification is introduced as $M_l = -\left(dX'/dX\right)$, which determines the sharpness of the focal spot. Differentiating the **Newton** formula (3.32) yields

$$XdX' + X'dX = 0 \tag{3.35}$$

Correspondingly,

$$M_l = -\frac{dX'}{dX} = \frac{X'}{X} \tag{3.36}$$

Using eqs. (3.32) and (3.34), one obtains

$$M_l = \frac{X'}{X} = \left(\frac{X'}{F}\right)^2 = M_t^2 \tag{3.37}$$

In addition, the optical system is characterized by angular magnification M_a, related to characteristic angles left (Φ) and right (Φ') of the lens (see Figure 3.18):

$$M_a = \frac{\tan \Phi'}{\tan \Phi} = \frac{Y}{F}\frac{F}{Y'} = \frac{Y}{Y'} = \frac{1}{M_t} \tag{3.38}$$

Therefore,

$$M_l \cdot M_a = M_t \tag{3.39}$$

Certainly, the system's magnification cannot be infinitely large. First, as we already mentioned, real physical objects have a finite thickness (see the vertical lines as in Figures 3.18 and 3.19). This means that the X-value cannot be zero. In practical terms, the more important circumstance is that the closer the object is to the focal point, the further the image moves away from the lens center. Therefore, to achieve very high magnification with a single-lens device, the latter's size should be unrealistically large. In Section 3.6, we discuss the two-lens design of a microscope, which provides rather high magnification compatible with a still compact design.

3.5 Limitations of Fermat's principle

The most fundamental factor limiting the resolving power of optical devices and, hence, their magnification, is light diffraction, which is discussed in detail later in this book. Here, we can say that the minimal size of the object, viewed through a conventional optical system, cannot be considerably less than the light wavelength λ. This is the consequence of the fundamental **Heisenberg** principle, stating that the product of uncertainties in determining some interrelated quantities, such as momentum P and size Y, is of the order of the **Planck** constant h:

$$\Delta P \cdot \Delta Y \approx h \tag{3.40}$$

Using the **de Broglie** linear relationship between momentum P and wave vector $k = \frac{2\pi}{\lambda}$,

$$P = \frac{h}{2\pi} k \tag{3.41}$$

eq. (3.40) transforms into

$$\Delta Y \approx \frac{h}{\Delta P} = \frac{2\pi}{\Delta k}$$

Since $\Delta k \leq k$ and $\Delta Y \leq Y$, one can decide that

$$Y \gtrsim \frac{2\pi}{k} = \lambda \tag{3.42}$$

We can support this conclusion by considering more carefully the foundations of **Fermat**'s principle – specifically, the statement that for light focusing, all trajectories should have equal optical paths. This statement, however, is completely true in geometrical optics only, that is, when wavelength $\lambda \longrightarrow 0$. For a finite wavelength $\lambda \neq 0$, these pathways may be not exactly equal, the differences being on the order of λ. If so, the objects that are smaller than some limiting height Y_l become "invisible" because the ray trajectories issuing from various object points differ in length by not more than λ. This principle is well illustrated in Figure 3.21. In fact, for small object located close to the left-side focal point F, we have two nearly equivalent ray paths, BC and AC, the difference being roughly equal to

$$AD = BC - AC \approx Y \sin \Theta \tag{3.43}$$

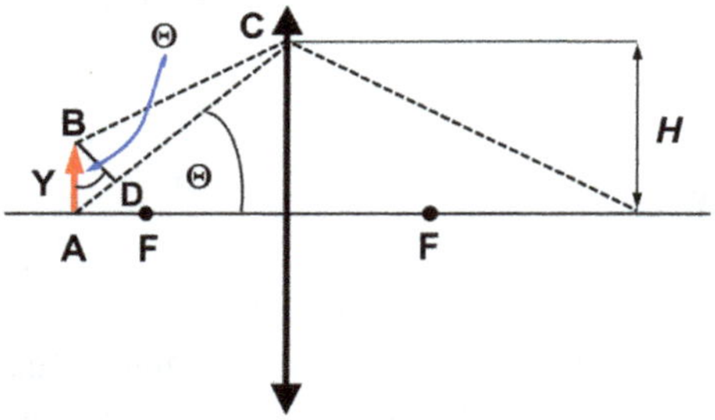

Figure 3.21: Sketch illustrating limitations of the **Fermat**'s principle.

On the other hand,

$$\tan \Theta \approx \frac{H}{F} \tag{3.44}$$

where H is half of the lens size. For small Θ, $\sin \Theta \approx \tan \Theta$, and then

$$AD \approx Y \frac{H}{F} \tag{3.45}$$

Setting $AD \approx \lambda$, one finally finds

$$Y_1 \approx \frac{F}{H}\lambda \tag{3.46}$$

We can evaluate Y_1 in another way, that is, considering the time difference $t_1 - t_2$ for light propagation along two closely related trajectories, BC and AC, in Figure 3.21. This quantity, taking into account eq. (3.43), equals

$$t_1 - t_2 = \frac{AD}{V_p} = n\frac{Y_1 \sin\Theta}{c} \tag{3.47}$$

To be more general, we consider in (3.47) light rays propagating with phase velocity V_p in homogeneous medium characterized by the refractive index n. On the other hand, two points of the object A and B become unresolved in the image plane, if the respective time difference $t_1 - t_2$ is reduced to the vibration period t_0 of the light wave (or below):

$$t_1 - t_2 \approx t_0 = \frac{\lambda_0}{c} \tag{3.48}$$

where λ_0 is the light wavelength in a vacuum. Combining eqs. (3.47) and (3.48) yields

$$Y_1 \approx \frac{\lambda_0}{n \sin\Theta} \tag{3.49}$$

The product $n \sin\Theta$ is called the numerical aperture and cannot be much larger than 1. Therefore, we again find that the smallest object features still resolvable with the aid of conventional optics are restricted by the light wavelength λ_0 (or some part of it).

More accurate estimations of the resolving power of optical instruments will be given in Section 10.5 after discussing the main diffraction phenomena.

3.6 Conventional light microscope

As already mentioned, to obtain rather high magnification (up to a few thousands) with a compact instrument, the latter in the simplest version should comprise at least two lenses, which are called objective and ocular (eyepiece) lenses. They are separated by the distance Δ, which is much larger than the focal lengths F_{ob} and F_{oc} of both lenses. Magnification of the microscope M_m is the product of magnification supplied by the objective M_{ob} and ocular M_{oc} lenses.

The principal scheme of the microscope is shown in Figure 3.22.

An object (Y) is placed close to the left focal point of the objective lens, which provides the magnified and inverted image (Y'). At this stage, the objective magnification M_{ob} is given by eq. (3.34):

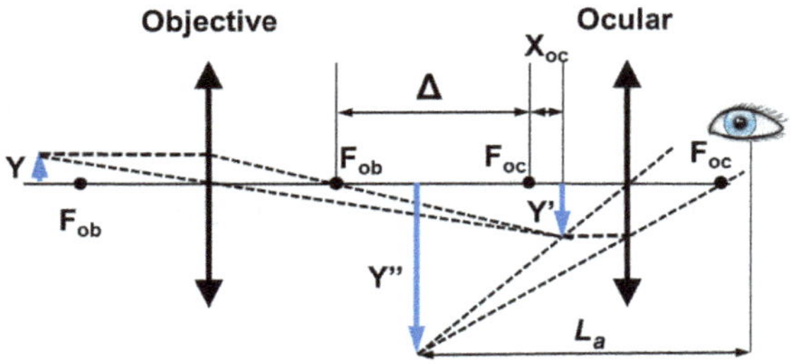

Figure 3.22: Scheme of a conventional light microscope.

$$M_{ob} = \frac{Y'}{Y} = \frac{X'}{F_{ob}} = \frac{\Delta + X_{oc}}{F_{ob}} \tag{3.50}$$

Image Y' is viewed via the ocular lens installed in such a way that Y' is on the right side and close to the ocular focus F_{oc} (see Figure 3.22). Therefore, the ocular lens works as a magnifying glass providing an enlarged image Y'', which is viewed by the eye from the best accommodation distance L_a. Practically, the eye is located near the back focus of the ocular lens. Magnification achieved at this stage is, again, calculated using eq. (3.34):

$$M_{oc} = \frac{Y''}{Y'} \approx \frac{L_a - F_{oc}}{F_{oc}} \tag{3.51}$$

Taking into account that $X_{oc} \ll \Delta$ and $F_{oc} \ll L_a$, we finally obtain the magnification of the microscope:

$$M_m = M_{ob} M_{oc} \approx \frac{\Delta}{F_{ob}} \cdot \frac{L_a}{F_{oc}} \tag{3.52}$$

The objective lens has a very short focal distance, about $F_{ob} \approx 1$ mm. Taking $\Delta = 100$ mm, we can achieve $M_{ob} \approx 100$. For adults, the best accommodation distance is $L_a \approx 250$ mm. Taking $F_{oc} = 10$ mm, we get $M_{oc} \approx 25$. These values give us an overall magnification of $M_m = M_{ob} M_{oc} \approx$ 2,500.

3.6.1 Köhler illumination and Abbe theory of image formation

In the preceding sections, we silently assumed that light is emanating from the object points, that is, an object is light emitting by itself. In most cases, however, this is not true, and to see the object the latter should be illuminated from an external source. At the end of the nineteenth century, mostly through the combined efforts of two German scientists, **August Köhler** and **Ernst Abbe**, an optimal illumination scheme using a condenser system was invented and the image formation theory was developed. Both **Köhler** and

Abbe worked closely with **Carl Zeiss** AG in Jena, which allowed this company to become a world leader in the optical industry.

Optimal illumination of the object by parallel beam using a condenser system provides two important benefits: an evenly illuminated field of view and coherent illumination of different object points. Note that a parallel beam can be along the optical axis or inclined to it. The term "coherence" will be explained in detail later in this book. For us, here, it means that the waves issuing from different points of the object are phase correlated. This can be achieved, for instance, when an object is illuminated by a plane (or nearly plane) wave. Image formation under such illumination is the essence of the microscope theory developed by **Ernst Abbe**.

Let us consider an object like diffraction grating (see Section 10.1), which is illuminated by a plane wave parallel to the optical axis. The overall wavefield $U(x,y)$, existing just after the object, depends on the local coordinates (x,y) in the object plane (see Figure 3.23).

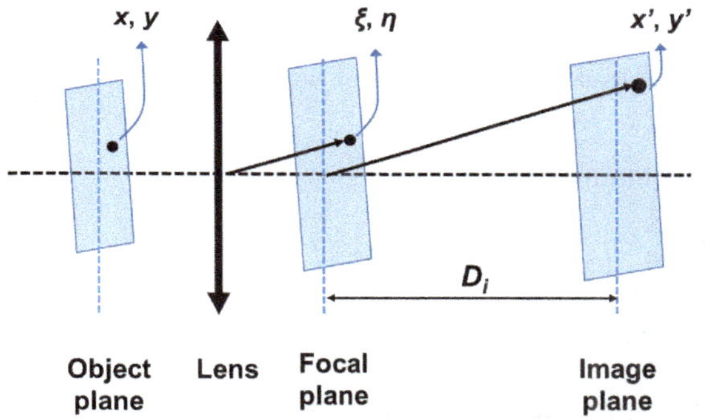

Figure 3.23: Illustration of the **Abbe** theory of image formation in a microscope.

In other words, function $U(x,y)$ is the transmission function of the object. Furthermore, the wavefields originating in individual points (x,y) are first transmitted to the points (ξ,η) in the system's focal plane, providing the overall wavefield here

$$V(\xi,\eta) = C_1 \int\int U(x,y)\exp[i(q_x x + q_y y)]dxdy \tag{3.53}$$

where C_1 is some constant and integration is over the entire object plane. Coordinates (ξ,η) are entering eq. (3.53) via projections of wave vectors $\boldsymbol{k}$ onto the focal plane

$$q_x \approx k\frac{\xi}{F}; \quad q_y \approx k\frac{\eta}{F} \tag{3.54}$$

with F being the focal length of the generalized optical system. Correspondingly,

$$V(\xi,\eta)=C_1\int\int U(x,y)\exp\left[ik\left(\frac{\xi}{F}x+\frac{\eta}{F}y\right)\right]dxdy \tag{3.55}$$

The wavefield $V(\xi,\eta)$ is then considered as a new source, which creates the distribution of light intensity in the image plane (x',y'):

$$W(x',y')=C_2\int\int V(\xi,\eta)\exp\left[i(q_\xi\xi+q_\eta\eta)\right]d\xi d\eta \tag{3.56}$$

where integration is performed over the aperture restricting the focal plane. Projections of wave vectors $\boldsymbol{k}$ onto the image plane (x',y') are

$$q_\xi\approx k\frac{x'}{D_i};\quad q_\eta\approx k\frac{y'}{D_i} \tag{3.57}$$

with D_i being the distance from the focal plane to image plane. Correspondingly,

$$W(x',y')=C_2\int\int V(\xi,\eta)\exp\left[ik\left(\frac{x'}{D_i}\xi+\frac{y'}{D_i}\eta\right)\right]d\xi d\eta \tag{3.58}$$

Combining eqs. (3.55) and (3.58), one obtains

$$W(x',y')=C_1C_2\int\int\int\int U(x,y)\cdot\exp\left[i\frac{k}{F}\left(x+\frac{x'F}{D_i}\right)\xi\right]\cdot\exp\left[i\frac{k}{F}\left(y+\frac{y'F}{D_i}\right)\right]dxdyd\xi d\eta \tag{3.59}$$

To further proceed we assume that the aperture is widely open and use the integral representation of the delta function:

$$\delta(t)=\frac{1}{2\pi}\int_{-\infty}^{\infty}e^{itz}dz \tag{3.60}$$

Combining eqs. (3.60) and (3.59) yields

$$W(x',y')=(2\pi)^2C_1C_2\int\int U(x,y)\cdot\delta\left[i\frac{k}{F}\left(x+\frac{x'F}{D_i}\right)\xi\right]\cdot\delta\left[i\frac{k}{F}\left(y+\frac{y'F}{D_i}\right)\right]dxdy \tag{3.61}$$

or

$$W(x',y')=(2\pi)^2C_1C_2\frac{F^2}{k^2}\int\int U(x,y)\cdot\delta\left[i\frac{k}{F}\left(x+\frac{x'F}{D_i}\right)\xi\right]\cdot\delta\left[i\frac{k}{F}\left(y+\frac{y'F}{D_i}\right)\right]d\left(\frac{k}{F}x\right)d\left(\frac{k}{F}y\right) \tag{3.62}$$

Recalling that $k = \frac{2\pi}{\lambda}$ and

$$\int f(t)\delta(t - t')dt = f(t') \tag{3.63}$$

we find

$$W(x', y') = C\, U\left(-\frac{F}{D_i}x', -\frac{F}{D_i}y'\right) \tag{3.64}$$

where the normalized constant is $C = C_1 C_2 \lambda^2 F^2$. Taking into account that $\frac{F}{D_i} = \frac{1}{M_t}$ (see eq. (3.34)) and the image inversion by the optical system, one can write

$$-\frac{F}{D_i}x' = x; \quad -\frac{F}{D_i}y' = y \tag{3.65}$$

and finally

$$W(x', y') = C\, U(x, y) \tag{3.66}$$

This is a very important result that shows that under coherent illumination by a parallel beam of light, the light distribution in the image plane $W(x', y')$ exactly reflects the one in the object plane. In simple words, an image of the sample is an exact enlarged copy of an object. Certainly, this is true if the integration in eq. (3.59) can be expanded to infinity, that is, when the microscopic aperture is wide enough.

Chapter 4
Refractive index in anisotropic crystals

In previous chapters we treated refractive index n as a number. This makes sense, however, for isotropic media only. In general case of anisotropic media, and certainly in crystals, refractive index is described by tensor of the same rank, as dielectric permittivity ε_{ik} and specific electrical conductivity σ_{ik}. Remember that both quantities enter the expression for refractive index in general case (see eq. (1.29)). Note that in three-dimensional space, indices i and k are 1, 2, and 3. We also remind the readers that in crystals, the dielectric permittivity ε_{ik} and specific electrical conductivity σ_{ik} are tensors of second rank since they interconnect between two vectors (i.e., tensors of rank 1). In case of dielectric permittivity these are the vectors of the electric displacement field D_i and external electric field E_k (see eq. (1.6)):

$$D_i = \varepsilon_{ik} E_k \tag{4.1}$$

whereas for conductivity, these are the vectors of the density of electric current J_i and electric field E_k (see eq. (1.8)),

$$J_i = \sigma_{ik} E_k \tag{4.2}$$

For simplicity, in further analysis we will focus on dielectric materials, in which refractive index n_{ik} depends on dielectric permittivity ε_{ik} lonely, via dielectric constant of a medium $(\varepsilon_m)_{ik} = \frac{\varepsilon_{ik}}{\varepsilon_0}$ (see eq. (1.30)). In its general form tensor of second rank is defined by $3^2 = 9$ parameters:

$$\varepsilon_{ik} = \begin{matrix} \varepsilon_{11} & \varepsilon_{12} & \varepsilon_{13} \\ \varepsilon_{21} & \varepsilon_{22} & \varepsilon_{23} \\ \varepsilon_{31} & \varepsilon_{32} & \varepsilon_{33} \end{matrix} \tag{4.3}$$

In most practical situations, tensors ε_{ik} and, hence, n_{ik} are symmetric, that is, $\varepsilon_{ik} = \varepsilon_{ki}$, and the number of independent parameters is reduced to six:

$$\varepsilon_{ik} = \begin{matrix} \varepsilon_{11} & \varepsilon_{12} & \varepsilon_{13} \\ \varepsilon_{12} & \varepsilon_{22} & \varepsilon_{23} \\ \varepsilon_{13} & \varepsilon_{23} & \varepsilon_{33} \end{matrix} \tag{4.4}$$

Further restrictions on tensor elements are imposed by crystal symmetry. To find tensor shapes for different symmetry systems, we must apply the transformation law for tensors of second rank

$$\varepsilon'_{ik} = f_{ip} f_{kq} \varepsilon_{pq} \tag{4.5}$$

https://doi.org/10.1515/9783111140100-005

where the transformation matrix f_{ik} describes the change of vector projections under rotation of the coordinate system, the rotation related to a certain symmetry element. Transformation matrix is determined by the angles Φ_{ik} between new vector projections (x'_1, x'_2, x'_3) and initial vector projections (x_1, x_2, x_3)

$$f_{ik} = \cos(\Phi_{ik}) \tag{4.6}$$

For example, inversion operation $((x_1, x_2, x_3) \to (-x_1, -x_2, -x_3))$ in its matrix form is

$$f_{ik} = \begin{matrix} -1 & 0 & 0 \\ 0 & -1 & 0 \\ 0 & 0 & -1 \end{matrix} \tag{4.7}$$

For a mirror symmetry plane perpendicular to the x_3-axis ($x_3 \to -x_3$), one obtains

$$f_{ik} = \begin{matrix} 1 & 0 & 0 \\ 0 & 1 & 0 \\ 0 & 0 & -1 \end{matrix} \tag{4.8}$$

while for the twofold rotation axis (rotation by 180°) parallel to the x_3-axis (i.e., changing $x_1 \to -x_1$ and $x_2 \to -x_2$), one finds the complementary matrix interrelated with (4.8) via inversion center (4.7):

$$f_{ik} = \begin{matrix} -1 & 0 & 0 \\ 0 & -1 & 0 \\ 0 & 0 & 1 \end{matrix} \tag{4.9}$$

In more general case of rotation about the x_3-axis by an arbitrary angle Φ, applying eq. (4.6) yields

$$f_{ik} = \begin{matrix} \cos\Phi & \sin\Phi & 0 \\ -\sin\Phi & \cos\Phi & 0 \\ 0 & 0 & 1 \end{matrix} \tag{4.10}$$

Using eq. (4.10), we find that for fourfold rotation axis (rotation by 90°) parallel to the x_3-axis, the transformation matrix has a nondiagonal form

$$f_{ik} = \begin{matrix} 0 & 1 & 0 \\ -1 & 0 & 0 \\ 0 & 0 & 1 \end{matrix} \tag{4.11}$$

Here we will also need the transformation matrices for threefold rotation axis (rotation by 120°) and sixfold rotation axis (rotation by 60°) parallel to the x_3-axis. The respective matrix forms are derived straightforwardly from eq. (4.10) as

$$f_{ik} = \begin{matrix} -1/2 & \sqrt{3}/2 & 0 \\ -\sqrt{3}/2 & -1/2 & 0 \\ 0 & 0 & 1 \end{matrix} \quad (4.12)$$

for threefold rotation axis and

$$f_{ik} = \begin{matrix} 1/2 & \sqrt{3}/2 & 0 \\ -\sqrt{3}/2 & 1/2 & 0 \\ 0 & 0 & 1 \end{matrix} \quad (4.13)$$

for sixfold rotation axis.

We stress that the transformation matrix f_{ik} enters eq. (4.5) twice, which has important implications on respective physical properties in crystals. For example, based only on this fact, we immediately realize that all physical properties described by tensors of second rank are not sensitive to the presence or absence of inversion center (4.7). Furthermore, the twofold rotation axis and horizontal (perpendicular to the axis) mirror plane, producing together the inversion center, impose equal constraints on tensors of second rank. These and similar considerations allow us to finally conclude that the second rank tensors will have akin shapes for all point groups within a given symmetry system. In other words, tensors of second rank are only sensitive to the type of symmetry system (triclinic, monoclinic, orthorhombic, etc.), to which the crystal belongs.

Let us analyze in more detail the symmetry constraints imposed on the second rank tensors in different symmetry systems. Since there is no sensitivity to inversion center, triclinic system comprising point groups 1 and $\bar{1}$ does not impose any additional constrain on symmetric tensor of second rank keeping it in the most general form with six independent constants (see eq. (4.4)).

In monoclinic system, the presence of the twofold rotation axis or mirror plane further reduces this amount down to four parameters. In fact, applying the transformation law (4.5) with transformation matrix (4.8) or (4.9), we find that $\varepsilon_{13} = \varepsilon_{23} = 0$. This follows from the **Neumann**'s principle, which states that the symmetry of any physical property in a crystal cannot be lower than the symmetry of the crystal itself. In practical words, if after the coordinate transformation induced by any symmetry element of the crystal point group, we find the discrepancy between the initial and transformed magnitudes of some tensor elements, it means that these elements should be equal to zero (in case of differing in sign) or be linear combinations of other tensor elements. Using these considerations, we find that tensor ε_{ik} in crystals of monoclinic symmetry has the following form:

$$\varepsilon_{ik} = \begin{matrix} \varepsilon_{11} & \varepsilon_{12} & 0 \\ \varepsilon_{12} & \varepsilon_{22} & 0 \\ 0 & 0 & \varepsilon_{33} \end{matrix} \tag{4.14}$$

At this point, let us address an important question: how the obtained tensor forms (4.4) and (4.14) match the general statement from linear algebra that every symmetric tensor of second rank can be expressed in the diagonal form? To answer this question, we recall that in monoclinic symmetry system, only one coordinate axis (say the x_3-axis) is physically fixed by crystal symmetry being parallel to the twofold rotation axis or perpendicular to a mirror plane. Two other mutually perpendicular axes (x_1 and x_2) can be rotated as a rigid frame about the x_3-axis by an arbitrary angle Φ. If so, we can find the specific angle Φ_s at which the single nondiagonal element in matrix (4.14) becomes zero, that is, (ε_{12})new = 0. By using the transformation law (4.5) and transformation matrix (4.10), describing the rotation mentioned, one finds that the condition (ε_{12})new = 0 satisfies, if

$$\tan(\Phi_s) = \frac{\varepsilon_{11} + \varepsilon_{22}}{2\varepsilon_{12}} \tag{4.15}$$

Therefore, it is possible, in fact, to diagonalize tensor (4.14), but the number of independent parameters for its complete determination remains to be four, that is, three diagonal tensor components and one specific angle Φ_s of the azimuthal rotation. Applying the same approach to tensor (4.4) for triclinic symmetry system, we can also reduce it to diagonal form, but when rotating our coordinate system by certain angles about all three coordinate axes. So, in this case, we still need to define six independent parameters, that is, three diagonal tensor elements and three angles of rotation.

In orthorhombic symmetry system, we already have the intrinsically built-in **Cartesian (Descartes)** coordinate system with mutually perpendicular axes related to the existing symmetry elements (twofold rotation axes and/or mirror planes). As a result, a tensor of second rank inherently has diagonal form with three independent parameters:

$$\varepsilon_{ik} = \begin{matrix} \varepsilon_{11} & 0 & 0 \\ 0 & \varepsilon_{22} & 0 \\ 0 & 0 & \varepsilon_{33} \end{matrix} \tag{4.16}$$

Tensor (4.16) can be directly obtained from (4.14) by applying to the latter the 180° rotation about the x_1- or x_2-axis (point group *222*), or mirror reflection in an extra symmetry plane for orthorhombic point groups (*mm2*) and (*mmm*).

Taking into account the above-mentioned diagonalization procedures in triclinic and monoclinic crystals, we can consider the tensor form (4.16), as characteristic for three symmetry systems: triclinic, monoclinic, and orthorhombic, that is, having three different nonzero diagonal elements. In crystals revealing these symmetries, by means of tensor (4.16) and eq. (4.1) we can find the magnitude D of the vector of the electric displacement field D_i, as a function of the projections of the applied electric field E_k:

$$D^2 = \varepsilon_{11}^2 E_1^2 + \varepsilon_{22}^2 E_2^2 + \varepsilon_{33}^2 E_3^2 \tag{4.17}$$

According to eq. (4.17), the representation surface for the ends of possible vectors $\boldsymbol{D}$ is triaxial ellipsoid with the axis lengths equal $\varepsilon_{11}E$, $\varepsilon_{22}E$, and $\varepsilon_{33}E$ (see Figure 4.1), where $E = \sqrt{E_1^2 + E_2^2 + E_2^2}$. Triaxial ellipsoid has two circular cross sections (shadowed in Figure 4.1), which are of great importance to light propagation in these crystals (see Chapter 5).

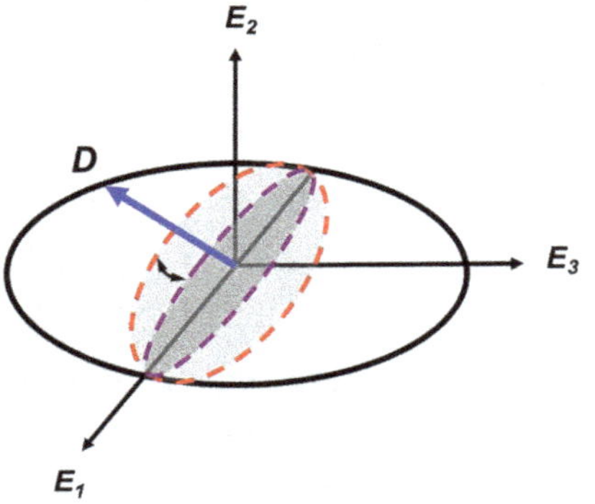

Figure 4.1: Representation ellipsoid (triaxial) for vectors of the electric displacement field ***D*** in crystals belonging to triclinic, monoclinic, or orthorhombic symmetry systems. Two circular cross sections (shadowed) of the ellipsoid are outlined by dashed lines.

For tetragonal symmetry system, the leading symmetry element is the fourfold rotation axis. Applying the transformation law (4.5) with transformation matrix (4.11) to tensor (4.16), we find that $\varepsilon_{11} = \varepsilon_{22}$, and the number of independent parameters is further reduced to two parameters. Correspondingly, the tensor of second rank for tetragonal symmetry system has the following form:

$$\varepsilon_{ik} = \begin{matrix} \varepsilon_{11} & 0 & 0 \\ 0 & \varepsilon_{11} & 0 \\ 0 & 0 & \varepsilon_{33} \end{matrix} \tag{4.18}$$

The equality of tensor elements $\varepsilon_{11} = \varepsilon_{22}$ in matrix (4.18) reveals the symmetry of tetragonal prism in the plane perpendicular to the x_3-axis.

The same form (4.18) with two independent constants, tensor of second rank has in crystals of rhombohedral and hexagonal symmetry, whenever tensor is expressed in the **Cartesian** coordinate system with the x_3-axis being parallel to the threefold or sixfold rotation axis. This can be directly obtained by applying the transformation law (4.5) with transformation matrices (4.12) or (4.13) to tensor (4.16). Also, for these symmetries, as for the fourfold symmetry, we recognize the isotropy of the respective physical properties within the plane perpendicular to the threefold or sixfold rotation axis.

According to the tensor form (4.18), the representation ellipsoid for vectors $\boldsymbol{D}$ in crystals belonging to these symmetry systems (tetragonal, rhombohedral, or hexagonal) is transformed into the ellipsoid of rotation about the x_3 (or E_3)-axis (see Figure 4.2):

$$D^2 = \varepsilon_{11}^2(E_1^2 + E_2^2) + \varepsilon_{33}^2 E_3^2 \tag{4.19}$$

with axes equal $\varepsilon_{11}E_1$, $\varepsilon_{11}E_2$, and $\varepsilon_{33}E_3$. It has lone circular cross section shadowed in Figure 4.2.

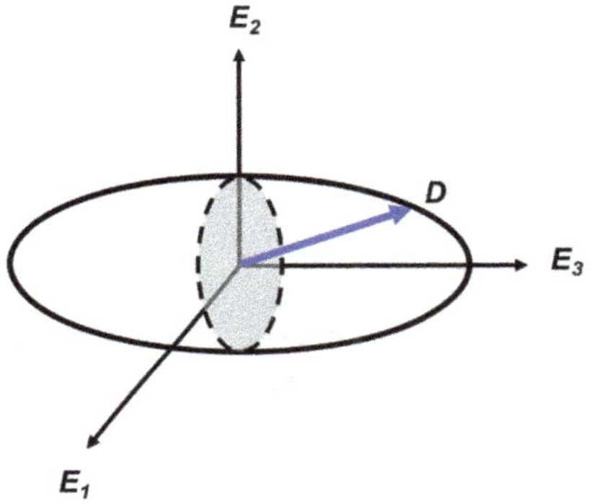

Figure 4.2: Representation ellipsoid (ellipsoid of rotation about the E_3-axis) for vectors of the electric displacement field $\boldsymbol{D}$ in crystals belonging to tetragonal, rhombohedral, or hexagonal symmetry systems. Circular cross section (shadowed) perpendicular to the E_3-axis is outlined by dashed line.

In cubic symmetry system, the coordinate axes, (x_1, x_2, x_3), can be chosen along the identical symmetry elements (three twofold rotation axes or three fourfold rotation axes along the cube edges) within cubic unit cell. Therefore, all three axes are equivalent and, hence, only one independent parameter, $\varepsilon = \varepsilon_{11} = \varepsilon_{22} = \varepsilon_{33}$, defines the shape of the tensor of second rank:

$$\varepsilon_{ik} = \begin{matrix} \varepsilon & 0 & 0 \\ 0 & \varepsilon & 0 \\ 0 & 0 & \varepsilon \end{matrix} \tag{4.20}$$

It means that in cubic crystals all physical properties, described by tensors of second rank, are isotropic, that is, do not depend on specific direction within a crystal. Representation ellipsoid for vectors $\boldsymbol{D}$ transforms to the sphere (see Figure 4.3):

$$D^2 = \varepsilon^2(E_1^2 + E_2^2 + E_3^2) = \varepsilon^2 E^2 \tag{4.21}$$

with radius equal εE. Certainly, all cross sections, passing through the center of the sphere, are circular ones. Note that spherical shape means that in cubic crystals, dielectric permittivity ε is the same along every crystallographic direction (x_1, x_2, x_3), and vectors $\boldsymbol{D}$ and $\boldsymbol{E}$ are parallel to each other. In crystals belonging to lower symmetry systems, vectors $\boldsymbol{D}$ and $\boldsymbol{E}$, in general, are not parallel.

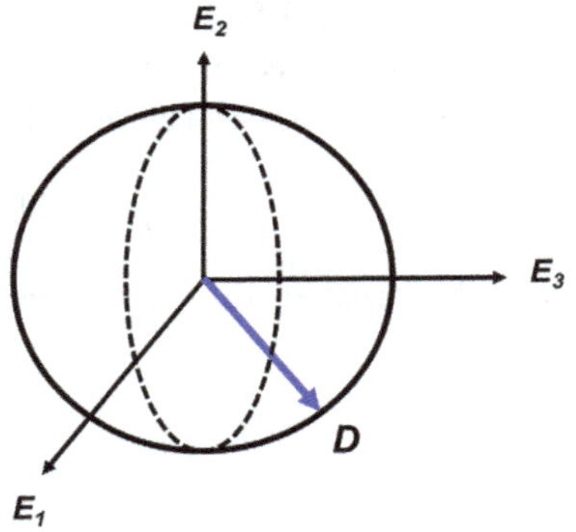

Figure 4.3: Representation ellipsoid (sphere) for vectors of the electric displacement field ***D*** in crystals belonging to cubic symmetry system.

In isotropic media, tensors of second rank remain to be in the most symmetric form, as for cubic crystals (see matrix (4.20)), and certainly two vectors, ***D*** and ***E***, interconnected by such a tensor, are parallel to each other.

Summarizing all mentioned above, we can draw an important conclusion: in isotropic materials and cubic crystals, light refraction and related phenomena are governed by single refractive index, which in dielectrics equals $n = \sqrt{\varepsilon/\varepsilon_0}$, where ε_0 is dielectric permittivity in a vacuum (see eqs. (1.6) and (1.30)). In crystals belonging to tetragonal, rhombohedral, or hexagonal symmetry systems, we need two refractive indices, $n_{11} = \sqrt{\varepsilon_{11}/\varepsilon_0}$ and $n_{33} = \sqrt{\varepsilon_{33}/\varepsilon_0}$, for this purpose, while in crystals of lower symmetry, this number is raised up to three, $n_{11} = \sqrt{\varepsilon_{11}/\varepsilon_0}$, $n_{22} = \sqrt{\varepsilon_{22}/\varepsilon_0}$, and $n_{33} = \sqrt{\varepsilon_{33}/\varepsilon_0}$.

In Chapter 5, we will further develop these issues toward analyzing the light polarization phenomena.

Chapter 5
Polarization, birefringence, and related phenomena

Let us recall again one of the **Maxwell** equations (1.4) adapted for use in dielectric materials ($\boldsymbol{J} = 0$)

$$\mathrm{rot}\,\boldsymbol{H} = \frac{\partial \boldsymbol{D}}{\partial t} \tag{5.1}$$

Substituting into eq. (5.1) plane waves, $\boldsymbol{H} = \boldsymbol{H}_0 \exp[i(\boldsymbol{kr} - \omega t)]$ and $\boldsymbol{D} = \boldsymbol{D}_0 \exp[i(\boldsymbol{kr} - \omega t)]$, yields an important relationship

$$\boldsymbol{k} \times \boldsymbol{H} = -\omega \boldsymbol{D} \tag{5.2}$$

which implies that in a material, the electric displacement field $\boldsymbol{D}$ is perpendicular to the wave vector $\boldsymbol{k}$. The same is valid for the electric field $\boldsymbol{E} = \frac{\boldsymbol{D}}{\varepsilon_0}$ in a vacuum

$$\boldsymbol{k} \times \boldsymbol{H} = -\omega \varepsilon_0 \boldsymbol{E} \tag{5.3}$$

In other words, electromagnetic waves (including light and X-rays) are transversely polarized with a polarization vector, indicating the direction of the electric displacement field in a material or electric field in vacuum. There is no longitudinal polarization component along vector $\boldsymbol{k}$; scientists from elementary particle physics say that it is because the rest mass of a photon is zero.

Since the polarization vector lies in the plane perpendicular to the wave vector $\boldsymbol{k}$, it, generally, is composed of two mutually orthogonal components: $\boldsymbol{D}_1$ and $\boldsymbol{D}_2$ (see Figure 5.1). For example, light emitted by an electric bulb or sealed X-ray tube is nonpolarized since it contains equal amounts of waves (or photons) having polarizations $\boldsymbol{D}_1$ and $\boldsymbol{D}_2$.

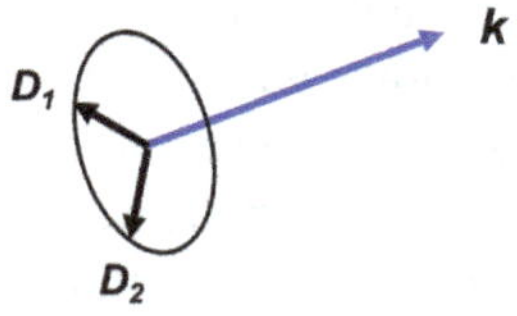

Figure 5.1: Arrangement of two mutually perpendicular linear polarizations (along the electric displacement field vectors $\boldsymbol{D}_1$ and $\boldsymbol{D}_2$) in the plane normal to the light wave vector $\boldsymbol{k}$.

5.1 The Brewster angle for visible light and X-rays

Let us consider the fate of differently polarized waves propagating in optically isotropic media like glass or cubic crystal. As we know, such materials have a sole refractive index n, which numerically is equivalent to the diagonal elements of the respective tensor:

https://doi.org/10.1515/9783111140100-006

$$n_{ik} = \begin{matrix} n & 0 & 0 \\ 0 & n & 0 \\ 0 & 0 & n \end{matrix} \tag{5.4}$$

Therefore, both waves propagate within a material with the same phase velocity $V_p = \frac{c}{n}$. The situation changes at the boundary between two media, having refractive indices n_1 and n_2. Because of the refraction phenomenon, we get a physically defined scattering plane, in which the wave vectors of the incident and refracted waves are situated (see Figure 5.2). In this case, the scattering plane also contains the wave vector of the reflected wave and the normal to the boundary. Now we can physically choose two polarizations, perpendicular to the scattering plane (out-of-plane polarization) and situated in that plane (the plane of Figure 5.2), which behave very differently upon refraction. The out-of-plane polarization is called σ-polarization, while the in-plane polarization is the π-polarization.

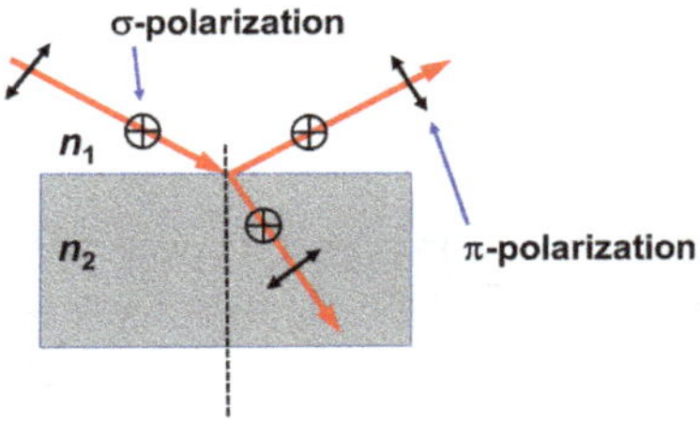

Figure 5.2: Illustration of the σ-polarization (out-of-plane) and π-polarization (in-plane) when an electromagnetic wave experiences refraction at the interface between two media with different refractive indices n_1 and n_2.

Essentially, the difference between these two polarizations follows from the fact that σ-polarization does not change its direction when a respective wave crosses the boundary between two media (i.e., its polarization remains perpendicular to the scattering plane), while π-polarization does (see Figure 5.2). Correspondingly, σ-polarization experiences refraction and reflection with no restrictions, while reflection of the π-polarized wave is suppressed by factor $\cos\delta$, where δ is the angle between polarization directions for the refracted and reflected rays (Figure 5.3). If $\delta = 90°$, $\cos\delta = 0$, and there is no π-polarized component in the reflected light, which becomes purely σ-polarized (out-of-plane polarization). This result is widely used to prepare linearly polarized electromagnetic waves. Note that the refracted light remains partially polarized with an excess of π-polarization.

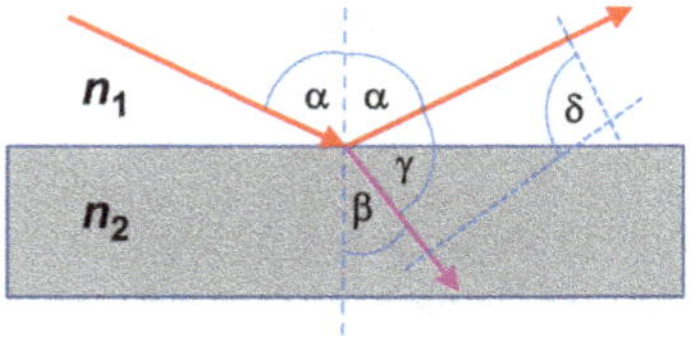

Figure 5.3: The **Brewster** angle α_B corresponds to the condition of $\delta = 90°$.

Using Figure 5.3, we can easily express this requirement ($\delta = 90°$) via experimental parameters, that is, via refractive indices n_1 and n_2 and incident angle α. In fact, $\delta = 360° - 90° - 90° - \gamma = 180° - \gamma$. On the other hand: $\gamma + \alpha + \beta = 180°$; therefore, $\delta = \alpha + \beta = 90°$. Recalling **Snell**'s law (2.23), one obtains $n_1 \sin\alpha = n_2 \sin\beta = n_2 \sin(90° - \alpha) = n_2 \cos\alpha$, and, finally, the so-called **Brewster** angle α_B for the incident light:

$$\tan\alpha_B = \frac{n_2}{n_1} \tag{5.5}$$

If the first medium is air and the second one is a material with refractive index n, then

$$\tan\alpha_B = n \tag{5.6}$$

For example, for glass with $n = 1.5$, $\alpha_B = 56.31°$.

As already mentioned in Section 2.3, the refractive index for X-rays is very close to one, $n \approx 1$. Therefore, according to eq. (5.6), the **Brewster** angle for X-rays is $\alpha_B = 45°$, irrespective of the material used. Note that X-rays interact weakly with materials, so such large scattering angles are available only in the X-ray diffraction experiments that we describe in Chapters 10, 12, and 15. In X-ray diffraction terminology, the angle between a diffracted beam and an incident beam (or a transmitted one since practically there is no refraction) is called the double **Bragg** angle $2\Theta_B$. Correspondingly, to achieve the σ-polarized X-rays, the **Bragg** angle Θ_B, counted from the air–material interface (see Figure 5.4), should be $\Theta_B = 90° - \alpha_B = 45°$. We will provide more details about polarization issues in the next sections of this chapter.

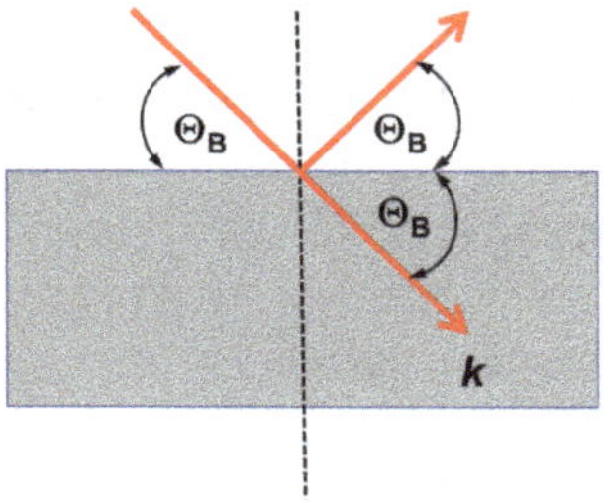

Figure 5.4: For X-rays, the **Brewster** angle equals $\alpha_B = 45°$ since the double **Bragg** angle $2\Theta_B = 90°$, and in this case, $\alpha_B = \Theta_B$.

5.2 Birefringence in anisotropic crystals

Even more interesting refraction phenomena are observed in uniaxial crystals belonging to the tetragonal, rhombohedral, or hexagonal symmetry systems. Like the tensor of dielectric permittivity (4.18), the tensor of the refractive index, in this case, comprises two dissimilar elements n_{11} and n_{33}:

$$n_{ik} = \begin{matrix} n_{11} = n_0 & 0 & 0 \\ 0 & n_{11} = n_0 & 0 \\ 0 & 0 & n_{33} = n_e \end{matrix} \tag{5.7}$$

This means that differently polarized refractive waves, in principle, can propagate along the same crystallographic direction (the same direction of wave vector $\boldsymbol{k}$) with different phase velocities $V_{p_1} = \frac{c}{n_{11}}$ or $V_{p_2} = \frac{c}{n_{33}}$. Furthermore, the existence of two refractive indices for two transverse polarizations situated in the plane perpendicular to the direction of light propagation implies that, in the general case, these two waves will spatially split when crossing the boundary between two media. This phenomenon is called *birefringence* (or double refraction) and was discovered in 1669 by **Erasmus Bartholinius**.

Let us investigate this issue in more detail. Suppose that light propagates in a uniaxial crystal along its main optic axis (the z-axis or x_3-axis), which per definition coincides with a high-order rotation axis (threefold, fourfold, or sixfold). In this case, two orthogonal polarizations will be parallel to the x_1- and x_2-axes, and hence, the respective waves will experience the same refractive index n_{11} and propagate with equal phase velocities $V_{p1} = \frac{c}{n_{11}}$ (see eq. (5.7)). Correspondingly, there is no birefringence, as in an isotropic body. For this reason, index n_{11} is called the ordinary refractive index $n_{11} = n_o$. Respectively, index n_{33}, offering the second phase velocity of light, $V_{p2} = \frac{c}{n_{33}}$ is called the extraordinary refractive index, $n_{33} = n_e$. Any other direction in uniaxial crystals is birefringent, allowing light propagation with two dissimilar velocities. It is worth noting that in the previously described light focusing system in brittle stars (see Section 3.2), individual spherical lenses made of rhombohedral calcite are arranged such that their main optic axis is vertical, that is, along the light flux coming from the upper water layers. Consequently, the birefringence, degrading the optical quality of the "eye" system, is avoided.

Birefringence in different materials can be analyzed using the dielectric ellipsoid shapes depicted in Figures 4.1–4.3 in Chapter 4. To avoid birefringence, we need to find specific directions of the light wave vector $\boldsymbol{k}$ for which perpendicular planes produce circular cross sections of the dielectric ellipsoid. In terms of vector $\boldsymbol{D}$, this means that within the plane perpendicular to vector $\boldsymbol{k}$, the projections D_1 and D_2 are equivalent, that is, they are characterized by the same elements of the tensor of dielectric permittivity ε_{ik} and, hence, the tensor of refractive index n_{ik}. Therefore, we must find appropriate planar cross sections of the representation ellipsoids in different symmetry systems, that is, the projections that cross the origin of the coordinate system and produce circular rims at the ellipsoid's surface. The normals to these circular cross sections will indicate the light propagation directions with no birefringence.

Starting with amorphs and crystals belonging to cubic symmetry system, we immediately recognize that all planar cross sections of the representation ellipsoid are circular since the representation ellipsoid is a sphere (see Figure 4.3). Hence, in these materials there is no birefringence effect in any direction of light propagation. Note that even in such highly symmetric materials, the birefringence can be intentionally induced (through the symmetry reduction, as explained below in Section 5.5) by applying stress, that is, via a photoelastic effect, or by applying an electric field, that is, via a quadratic electro-optic effect or linear electro-optic effect, the latter only in cubic crystals belonging to point groups *23* and $\bar{4}3$ m.

In uniaxial crystals, the dielectric ellipsoid is the ellipsoid of rotation (see Figure 4.2). Correspondingly, the tensor of refractive indices has two dissimilar diagonal elements n_o and n_e (see eq. (5.7)). In the case of ellipsoid of rotation, there is only one circular cross section (shadowed in Figure 4.2), which is perpendicular to the main optic axis (the x_3-axis). We recall that in uniaxial crystals, the main optic axis always coincides with the sole high-order rotation axis, that is, the fourfold, threefold, or sixfold one. In other words, in uniaxial crystals there is only one direction of light propagation with no birefringence, namely, along the high-order rotation axis. In that case, both waves with polarizations D_1 or D_2 will encounter the same refractive index (the ordinary refractive index), $n_o = n_{11}$. In any other direction the birefringence does occur, the maximum effect being for light propagating parallel to the planar circular cross-sectional area (shadowed in Figure 4.2).

For any direction of propagation within the shadowed area, one wave (the ordinary one) will be polarized in this plane and, hence, will "feel" the ordinary refractive index n_o, while the second wave (the extraordinary one) will be polarized along the main optic axis and, therefore, will be affected by the extraordinary refractive index, $n_e = n_{33}$. We stress that notations "ordinary wave (ray)" and "extraordinary wave (ray)" refer to the polarization orientation with respect to the main optic axis in a crystal, whereas the terms "σ-polarization" and "π-polarization" describe the polarization direction with respect to the scattering plane (the plane outlined by the respective wave vectors).

The birefringence effect in uniaxial crystals may be very strong. It is characterized by the ratio

$$\eta = \frac{2\Delta n}{n_e + n_o} = 2\frac{n_e - n_o}{n_e + n_o} \tag{5.8}$$

which in some crystals is more than 10%. For example, for tetragonal rutile (TiO_2), $n_e = 2.903$, $n_o = 2.616$, and $\eta = 10.4\%$, whereas for rhombohedral calcite ($CaCO_3$), $n_e = 1.4864$, $n_o = 1.6584$, and $\eta = -10.9\%$. Birefringence in calcite is illustrated in Figure 5.5. Some investigators claim that owing to the huge birefringence effect, large natural calcite crystals (so-called Iceland spar) were used by Vikings for navigation before the invention of the magnetic compass by Chinese sailors. According to these researchers, calcite crystals allowed for sun positioning (via crude sunlight polarimetry) even in partly cloudy sky conditions by monitoring the birefringent shadow of the slit fabricated at one of the crystal faces.

In crystals belonging to orthorhombic, monoclinic, or triclinic symmetry systems, the representation ellipsoid is triaxial (see Figure 4.1). Correspondingly, the tensor of refractive indices has the following shape:

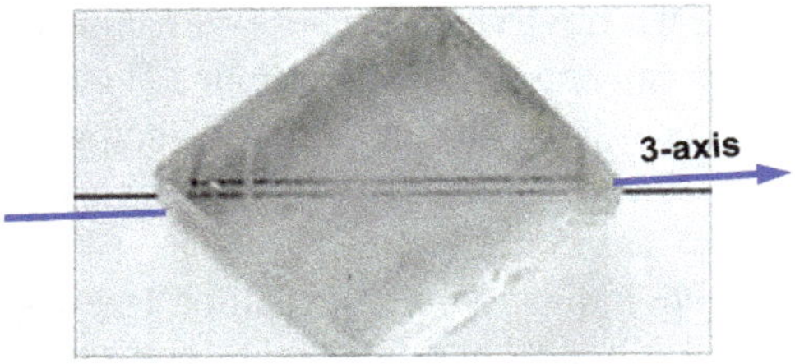

Figure 5.5: Birefringence in a single crystal of calcite.

$$n_{ik} = \begin{matrix} n_{11} & 0 & 0 \\ 0 & n_{22} & 0 \\ 0 & 0 & n_{33} \end{matrix} \quad (5.9)$$

As was mentioned in Chapter 4, in triaxial ellipsoids there are two circular planar cross sections (shadowed in Figure 4.1), inclined with respect to each other by some characteristic angle, which is defined (as is seen below) by the combination of tensor elements (5.9). Both these cross sections have radii (in terms of refractive indices) equal to the intermediate axis length (i.e., in between the lengths of the largest and the smallest ellipsoid axes).

Let us suppose, for the sake of convenience, that $n_{11} < n_{22} < n_{33}$. Accordingly, we can find the corresponding inclination angles by transforming the tensor (5.9) from the initial axes (x_1, x_2, x_3), which coincide with projections E_1, E_2, E_3 in Figure 4.1, to new axes (x'_1, x'_2, x'_3) rotated by angles α, β, γ with respect to the initial ones. If the inclination angle Θ is measured between the inclined circular projection and the old axis E_3 (see Figure 5.6), while angles α, β, γ are measured between the trace of this planar projection (blue line) in Figure 5.6 and the axes E_1, E_2, and E_3, then $\alpha = 90° - \Theta$, $\beta = 90°$, and $\gamma = \Theta$.

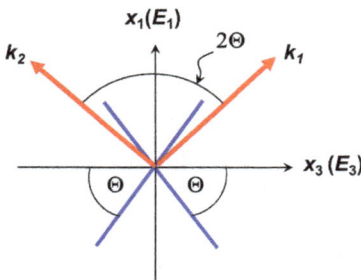

Figure 5.6: Angular relationships between the traces of the planar circular cross sections (blue solid lines) and directions of light propagation ($\boldsymbol{k}_1$ and $\boldsymbol{k}_2$) with no birefringence (red arrows) in biaxial crystals. The x_2-axis (parallel to E_2) is normal to the drawing.

The next step is transforming the tensor (5.9) to new axes (x'_1, x'_2, x'_3). For this purpose, we use the transformation law for tensors of the second rank (4.5). In our case, the transformation matrix f_{ik} is

$$f_{ik} = \begin{matrix} \cos\alpha & \cos\beta & \cos\gamma \\ f_{21} & f_{22} & f_{23} \\ f_{31} & f_{32} & f_{33} \end{matrix} \tag{5.10}$$

Combining eqs. (4.5) and (5.10) yields the modified refractive index n'_{11}:

$$n'_{11} = f_{1p} f_{1q} n_{pq} = f_{11}^2 n_{11} + f_{12}^2 n_{22} + f_{13}^2 n_{33} = n_{11}\cos^2\alpha + n_{22}\cos^2\beta + n_{33}\cos^2\gamma \tag{5.11}$$

Substituting the above-found angles α, β, γ (via the angle Θ) into eq. (5.11), one obtains the modified refractive index n'_{11} within the circular cross section:

$$n'_{11} = n_{11}\sin^2\Theta + n_{33}\cos^2\Theta \tag{5.12}$$

To avoid birefringence, n'_{11} should be equal to the intermediate refractive index n_{22}:

$$n'_{11} = n_{11}\sin^2\Theta + n_{33}\cos^2\Theta = n_{22} \tag{5.13}$$

Solving eq. (5.13) yields

$$\cos\Theta = \sqrt{\frac{n_{22} - n_{11}}{n_{33} - n_{11}}} \tag{5.14}$$

Respectively, this solution provides two inclination angles, Θ_1 and Θ_2, for two planar circular cross sections of the dielectric ellipsoid:

$$\Theta_1 = \arccos\sqrt{\frac{n_{22} - n_{11}}{n_{33} - n_{11}}} = \Theta \tag{5.15}$$

$$\Theta_2 = 180° - \Theta \tag{5.16}$$

Therefore, for crystals belonging to orthorhombic, monoclinic, or triclinic symmetry systems, there are two directions of wave propagation, inclined by an angle 2Θ with respect to each other (red arrows in Figure 5.6), the directions in which the birefringence effect does not occur. For this reason, such crystals are called biaxial crystals. In other directions of light propagation, a biaxial crystal is birefringent. Note that the magnitudes of the birefringence effect in biaxial crystals are much smaller than in uniaxial crystals. For example, orthorhombic topaz crystal ($Al_2SiO_4(F,OH)_3$) is characterized by $n_{11} = 1.618$, $n_{22} = 1.620$, and $n_{33} = 1.627$ that provides the largest relative difference of about 0.5% between the refractive indices (see, eq. (5.8)), that is, 20 times lower than in calcite.

5.3 Producing polarized light

In addition to what was discussed in Section 5.1, there are several other ways to generate polarized light besides using the **Brewster** angle. Polarized light, for example, can be produced employing materials in which a certain polarization type is strongly absorbed, while another one (orthogonal to the first) is not. This phenomenon of selective polarization-dependent light absorption is called light dichroism. Perhaps the most famous crystalline material of this kind is tourmaline, which is a borosilicate with additions of elements such as aluminum, iron, magnesium, sodium, lithium, or potassium. The uniaxial tourmaline crystals have a rhombohedral (trigonal) structure with a threefold rotation axis as the main optic axis. They produce a birefringence effect (see eq. (5.8)) of about 2%. When passing through a tourmaline crystal, an ordinary ray is absorbed on the length of about 1 mm, providing a pure extraordinary ray at the exit.

A much shorter absorption length (about 0.1 mm) is achieved with polaroid films, which were invented in 1928 by **Edwin Land** and led to the launching of the Polaroid company. In these films, **Land** used herapathite (iodoquinine sulfate) crystals with the complicated composition of $4QH_2^{2+} \cdot 3SO_4^{2-} \cdot 2I_3^- \cdot 6H_2O \cdot CH_3COOH$, where Q is quinine ($C_{20}H_{24}N_2O_2$). The I_3^- molecules are the light-absorbing entities and run in a zigzag chain along the absorption axis. These crystals have an orthorhombic structure, that is, they are biaxial from a polarization point of view. **Land** grounded the herapathite crystals (using ball milling) down to sub-micron sizes and then extruded them in polymers, such as nitrocellulose, to fabricate polarizing films for sunglasses and photographic filters. During the fabrication process of the film, the needle-like crystals were aligned by stretching or by applying electric or magnetic fields. With the crystals aligned, the film sheet becomes dichroic: it absorbs light that is polarized along the direction of crystal alignment and transmits light polarized perpendicular to it.

Most useful, however, are methods in which eliminating one sort of light polarization is based on total internal reflection; the latter described in Section 2.2. Such optical elements, polarizers and analyzers, are often made of calcite because it offers two essential benefits: a huge birefringence effect of nearly 11% (see eq. (5.8)) and availability of large size, pure natural crystals (such as Iceland spar calcite) of high optical quality. The most popular device based on this principle is called **Nicol** prism (or simply Nicol) after **William Nicol** who invented it in 1828.

A **Nicol** prism is produced from calcite scalenohedron formed by the {214}-type (or $\{21\bar{3}4\}$-type in the four-digit setting) crystallographic planes making a dihedral angle of 72° (see Figure 5.7).

Using appropriate cleavage and grinding procedures, a parallelepiped calcite block with an angle between two adjacent faces nearly equal to 68° is fabricated (see Figure 5.8). After that, this crystal block is cut into two pieces along its diagonal (see Figure 5.8) and then these two parts are bonded back using Canada balsam as a glue. The latter has a refractive index, $n_{Cb} = 1.54–1.55$, which is in between the extraordinary refractive index $n_e = 1.4864$ and ordinary refractive index $n_o = 1.6584$ of calcite, that is, $n_e < n_{Cb} < n_o$.

Figure 5.7: Calcite scalenohedron formed by the {214}-type crystallographic planes (or the {21$\bar{3}$4}-planes in the four-digit indexing system).

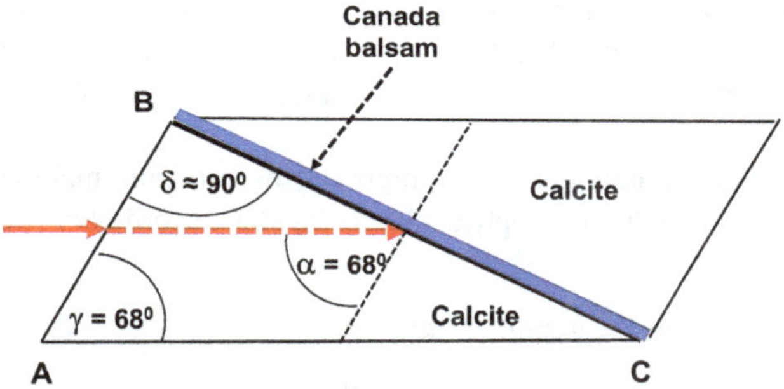

Figure 5.8: Fabricating a **Nicol** prism.

Therefore, for the ordinary ray (O-ray), Canada balsam is optically less dense than calcite, and it is possible to organize the total internal reflection of the O-ray at the calcite/balsam interface. If the exit surface for the O-ray is painted black, the σ-polarized O-ray will be strongly absorbed. As a result, only linearly polarized (π-polarized), extraordinary ray (E-ray) will exit from the prism close to the direction of the incident nonpolarized light (Figure 5.9).

Using eq. (2.27), we can calculate the critical angle of total internal reflection in this case as

$$\sin\alpha_c = \frac{n_{Cb}}{n_o} \tag{5.17}$$

which yields $\alpha_c = 68° - 69°$.

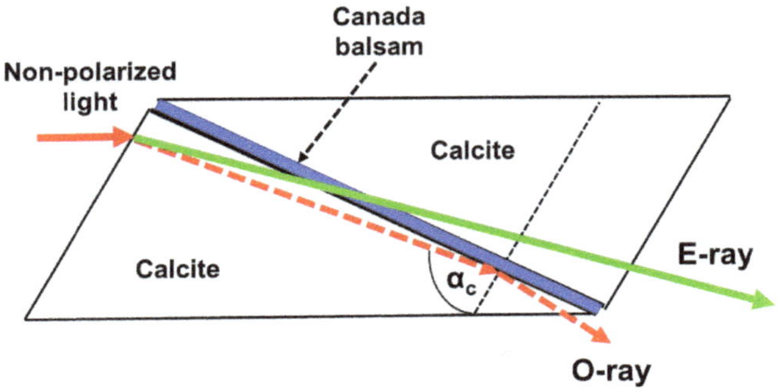

Figure 5.9: Working principle of a **Nicol** prism.

To create total reflection conditions, a calcite block, as noted above, is cut to have one angle of the parallelogram equal $\gamma = 68°$, as shown in Figure 5.8. Further, the length of one edge, *AC*, is chosen to be approximately three times larger than the length of edge *AB*. In this case, the angle $\angle ABC$ is nearly equal to 90°, that is, $\delta \approx 90°$ (see Figure 5.8). If incident light is entering the **Nicol** prism along the *AC* edge and there is no refraction on the air–calcite boundary, then looking at Figure 5.8 we immediately understand that the incident angle α at the calcite–balsam interface is $\alpha = \gamma = 68°$, that is, practically equals α_C.

Light refraction at the air–calcite boundary even improves the situation, making angle α larger than α_C (see Figure 5.10). In fact, applying **Snell**'s law for the ordinary ray here

$$\sin\alpha_0 = n_0 \sin\beta_0 \tag{5.18}$$

we find $\beta_0 = 13°$ for $\alpha_o \simeq 90° - 68° = 22°$. Considering triangle *NKM* in Figure 5.10, one obtains the incident angle at the calcite/balsam interface (point *M*) $\alpha \simeq \delta - \beta_0 \simeq 90° - 13° = 77° > \alpha_C$.

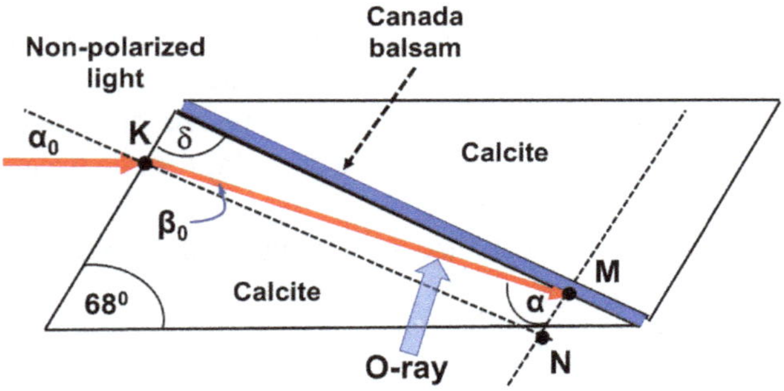

Figure 5.10: Angle calculation at the calcite/balsam interface in a **Nicol** prism.

There exist several other prism variants aimed at producing polarized light beams. All of them utilize dissimilar refraction of the ordinary and extraordinary rays at the boundary between two calcite blocks. One option is called a **Wollaston** prism (after **William Hyde Wollaston**), which consists of two right-angle prisms having a triangle-shaped base and glued together by Canada balsam (see Figure 5.11). The prisms retain mutually orthogonal optic axes, as depicted in Figure 5.11, the axes being also perpendicular to the direction of the incident light beam. The latter enters a **Wollaston** prism normally to one of the faces of the first prism (and perpendicular to the main optic axis in there) and transforms into two beams, the ordinary and extraordinary ones. Both beams propagate up to the Canada balsam layer in the spatially unsplit mode, but with different velocities dictated by the corresponding refractive indices, n_o and n_e.

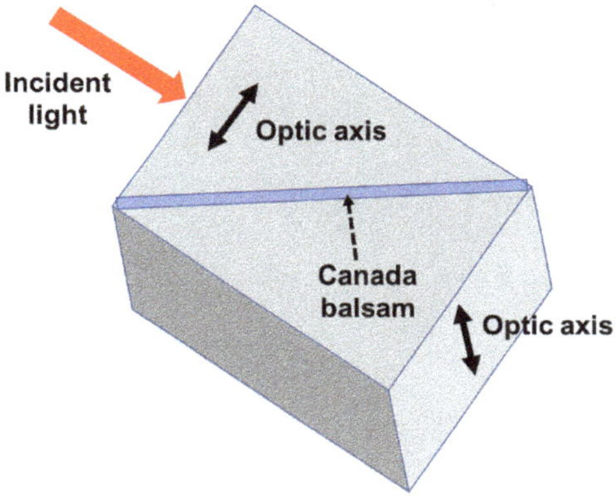

Figure 5.11: Principal design of a **Wollaston** prism.

The σ- and π-polarizations are, respectively, perpendicular and parallel to the main optic axis here (see Figure 5.12). Spatial spitting between two linearly polarized rays (σ-polarized and π-polarized rays) occurs at the calcite/balsam boundary due to the differences in the refractive indices. Separated light beams, that is, σ-polarized E-ray and π-polarized O-ray (both with respect to the orientation of the optic axis in the second calcite block), propagate until the end of the prism (see Figure 5.12). Here they escape the prism at different exit angles, the angular divergence (up to about 45°) being determined by the wedge angle of the base triangle and the wavelength of the light. Because of this feature, **Wollaston** prisms are commonly used as beam splitters and deflectors in different optical devices and systems.

A **Rochon** prism (invented by **Abbé Rochon**) is similar to the **Wollaston** prism, but the incident light beam is parallel to the main optic axis in the first calcite block (see Figure 5.13). As a result, one ray (being the ordinary one in both blocks) passes through the prism with no refraction. Therefore, at the exit we have the π-polarized O-ray propagating along the direction of the incident light. The second ray, which

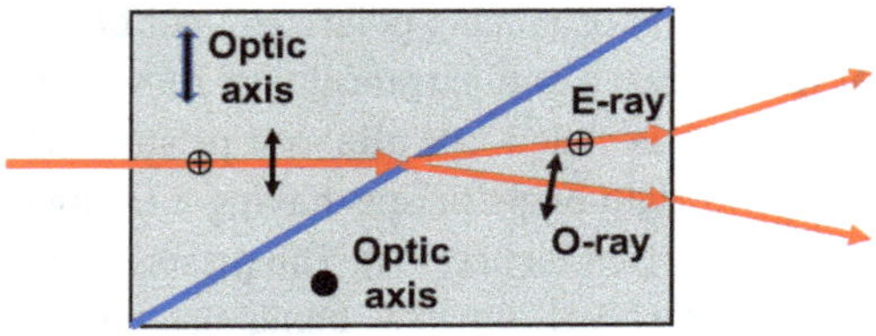

Figure 5.12: How a **Wollaston** prism operates.

propagates as the σ-polarized O-ray within the first block, with polarization perpendicular to the main optic axis in there, transforms, after refraction at the calcite/balsam interface, into the σ-polarized E-ray with polarization parallel to the main optic axis in the second calcite block. For this reason, this ray substantially deviates from the direction of the incident beam (see Figure 5.13).

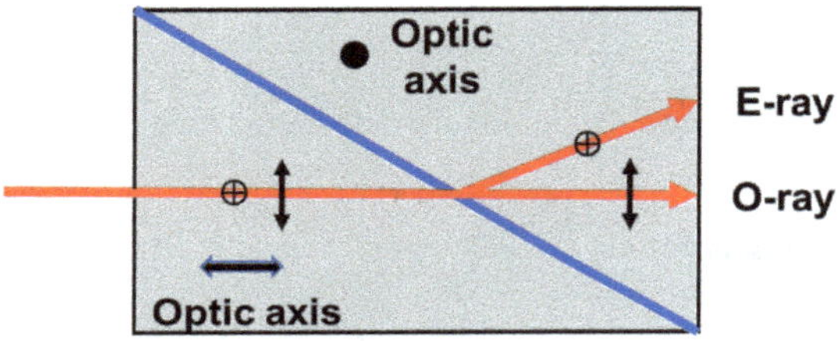

Figure 5.13: A **Rochon** prism in action.

In a **Glan–Thompson** prism, the main optic axes in both calcite blocks are parallel to the entrance crystal face and perpendicular to the plane of drawing (Figure 5.14). Also in this design, the O- and E-rays propagate unsplit in the first block and split into separate rays after refraction at the calcite/balsam interface. At the exit, we have a nondeviated, σ-polarized E-ray propagating in the direction of the incident beam since it remains extraordinary in both calcite blocks.

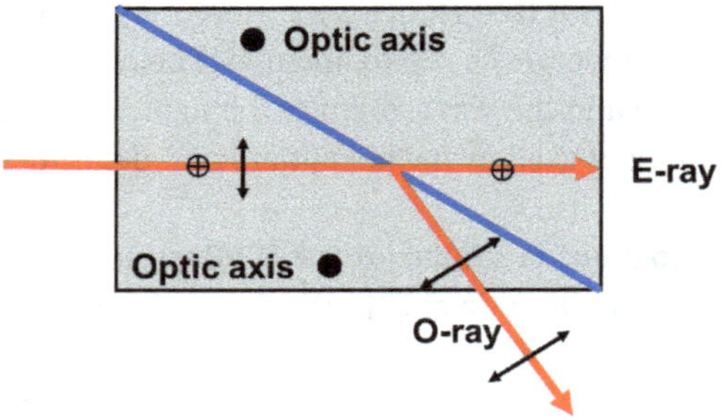

Figure 5.14: A **Glan–Thompson** prism in action.

The **Glan–Foucault** prism, named after **Leon Foucault** (Figure 5.15) is very similar to the **Glan–Thompson** prism with one exception: the two calcite prisms are separated by an air gap instead of Canada balsam. Again, only the σ-polarized E-ray is transmitted straightforwardly through the prism couple; the π-polarized O-ray is rejected by the total internal reflection at the calcite/air boundary.

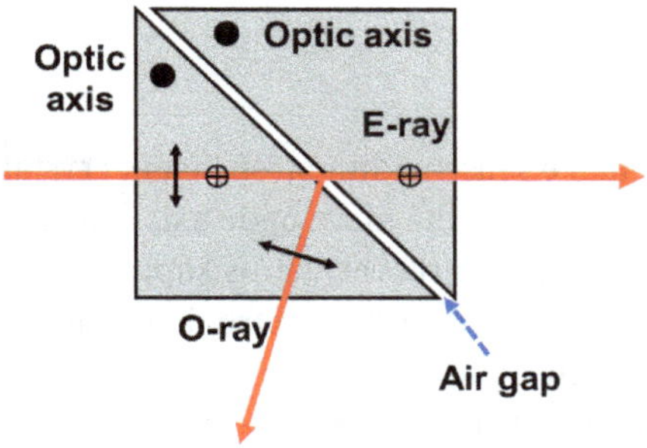

Figure 5.15: A **Glan–Foucault** prism in action.

A twist was introduced into this design by **John Archard** and **A. M. Taylor** who invented the **Glan–Taylor** prism (Figure 5.16). In their design, the main optic axes in both calcite parts are parallel to the entrance calcite face and perpendicular to the incident light direction. Correspondingly, at the exit one obtains the nondeviated, π-polarized E-ray propagating in the direction of the incident beam.

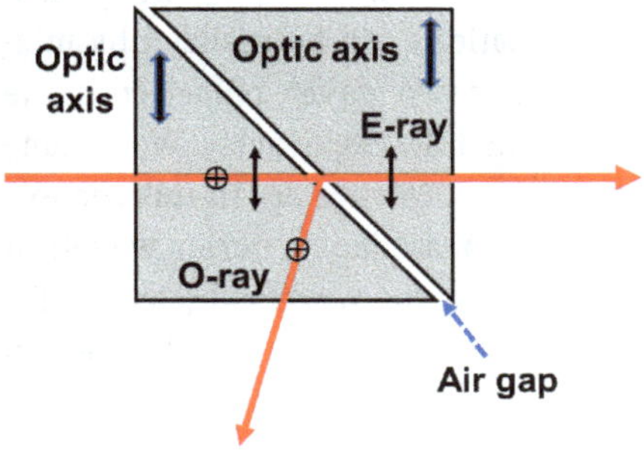

Figure 5.16: A **Glan–Taylor** prism in action.

It is also worth mentioning another modification of a **Wollaston** prism, the so-called **Nomarski** prism invented by **Georges Nomarski** for use in a microscope named after him (see Section 9.2). Similar to a **Wollaston** prism, the **Nomarski** prism is composed of two calcite blocks glued together with Canada balsam (see Figure 5.17). In the second block, the optic axis is perpendicular to the ray propagation (and normal to the drawing plane in Figure 5.17). In the first block, the oblique optic axis is oriented in the plane of the drawing. The **Nomarski** prism acts as a focusing element for O- and E-rays at the exit (as shown in Figure 5.17). This feature is used in the **Nomarski** microscope design (for details see Section 9.2).

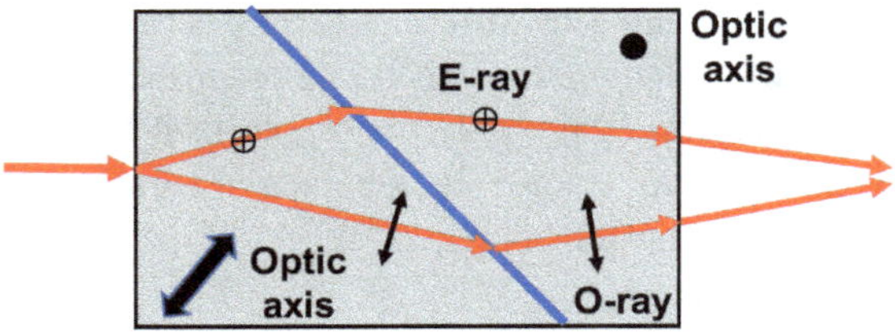

Figure 5.17: Working principle of a **Nomarski** prism.

We stress that the terms O-rays and E-rays make sense within birefringent crystal only since only here can their polarizations be related to the main optic axis. Outside a crystal, these terms become physically meaningless. Moreover, let us suppose that the O-ray exits a crystal and, after traveling in air (or some other isotropic medium), enters another birefringent crystal with the main optic axis being rotated by 90° (about the wave vector direction) with respect to that in the first one. In the second crystal, this ray will behave as the E-ray, with polarization parallel to the optic axis in there, as we saw, for example, in a **Rochon** prism (see Figure 5.13). This fact is widely used in the design of crossed polarizers for polarization microscopy (see Section 5.7).

5.4 Circular polarization and rotation of the polarization plane

Earlier we only spoke about linear polarization (sometimes also called plane polarization), but, in principle, more complex light polarizations can be realized by mixing two orthogonal linear polarizations. Let us consider two waves, polarized in the x- and y-directions and having different amplitudes A_1 and A_2, respectively. We assume that these waves are coherent possessing a steady phase difference $\Delta\varphi$. In this context, the term "coherence" means that there are no stochastic phase jumps during wavefield propagation (see also Chapter 7). Consequently, to calculate the wavefield (electric displacement field $\boldsymbol{D}$) in a medium, we must summate two waves, which are changing in time as:

$$D_x = A_1 \cos(\omega t) \tag{5.19}$$

$$D_y = A_2 \cos(\omega t + \Delta\varphi) \tag{5.20}$$

Equations (5.19) and (5.20) provide a parametric description of the shape of the curve in the (x,y)-plane indicating the ends of vectors $\boldsymbol{D}(D_x, D_y)$ (see Figure 5.18). If the phase difference $\Delta\varphi = 0$, then $\frac{D_x}{A_1} = \frac{D_y}{A_2}$ and one obtains linear polarization

$$D_y = \frac{A_2}{A_1} D_x \tag{5.21}$$

inclined with respect to the x- and y-axes, the inclination angle being defined by the amplitude ratio $\frac{A_2}{A_1}$ (see Figure 5.18a).

If the phase difference $\Delta\varphi = 90°$, then

$$D_x = A_1 \cos(\omega t) \quad (5.22)$$

$$D_y = -A_2 \sin(\omega t) \quad (5.23)$$

and the shape of the curve $\boldsymbol{D}(D_x, D_y)$ is elliptic

$$\frac{D_x^{\ 2}}{A_1^{\ 2}} + \frac{D_y^{\ 2}}{A_2^{\ 2}} = \cos^2(\omega t) + \sin^2(\omega t) = 1 \quad (5.24)$$

that is, one obtains elliptic light polarization (see Figure 5.18b). If, in addition, the wave amplitudes are equal, that is, $A_1 = A_2 = A$, the ellipse transforms into the circle

$$D_x^{\ 2} + D_y^{\ 2} = A^2 \quad (5.25)$$

and the light polarization becomes circular (see Figure 5.18c). Even for equal amplitudes $A_1 = A_2$, but $\Delta\varphi$ differing from $\Delta\varphi = 90°$ and $\Delta\varphi = 0$, we get elliptic polarization.

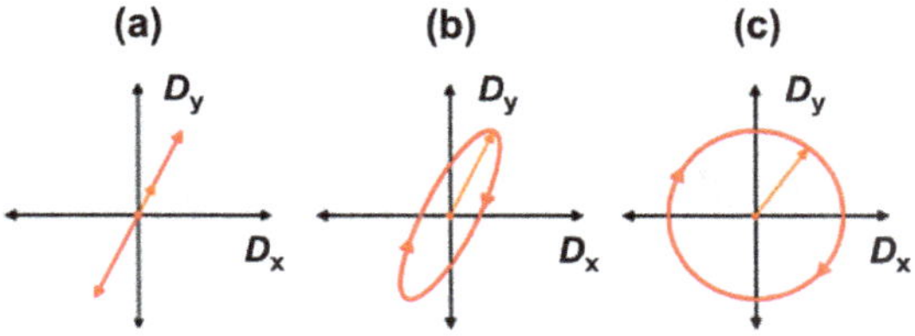

Figure 5.18: Summation of two orthogonal linear polarizations providing: (a) linear polarization; (b) elliptic polarization; and (c) circular polarization.

In fact, in the general case, we can rewrite (5.20) as

$$D_y = A_2[\cos(\omega t)\cos(\Delta\varphi) - \sin(\omega t)\sin(\Delta\varphi)] = A_2[\cos(\omega t)\cos(\Delta\varphi) - \sqrt{1 - \cos^2(\omega t)}\sin(\Delta\varphi)] \quad (5.26)$$

Substituting eq. (5.19) into eq. (5.26) yields

$$\frac{D_y}{A_2} = \frac{D_x}{A_1}\cos(\Delta\varphi) - \sqrt{1 - \left(\frac{D_x}{A_1}\right)^2}\sin(\Delta\varphi) \quad (5.27)$$

or

$$\sqrt{1 - \left(\frac{D_x}{A_1}\right)^2}\sin(\Delta\varphi) = \frac{D_x}{A_1}\cos(\Delta\varphi) - \frac{D_y}{A_2} \quad (5.28)$$

After squaring both parts of eq. (5.28), one obtains

$$\left(\frac{D_x}{A_1}\right)^2\cos^2(\Delta\varphi)+\left(\frac{D_x}{A_1}\right)^2\sin^2(\Delta\varphi)+\left(\frac{D_y}{A_2}\right)^2-2\frac{D_x}{A_1}\cdot\frac{D_y}{A_2}\cos(\Delta\varphi)=\sin^2(\Delta\varphi)$$

and finally,

$$\left(\frac{D_x}{A_1}\right)^2+\left(\frac{D_y}{A_2}\right)^2-2\frac{D_x}{A_1}\cdot\frac{D_y}{A_2}\cos(\Delta\varphi)=\sin^2(\Delta\varphi) \tag{5.29}$$

If the amplitudes are equal, that is, $A_1 = A_2 = A$, then

$$(D_y - D_x)^2 + 4D_x \cdot D_y \sin^2\left(\frac{\Delta\varphi}{2}\right) = A^2 \sin^2(\Delta\varphi) \tag{5.30}$$

For all $\Delta\varphi$-values differing from $\Delta\varphi = 0$ or $\Delta\varphi = 90°$, this is the ellipse equation. To prove it, let us compose new polarization states

$$\begin{aligned} P_1 &= D_y - D_x \\ P_2 &= D_y + D_x \end{aligned} \tag{5.31}$$

Correspondingly

$$\begin{aligned} D_y &= \frac{P_2 + P_1}{2} \\ D_x &= \frac{P_2 - P_1}{2} \end{aligned} \tag{5.32}$$

Substituting definition (5.32) into eq. (5.30), we find the ellipse equation in the new axes (P_1, P_2) as

$$\left[\frac{P_1}{2A\sin\left(\frac{\Delta\varphi}{2}\right)}\right]^2+\left[\frac{P_2}{2A\cos\left(\frac{\Delta\varphi}{2}\right)}\right]^2=1 \tag{5.33}$$

When $\Delta\varphi = 90°$, eq. (5.33) converts into a purely circular form

$$\frac{P_1^2}{2}+\frac{P_2^2}{2}=1 \tag{5.34}$$

or substituting back the definitions (5.31) to circular polarization in terms of D_x and D_y (see eq. (5.25)).

It is worth noting that it is possible not only to represent circular (elliptic) polarization as the sum of two orthogonal linear polarizations, but in reverse to separate a linear polarization into two circular ones, rotated clockwise and counterclockwise:

$$P_1 = A[\cos(\omega t) + \sin(\omega t)]$$
$$P_2 = A[\cos(\omega t) - \sin(\omega t)] \tag{5.35}$$

The respective linear polarizations are

$$D_1 = \frac{P_1 + P_2}{2} = A\cos(\omega t)$$
$$D_2 = = \frac{P_1 - P_2}{2} = A\sin(\omega t) \tag{5.36}$$

We will use this presentation in Section 5.6, considering the rotation of the polarization plane.

All this leads to the important question of how to arrange the phase difference $\Delta\varphi = 90°$ necessary for producing circular polarization. This can be done using the differences between phase velocities of light propagation in birefringent crystals. Let us suppose that the wave propagation direction in a uniaxial crystal is in the plane perpendicular to its main optic axis. As we know, in this case, light splits into ordinary and extraordinary waves propagating with phase velocities $V_{p1} = \frac{c}{n_o}$ and $V_{p2} = \frac{c}{n_e}$, respectively. If the crystalline plate has thickness T, then at the exit, the accumulated phase difference $\Delta\varphi$ between these two waves is

$$\Delta\varphi = \omega\left(\frac{T}{V_{p1}} - \frac{T}{V_{p2}}\right) = \omega T\frac{n_o - n_e}{c} = 2\pi T\frac{n_o - n_e}{\lambda_o} \tag{5.37}$$

where λ_o is the light wavelength in a vacuum. To provide phase difference $\Delta\varphi = \frac{\pi}{2} = 90°$, the plate thickness should be

$$T_{\frac{1}{4}} = \frac{\lambda_o}{4(n_o - n_e)} \tag{5.38}$$

Such a crystalline plate is called a quarter-wave plate (or quarter-wave phase retarder).

Half-wave plates (phase retarders) are also commonly used. They produce the phase difference between the ordinary and extraordinary waves $\Delta\varphi = \pi = 180°$. Respectively, the thickness of a half-wave plate is

$$T_{\frac{1}{2}} = \frac{\lambda_o}{2(n_o - n_e)} \tag{5.39}$$

If polarization vector $\boldsymbol{P}_1$ is initially inclined by angle Ω to the optic axis, then after passing through a half-wave plate, the polarization $\boldsymbol{P}_2$ will be rotated by 180° about the optic axis. This means that the angular difference between final and initial polarization directions is 2Ω (see Figure 5.19). Setting the optic axis of a half-wave retarder at 45° to the initial polarization plane (polarization direction) results in a polarization rotation of 90°.

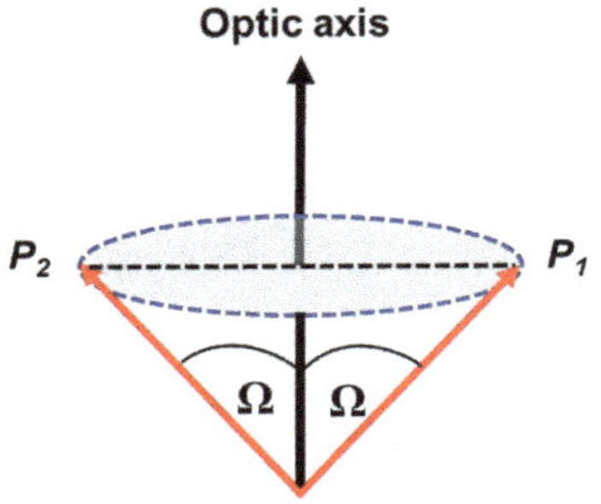

Figure 5.19: Rotation of the inclined polarization vector by 90° with the aid of the half-wave plate.

Practically, to create or compensate for the desired phase difference, two sliding wedges made of uniaxial crystals are often used (see Figure 5.20). In this way, the required value of the plate's effective thickness can be achieved, which is thereafter converted to the phase difference according to eq. (5.37).

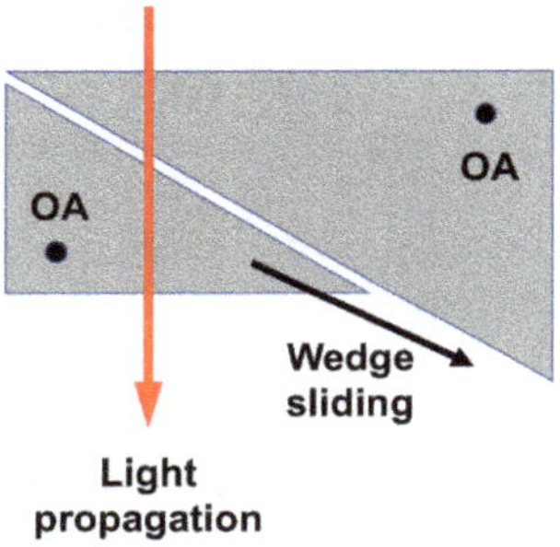

Figure 5.20: Changing the effective plate thickness by wedge sliding. Abbreviation OA stands for optic axis.

5.5 Induced birefringence

We can play with birefringence by applying an external electric field $\boldsymbol{E}$, which, in general, reduces the overall crystal symmetry. As already mentioned, there exist two electro-optic effects: the linear one (the **Pockels** effect) and quadratic one. Both effects define the change Δn_{ik} of the refractive tensor n_{ik}, as a function of the electric field components or its quadratic form. Correspondingly, the **Pockels** effect is described by the tensor of third rank r_{ikl} that connects vector E_i and the tensor of second rank Δn_{ik}

$$\Delta n_{ik} = r_{ikl} E_l \tag{5.40}$$

Respectively, the quadratic electro-optic effect is described by the tensor of fourth rank g_{iklm} connecting the tensor of second rank Δn_{ik} and two vectors

$$\Delta n_{ik} = g_{iklm} E_l E_m \tag{5.41}$$

At this point, we remind the reader about the earlier mentioned transformation law for tensors of second rank (eq. (4.5)), in which the transformation matrix f_{ik} enters twice. Correspondingly, in the transformation law for tensors of third rank, the transformation matrix f_{ik} enters three times:

$$r'_{ikl} = f_{ip} f_{kq} f_{ls} r_{pqs} \tag{5.42}$$

whereas in the case of tensors of fourth rank, it enters four times:

$$g'_{iklm} = f_{ip} f_{kq} f_{lr} f_{ms} g_{pqrs} \tag{5.43}$$

This implies that under an inversion operation (4.7), all components of the tensor of third rank do not "survive" and hence should be zero. Here, this means that the **Pockels** effect does not exist in center-symmetric crystals and isotropic media.

In contrast, the quadratic electro-optic effect is insensitive to the presence or absence of an inversion center. Practically, birefringence can be induced even in amorphous materials or liquids, which under an electric field pick up some properties of uniaxial crystals with the optic axis situated along the applied electric field. Such induced birefringence is called the **Kerr** effect. In this case,

$$n_o - n_e = KE^2 \tag{5.44}$$

where K is the proportionality factor. It is also convenient to calculate the phase difference $\Delta\varphi$ accumulated over the propagation path l

$$\Delta\varphi = \frac{2\pi}{\lambda_0}(n_o - n_e)l = \frac{2\pi}{\lambda_0} KlE^2 \tag{5.45}$$

The latter equation can be rewritten as

$$\Delta\varphi = 2\pi BlE^2 \tag{5.46}$$

where B is the wavelength-dependent **Kerr** constant. In some polar liquids, the value of B can be rather high, about 10^{-12} m/V^2.

Under applied electric field $\boldsymbol{E}$, polar molecules try to orient themselves to have dipole moments along vector $\boldsymbol{E}$, and such liquid becomes strongly anisotropic. The preferred orientation of polar molecules acts against thermal molecular movements. For this reason, the **Kerr** constant diminishes as the temperature increases. Note that the molecular reorganization upon application of an electric field is very fast, characteristic time being on the order of 0.1 ns. Consequently, **Kerr** cells are used for fast manipulation with light beams.

5.6 Optical activity

This term refers to circular birefringence and/or circular dichroism, which in turn implies the difference, respectively, in phase velocities and/or light absorption for the right-hand (clockwise-rotated) or left-hand (counterclockwise-rotated) circular polarizations. As was proposed by **Fresnel**, different velocities lead to the new phenomenon of the rotation of linear polarization (sometimes called rotation of the polarization plane) during light propagation. This phenomenon was first observed in quartz in 1811 by **François Arago**. Later, it turned out that to be optically active, a material must be composed of asymmetric (chiral) molecules or asymmetric atomic groups. For example, optical activity may be observed in so-called enantiomorphic crystals, such as quartz, existing in the left-hand and right-hand chiral forms, which are related to each other by mirror symmetry. Note that enantiomorphic point groups are 1 (triclinic symmetry system), 2 (monoclinic system), 222 (orthorhombic system), 23 (trigonal system), 622 (hexagonal system), and 432 (cubic system). Namely, using the existence of the two natural enantiomorphs of quartz crystals (point group 23), in 1820 **John Herschel** discovered that these enantiomorphic crystalline forms rotate linear polarization by equal amounts but in opposite directions.

To understand more deeply the **Fresnel**'s splendid idea, we recall that linear polarization can be presented as the superposition of two circular polarizations with equal amplitudes, the right-hand and left-hand rotated ones (see eqs. (5.35) and (5.36)). In an inactive material, the polarization vectors $\boldsymbol{P}_1$ and $\boldsymbol{P}_2$ in these waves are rotated, respectively, clockwise and counterclockwise, but with equal phase velocities (see Figure 5.21). As a result, the sum $\boldsymbol{P}$ of vectors $\boldsymbol{P}_1$ and $\boldsymbol{P}_2$ will remain oriented in the same plane m. In contrast, in an active material, these velocities are different, and we will find the sum $\boldsymbol{P}$ of vectors $\boldsymbol{P}_1$ and $\boldsymbol{P}_2$ in plane m^* being rotated with respect to the initial plane m (see Figure 5.21). The rotation angle $\Delta\Theta$ is proportional to the light path l in an active material

$$\Delta\Theta = a_r l \tag{5.47}$$

In the case of light propagation along the main optic axis in quartz, the proportionality factor equals $a_r = 21.7°$ per mm, for yellow light with $\lambda_o = 589$ nm, and $a_r = 48.9°$ per mm, for purple light with $\lambda_o = 404.7$ nm.

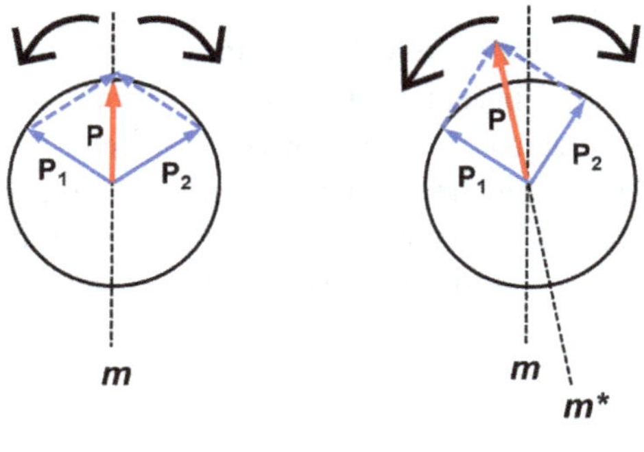

Figure 5.21: Rotation of the polarization plane.

Liquids and gases can be optically active as well, if they contain chiral molecules. Note that most organic molecules are, in fact, chiral, prominent examples being amino acids and sugars. In chemistry, the right-hand (clockwise) rotating enantiomorphic molecules are called dextrorotary ones, whereas their left-hand (counterclockwise) rotating enantiomorphs are called levorotary ones. In the laboratory, one can synthesize both enantiomorphs;

the mixture containing equal amounts of them would be optically inactive. Nature, however, often prefers the single enantiomorphic form, as in the case of simple sugars, which naturally consist of only one stereoisomer. For example, glucose is dextrorotary, while fructose is levorotary. Therefore, by measuring the polarization rotation of light passing through a cuvette filled with a specific sugar solution, one can find the sugar concentration c_s, according to the following relationship

$$\Delta\Theta = [a_r]c_s l \qquad (5.48)$$

where $[a_r]$ is the specific rotation coefficient and l is the cuvette length.

5.6.1 Faraday effect

Materials (even optically nonactive) may acquire the ability to rotate the polarization plane upon application of a magnetic field. This magneto-optic effect was discovered by **Michael Faraday** in 1845 and is considered to be the first direct proof that light and electromagnetism are strongly interrelated. Polarization rotation is observed when light propagates along the vector of magnetic induction $\boldsymbol{B}$. The rotation angle $\Delta\Theta$ is proportional to the B-magnitude and the light path length l in a material:

$$\Delta\Theta = V_H lB \qquad (5.49)$$

The wavelength-dependent proportionality factor V_H is called the **Verdet** constant and is measured in rad/(T m). It reaches its highest values in materials containing paramagnetic ions such as terbium.

The physics behind the **Faraday** effect is related to the precession of electron orbital movements upon magnetic field application. As a result, the velocities of the left-hand and right-hand circularly polarized waves become different, which as we already know, leads to the rotation of the polarization plane. **Faraday** rotators are used, for instance, as optical "isolators" (i.e., devices permitting only straightforward light propagation in one direction and not backward) for optical telecommunications and specific laser applications.

5.6.2 Magnetic circular dichroism and circular polarization of X-rays

A phenomenon closely related to the **Faraday** effect is magnetic circular dichroism (MCD). This means dissimilar absorption of the right-hand and left-hand polarized light in materials upon application of an external magnetic field. As in the case of the **Faraday** effect, MCD is observed when a magnetic field is situated along the light propagation direction. MCD is used to investigate the fine structure (e.g., spin–orbit mixing) of the electron energy shells in magnetic ions. Using X-rays, deeper energy states can be probed. For XMCD, the soft X-rays with energy below 1 keV are commonly used.

Since in the X-ray range, the refractive index is practically equal $n \approx 1$, producing circular polarization is not an easy task, and therefore, special methods, differing from those known for visible light, have been developed. One method is based on the manipulation of an electron beam circulating within the synchrotron storage ring. To obtain high X-ray intensity, special devices – wigglers and undulators – are inserted to produce additional (lateral) acceleration of electrons moving with velocity being very close to the speed of light. These devices consist of periodic arrays of magnets, causing the electrons to oscillate around the average circular trajectory. In an elliptical polarization undulator, the magnetic field vector rotates as the electrons pass through the device, resulting in a spiral electron trajectory about the central axis and, hence, circular (or elliptic) polarization of the issuing X-rays.

Another method employs perfect diamond crystals acting as quarter-wave plates. To attain the quarter-wave plate characteristics in the X-ray energy range and to produce circular X-ray polarization, dynamical X-ray diffraction is used. This issue will be discussed in more detail in Section 15.2.

5.7 Polarization microscopy

Polarization microscopy entails microscopic observations using polarized light. Respective microscopes are equipped with polarizers and analyzers that create polarized light and analyze it at the exit of an optical system. A rather easier solution is to use so-called wire grid polarizers installed perpendicular to the light beams (see Figure 5.22). This type of polarizer comprises a system of evenly oriented metallic wires, in which an electric field $\boldsymbol{E}$ of the incoming light creates the forced vibrations of electrons, if vector $\boldsymbol{E}$ is parallel to the wires ($E_{||}$). In turn, the electron vibrations generate the secondary electromagnetic waves, which cancel out the incoming wave and, hence, stop its transmission. If vector $\boldsymbol{E}$ is perpendicular to the wires ($E_{\perp}$), there are no electron vibrations, and the light is transmitted through the system.

Birefringent crystals and devices (prisms), fabricated on their basis, are most widely used as polarizers (see Section 5.3). To analyze the polarization state of the light behind the sample, the same kind of prisms is utilized with an option of rotating the optic axis about the direction of the propagating light. For example, if at the entrance we have linearly (plane) polarized light and the sample does not change it, then rotating the analyzer, the light extinction (vanishing) will occur twice over the rotation period, that is, when the optic axes in the polarizer and analyzer are orthogonal to each other. In other words, we will see a dark field of view in our microscope when polarizers/analyzers are crossed. More specifically, if the polarizer rejects, let us say, the O-ray and transmits an E-ray, the latter in the second prism with the optic axis rotated by 90° becomes the O-ray and will be rejected in there.

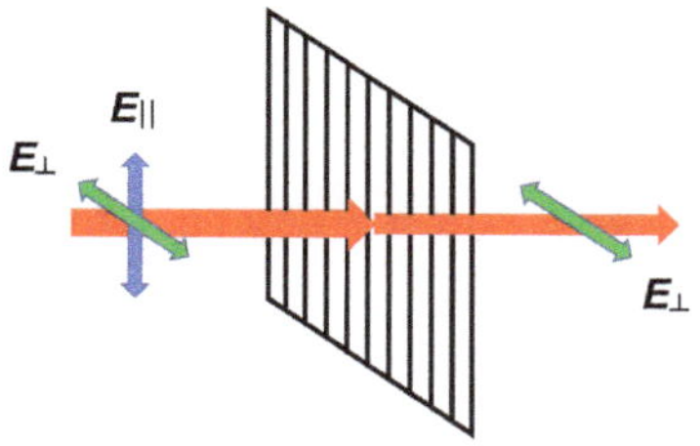

Figure 5.22: A grid polarizer in action.

For circularly polarized light, its intensity does not change upon analyzer rotation. The same behavior will be observed for nonpolarized light. To distinguish between these two cases, the light is first passed through a quarter-wave plate, which transforms a circularly polarized light into a linearly polarized light. The latter, as we know, does reveal extinctions upon the analyzer's rotation. In contrast, a nonpolarized light remains nonpolarized after passing through a quarter-wave plate, and therefore, there is no intensity change upon the analyzer's rotation.

The situation becomes more complicated (and more fascinating) if a uniaxial crystalline specimen is placed between the polarizer and analyzer (Figure 5.23). As we know, linearly polarized light after passing through such a specimen, in general, transforms into an elliptically polarized one. Correspondingly, the light intensity behind analyzer will depend on three angles (see Figure 5.23): angle Ω_p between the initial polarization plane (P) and optic axis (OA) of the specimen; angle Ω_a between the polarization planes in the polarizer (P) and analyzer (A); and the phase difference angle $\Delta\varphi = \frac{2\pi}{\lambda_0}(n_o - n_e)T$ between the ordinary (O) and extraordinary (E) rays.

The phase difference angle $\Delta\varphi$ is created when the O- and E-rays pass through the specimen's thickness T (see eq. (5.37)).

In fact, the light coming from the polarizer has polarization vector $\boldsymbol{D}$ along the polarizer plane, marked as P in Figure 5.23. Passing through the specimen, it splits into two components, the ordinary and extraordinary ones, with projections D_o and D_e, to be, respectively, perpendicular and parallel to the optic axis (OA) of the specimen. Correspondingly

$$D_e = D\cos\Omega_p$$
$$D_o = D\sin\Omega_p \tag{5.50}$$

Projections of the amplitudes D_e and D_o, on the polarization plane (A) of the analyzer (see Figure 5.23) will pass through the latter producing the following wave components

$$D_{ea} = D\cos\Omega_p\cos(\Omega_p + \Omega_a)$$
$$D_{oa} = D\sin\Omega_p\sin(\Omega_p + \Omega_a) \tag{5.51}$$

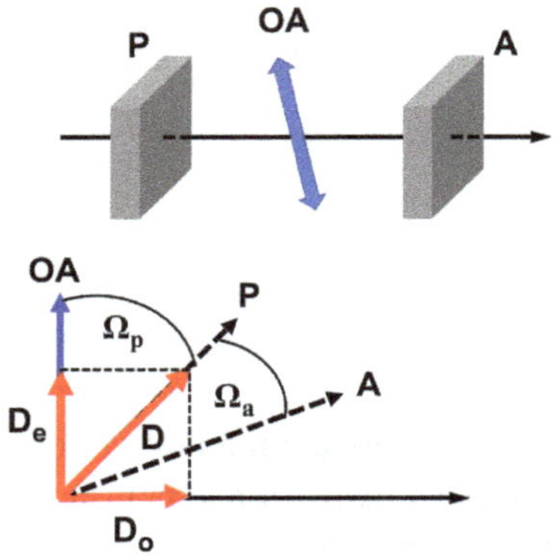

Figure 5.23: Principal scheme and action of a polarization microscope. P, A, and OA stand, respectively, for the polarizer, analyzer, and optic axis of the investigated specimen.

At the exit, these components will mix, creating a wavefield D_a, which also depends on the phase difference angle $\Delta\varphi = \frac{2\pi}{\lambda_0}(n_o - n_e)T$:

$$D_a = D_{oa} + D_{ea}e^{i\Delta\varphi} = D \sin\Omega_p \sin(\Omega_p + \Omega_a) + e^{i\Delta\varphi} D \cos\Omega_p \cos(\Omega_p + \Omega_a) \tag{5.52}$$

Respectively, the transmitted intensity I_a equals

$$\begin{aligned} I_a = D^*{}_a D_a = D^2 [&\sin^2(\Omega_p)\sin^2(\Omega_p + \Omega_a) + \cos^2(\Omega_p)\cos^2(\Omega_p + \Omega_a) \\ &+ 2\sin\Omega_p \sin(\Omega_p + \Omega_a)\cos\Omega_p \cos(\Omega_p + \Omega_a)\cos(\Delta\varphi)] \end{aligned} \tag{5.53}$$

Using standard trigonometric relationships

$$\cos(\Delta\varphi) = 1 - 2\sin^2\frac{\Delta\varphi}{2}$$

and

$$\sin\Omega_p \sin(\Omega_p + \Omega_a) + \cos\Omega_p \cos(\Omega_p + \Omega_a) = \cos(\Omega_p + \Omega_a - \Omega_p) = \cos(\Omega_a)$$

we can rewrite eq. (5.53) as

$$I_a = D^2\left\{\cos^2(\Omega_a) - \sin(2\Omega_p)\sin[2(\Omega_p + \Omega_a)]\sin^2\left(\frac{\Delta\varphi}{2}\right)\right\} \tag{5.54}$$

Let us analyze the key equation (5.54). With no specimen ($\Delta\varphi = 0$), we get

$$I_a = D^2\cos^2(\Omega_a) \tag{5.55}$$

This is the well-known **Malus** law, named after **Étienne-Louis Malus**, who discovered it at the beginning of the nineteenth century. This law describes the intensity of the linearly polarized light passing through the second polarizer/analyzer. If the polarization plane of the incoming light coincides with the polarization plane in the analyzer ($\Omega_a = 0$), the transmitted intensity is maximal, $I_a = D^2$. For $\Omega_a = 90°$, intensity $I_a = 0$.

Placing a specimen between the polarizer and analyzer, we generally introduce a phase shift $\Delta\varphi \neq 0$. If the polarizer (P) and analyzer (A) polarization planes are parallel, that is, $\Omega_a = 0$, eq. (5.54) converts into

$$I_{||} = D^2\left[1 - \sin^2(2\Omega_p)\sin^2\left(\frac{\Delta\varphi}{2}\right)\right] \tag{5.56}$$

For the crossed polarizer/analyzer geometry $\Omega_a = 90°$, one obtains the complimentary expression

$$I_{\perp} = D^2\left[\sin^2(2\Omega_p)\sin^2\left(\frac{\Delta\varphi}{2}\right)\right] \tag{5.57}$$

If the optic axis of the specimen is inclined by 45° with respect to the polarizer plane (P), that is, $\Omega_p = 45°$, then

$$I_{||} = D^2\left[1 - \sin^2\left(\frac{\Delta\varphi}{2}\right)\right] = D^2\cos^2\left(\frac{\Delta\varphi}{2}\right) \tag{5.58}$$

$$I_{\perp} = D^2\sin^2\left(\frac{\Delta\varphi}{2}\right) \tag{5.59}$$

We see that in this case, the transmitted intensities (5.58) and (5.59) are directly determined by the phase angle (5.37)

$$\Delta\varphi = \frac{2\pi}{\lambda_o}(n_o - n_e)T \tag{5.60}$$

which contains information on the thickness T of the specimen and the difference between the refractive indices, $\Delta n = n_o - n_e$. Note that the difference Δn is wavelength-dependent. In addition, light wavelength λ_o straightforwardly enters the expression of $\Delta\varphi$ (see eq. (5.60)). This means that under white light illumination, the colors in the exit picture will change continuously upon analyzer rotation. Disappearance of certain colors can be used for determining the birefringence Δn, if the crystal thickness T is known.

For polycrystalline samples, the exit image will be more complicated since in individual crystallites the optic axes are differently oriented. For this reason, even under monochromatic light illumination, individual crystallites will appear or disappear at different angles of the analyzer rotation, which allows us to interpret the crystallite orientation.

Appendix 5.A Fresnel equations

Let us consider more carefully the polarization changes at the boundary between two isotropic transparent media with refractive indices n_1 and n_2. Here the term "transparent" means that the magnitudes of the involved wave vectors $\boldsymbol{k}$ and refractive indices n_1 and n_2 are real. We will use **Maxwell** equations to follow the amplitudes of electric fields E in two media for the σ- and π-polarization components. Let us mark the amplitudes of the electric and magnetic fields, as well as the corresponding wave vectors, by indices 1, 2, and r for the incident, refracted, and reflected waves, respectively. Using eq. (1.25), we find that

$$k_1 = n_1 k_0$$
$$k_2 = n_2 k_0 \tag{5.A.1}$$
$$k_r = n_1 k_0$$

where k_0 is the light wave vector in a vacuum. The angles ($\wedge$) between the respective wave vectors and the unit vector $\hat{\mathbf{z}}$, normal to the interface, are indicated in Figure 5.24:

$$\boldsymbol{k}_1 \wedge \hat{\mathbf{z}} = \boldsymbol{k}_r \wedge \hat{\mathbf{z}} = \alpha \tag{5.A.2}$$

$$\boldsymbol{k}_2 \wedge \hat{\mathbf{z}} = \beta \tag{5.A.3}$$

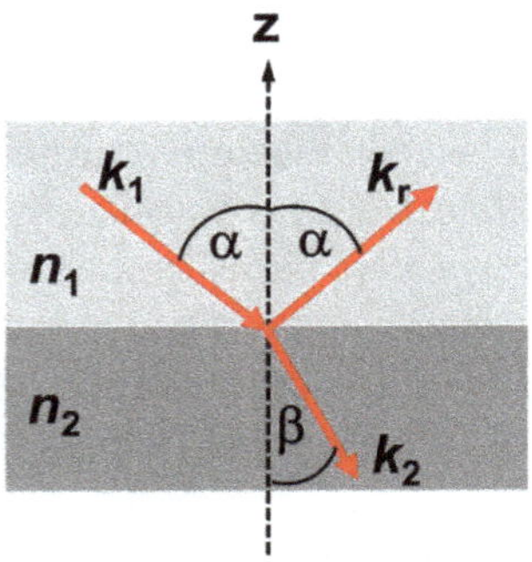

Figure 5.24: Light reflection and refraction at the interface between two homogeneous media with refractive indices n_1 and n_2.

In eq. (5.A.2), we used the conservation law of the wave vector component along the interface between two media. The same considerations provide **Snell**'s law (see Section 2.1)

$$n_1 \sin\alpha = n_2 \sin\beta \tag{5.A.4}$$

Relationships between the wavefield components are deduced from the boundary conditions, that is, the continuity of the tangential (t) components of vectors $\boldsymbol{E}$ and $\boldsymbol{H}$ at the interface between two media:

$$(\boldsymbol{E}_1)_t + (\boldsymbol{E}_r)_t = (\boldsymbol{E}_2)_t \tag{5.A.5}$$

$$(\boldsymbol{H}_1)_t + (\boldsymbol{H}_r)_t = (\boldsymbol{H}_2)_t \tag{5.A.6}$$

For σ-polarization, eq. (5.A.5) simply yields

$$E_1 + E_r = E_2 \tag{5.A.7}$$

To express the second boundary condition (5.A.6) via electric field components, we have, first, to recall one of the **Maxwell** eqs. (1.3) and the relationship between magnetic induction $\boldsymbol{B}$ and magnetic field $\boldsymbol{H}$ (see eq. (1.7)). Substituting plane waves, $\boldsymbol{B}\exp[i(\boldsymbol{kr} - \omega t)]$ and $\boldsymbol{E}\exp[i(\boldsymbol{kr} - \omega t)]$, into eq. (1.3) and using eq. (1.7) with $\mu_m = 1$, one obtains

$$\boldsymbol{H} = \frac{\boldsymbol{k} \times \boldsymbol{E}}{\mu_0 \omega} \tag{5.A.8}$$

It implies that vector $\boldsymbol{H}$ is perpendicular to both vectors $\boldsymbol{k}$ and $\boldsymbol{E}$. Correspondingly, for σ-polarization, eq. (5.A.6) becomes

$$\frac{k_1}{\mu_0 \omega} E_1 \cos\alpha + \frac{k_r}{\mu_0 \omega} E_r \cos(180° - \alpha) = \frac{k_2}{\mu_0 \omega} E_2 \cos\beta \tag{5.A.9}$$

Since the magnitude of the wave vector in a medium $\boldsymbol{k}$ is related to that in vacuum k_0 via refractive index n, as $k = nk_0$ (see eq. (5.A.1)), then

$$n_1 \cos\alpha (E_1 - E_r) = n_2 E_2 \cos\beta \tag{5.A.10}$$

With the aid of **Snell**'s law (5.A.4), the second boundary condition (5.A.10) can be rewritten as follows:

$$E_1 - E_r = \frac{\tan\alpha}{\tan\beta} E_2 \tag{5.A.11}$$

Solving the system of eqs. (5.A.7) and (5.A.11) yields

$$\frac{E_2}{E_1} = 2\frac{\tan\beta}{\tan\alpha + \tan\beta} = 2\frac{\sin\beta \cos\alpha}{\sin(\alpha + \beta)} \tag{5.A.12}$$

$$r_\sigma = \frac{E_r}{E_1} = \frac{\tan\beta - \tan\alpha}{\tan\alpha + \tan\beta} = -\frac{\sin(\alpha - \beta)}{\sin(\alpha + \beta)} \tag{5.A.13}$$

Equations (5.A.7) and (5.A.11) comprise the first pair of the so-called **Fresnel** equations, which provide the reflection coefficient r_σ given by eq. (5.A.13). If $\alpha > \beta$, which represents the wave reflection from the optically denser medium ($n_2 > n_1$), then the reflected wave E_r has a negative sign (see eq. (5.A.13)), that is, is in antiphase (phase shift by π) with the incident wave E_1. We will use this important result when discussing interference phenomena in Chapter 7. Note that eq. (5.A.13) may be rewritten in an alternative form that is also often used in the literature:

$$r_\sigma = \frac{n_1 \cos\alpha - n_2 \cos\beta}{n_1 \cos\alpha + n_2 \cos\beta} \tag{5.A.14}$$

For π-polarization, the boundary condition (5.A.5) transforms into

$$E_1 \cos\alpha + E_\mathrm{r} \cos(180° - \alpha) = E_2 \cos\beta$$

or

$$E_1 - E_\mathrm{r} = E_2 \frac{\cos\beta}{\cos\alpha} \tag{5.A.15}$$

For the boundary condition (5.A.6), we obtain (with the aid of eq. (5.A.8))

$$(E_1 + E_\mathrm{r})n_1 = E_2 n_2 \tag{5.A.16}$$

and hence

$$E_1 + E_\mathrm{r} = E_2 \frac{\sin\alpha}{\sin\beta} \tag{5.A.17}$$

Solving the system of eqs. (5.A.15) and (5.A.17) yields

$$\frac{E_2}{E_1} = 2\frac{\sin\beta\cos\alpha}{\cos\alpha\sin\alpha + \cos\beta\sin\beta} = 2\frac{\sin\beta\cos\alpha}{\sin(\alpha+\beta)\cos(\alpha-\beta)} \tag{5.A.18}$$

$$r_\pi = \frac{E_r}{E_1} = \frac{\sin\alpha\cos\alpha - \cos\beta\sin\beta}{\cos\alpha\sin\alpha + \cos\beta\sin\beta} = \frac{\sin(\alpha-\beta)\cos(\alpha+\beta)}{\sin(\alpha+\beta)\cos(\alpha-\beta)} = \frac{\tan(\alpha-\beta)}{\tan(\alpha+\beta)} \tag{5.A.19}$$

Equations (5.A.15) and (5.A.17) comprise the second pair of **Fresnel** equations, which provide the reflection coefficient r_π given by eq. (5.A.19). The latter is often used in a somewhat different form:

$$r_\pi = \frac{n_2 \cos\alpha - n_1 \cos\beta}{n_2 \cos\alpha + n_1 \cos\beta} \tag{5.A.20}$$

Note that if

$$\alpha + \beta = 90° \tag{5.A.21}$$

then the denominator of eq. (5.A.19) tends to infinity. This means that there is no π-polarization (in-plane polarization) in the reflected beam, which becomes fully σ-polarized. In other words, the **Brewster** condition (i.e., eq. (5.A.21), see also Section 5.1) follows directly from the **Fresnel** equations.

At normal incidence or close to it, $\cos\alpha \simeq \cos\beta$, the absolute values of both reflection coefficients become equal. In fact, using eqs. (5.A.14) and (5.A.20), we find

$$r_\sigma = \frac{n_1 - n_2}{n_1 + n_2} \tag{5.A.22}$$

$$r_\pi = \frac{n_2 - n_1}{n_2 + n_1} \tag{5.A.23}$$

In the case of nonpolarized incoming light, one can consider one half of the photons as σ-polarized, with the other half as π-polarized. Since for the initially nonpolarized light there is no coherence between two linearly (planar) polarized components, the ratio $\mathcal{R}$ of the reflected intensity I_r to the incident intensity I_0 is

$$\mathcal{R} = \frac{I_r}{I_0} = \frac{1}{2}r_\sigma^2 + \frac{1}{2}r_\pi^2 = \left(\frac{n_2 - n_1}{n_2 + n_1}\right)^2 \tag{5.A.24}$$

If the first medium is a vacuum or air ($n_1 \simeq 1$) and the second medium has refractive index $n_2 = n$, then

$$\mathcal{R} = \left(\frac{n-1}{n+1}\right)^2 \tag{5.A.25}$$

For example, most glasses have a refractive index of about $n = 1.5$; therefore, at normal incidence (or close to it), the surface of a glass plate reflects nearly $\mathcal{R} = 4\%$ of the incoming light. In Chapter 6, we calculate the reflectance of metal surfaces and will see that it is practically $\mathcal{R} \approx 100\%$.

Appendix 5.B Light ellipsometry and X-ray reflectivity

It turns out that by measuring the polarization state of the reflected light, it is possible to carefully determine the refractive index and thickness of thin films deposited on the substrate, even if the film thickness is much smaller than the light wavelength. This is a very important practical problem, especially for nanotechnology. The optical measurement technique solving this problem is called light ellipsometry.

At the end of nineteenth century, **Paul Drude** laid the theoretical foundations of ellipsometry when he calculated the polarization change introduced by a thin film deposited on a metal surface. In 1945, the first ellipsometer was constructed by **Alexandre Rothen**, who also coined the name of the method. His apparatus enabled film thickness to be measured with sub-nanometer precision, based on the change in elliptic polarization after coating a substrate with a thin film.

To better understand the principles of the method, we briefly show the calculation scheme for the polarization state of the reflected light, taking into account light reflection at two boundaries, that is, between the first medium (1) with refractive index n_1 and the film (2) with refractive index n_2 as well as between the film (2) and the substrate (3) having refractive index n_3. We further assume that the first medium (1), from which light enters a whole ensemble, is an air.

Let us indicate an electric field in the incident wave as $\boldsymbol{E}_1$, while the electric fields in the refracted waves within media (2) and (3), as $\boldsymbol{E}_2$ and $\boldsymbol{E}_3$, respectively. The wavefields reflected at the interfaces 1–2 and 2–3 are marked as $\boldsymbol{E}_{1r}$ and $\boldsymbol{E}_{2r}$, respectively. Similar indices are attributed to the corresponding wave vectors (see Figure 5.25). Let us first calculate the reflection coefficient r_σ for σ-polarization.

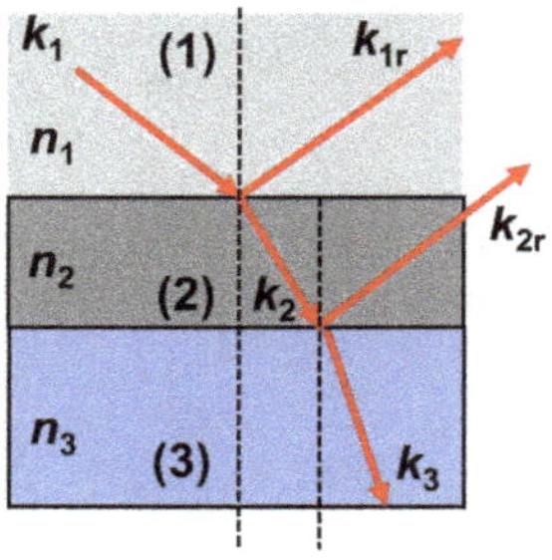

Figure 5.25: Calculating reflection coefficients from multilayers.

The electric field in the film $\boldsymbol{E}_f$ is the sum of the refracted wave $\boldsymbol{E}_2$ and the reflected wave $\boldsymbol{E}_{2r}$ from the boundary 2–3

$$\boldsymbol{E}_f = \boldsymbol{E}_2 + \boldsymbol{E}_{2r} \tag{5.B.1}$$

The boundary condition at the 1–2 interface is

$$E_{2r} = A_b(E_{1r} - K_b E_1) \tag{5.B.2}$$

where A_b and K_b are some constants. In eq. (5.B.2), we consider that E_2 is a certain fraction of E_1. If the film thickness T tends to infinity, then E_{2r} disappears, and eq. (5.B.2) yields

$$K_b = r_{12} = \frac{E_{1r}}{E_1} \tag{5.B.3}$$

In other words, parameter K_b is the reflection coefficient r_{12} at the 1–2 boundary, if the third layer does not exist. Correspondingly, eq. (5.B.2) converts into

$$E_{2r} = A_b(E_{1r} - r_{12}E_1) \tag{5.B.4}$$

The second equation is obtained by changing the sign of the z-component of the wave vectors in eq. (5.B.4), that is, by replacing E_{2r} by E_2 and switching the places of E_{1r} and E_1 in the right-hand side of eq. (5.B.4)

$$E_2 = A_b(E_1 - r_{12}E_{1r}) \tag{5.B.5}$$

In the substrate (3), there exists only one wave E_3 for which we can write the following relationships

$$E_2 e^{i\psi} = A_b E_3 \tag{5.B.6}$$

$$E_{2r} e^{-i\psi} = A_b r_{23} E_3$$

where r_{23} is the reflection coefficient at the 2–3 interface, if only two media (2) and (3) exist. Phase factor $e^{i\psi}$ reflects the phase change ψ accumulated during the wavefield propagation over the film thickness T

$$\psi = kT\cos\beta = 2\pi\frac{n_2 T}{\lambda_o}\sqrt{1-\sin^2\beta} = 2\pi\frac{T}{\lambda_o}\sqrt{n_2^2-\sin^2\alpha} \tag{5.B.7}$$

where λ_o again is the light wavelength in a vacuum. Such expression for phase change ψ is justified in Section 7.2.

Excluding E_3 from eqs. (5.B.6), one obtains

$$E_{2r} e^{-i\psi} = r_{23} E_2 e^{i\psi} \tag{5.B.8}$$

Combining eqs. (5.B.4), (5.B.5), and (5.B.8), we find the reflection coefficient r_σ from the film deposited on the substrate for the σ-polarization of the incoming light

$$r_\sigma = \frac{E_{1r}}{E_1} = \left[\frac{r_{12}e^{-2i\psi} + r_{23}}{e^{-2i\psi} + r_{12}r_{23}}\right]_\sigma \tag{5.B.9}$$

Index σ in the right-hand part of this equation means that the partial reflectivity coefficients at the 1–2 (r_{12}) and 2–3 (r_{23}) interfaces are also taken for σ-polarization (see eq. (5.A.14)). In the case of π-polarization, we receive the similar expression

$$r_\pi = \frac{E_{1r}}{E_1} = \left[\frac{r_{12}e^{-2i\psi} + r_{23}}{e^{-2i\psi} + r_{12}r_{23}}\right]_\pi \tag{5.B.10}$$

with partial reflectivity coefficients given by eq. (5.A.20).

We stress that unique sensitivity of the ellipsometry methods in measuring the film thickness is related to the phase factor $e^{-2i\psi}$, which can be altered by changing the incident angle α (see eq. (5.B.7)). The sensitivity to polarization change is higher when working close to the **Brewster** angle α_B at which π-polarization is eliminated. Substituting the **Brewster** condition (eq. (5.5)), that is, $\tan(\alpha_B) = n_2$, into eq. (5.B.7), we find

$$\psi = 2\pi\frac{T}{\lambda_o}\sqrt{n_2^2 - \sin^2\alpha_B} = 2\pi\frac{T}{\lambda_o}\frac{{n_2}^2}{\sqrt{1+{n_2}^2}} \tag{5.B.11}$$

The required equipment is very minimal and consists of a light source, polarizer, analyzer, and light detector as well as a simple goniometer allowing the incident angle of the incoming light to be continuously changed (see Figure 5.26).

In the beginning of the 1970s, spectroscopic ellipsometry (SE) was invented. In contrast to standard ellipsometry, which uses monochromatized laser radiation, SE employs wide-band light sources, working in the infrared, visible, or ultraviolet spectral

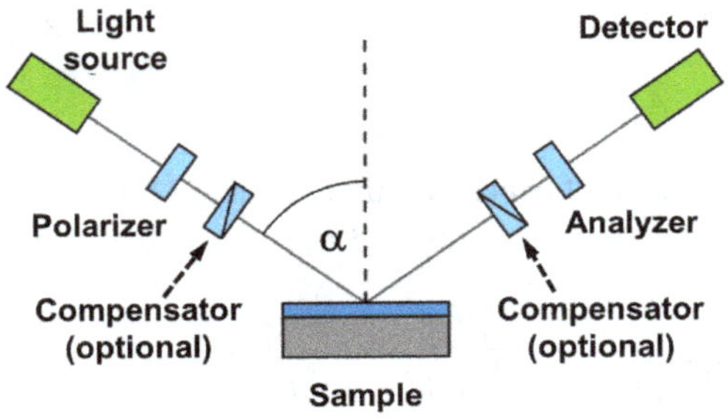

Figure 5.26: Principal scheme of an ellipsometer.

regions. In this way, the complex refractive index or the dielectric permittivity tensor in the spectral region of interest can be obtained, allowing a number of fundamental physical characteristics to be extracted. For example, infrared SE can probe lattice vibrational (phonon) and free charge carrier (plasmon) properties. SE in the near infrared, visible and up to ultraviolet spectral regions is able to provide the refractive index in the transparency regions or below bandgap regions and, hence, to probe electron properties, related, for example, to band-to-band transitions and exciton formation.

Therefore, in addition to the leading role of ellipsometry in determining film thickness, SE offers wide capabilities for measuring various properties of homogeneous and inhomogeneous materials in bulk or thin film form including characteristics of surfaces and interfaces.

Alternative X-ray-based techniques allowing very accurate measurement of film thickness and physical characteristics in the near surface material layers are also in common use. For single-crystalline films, high-resolution X-ray diffraction provides important structural information, as discussed in Section 10.2. In addition, the nondiffractive technique, called X-ray reflectivity, which can be applied equally to noncrystalline materials (and even to study liquid layers), is widely exploited.

In this method, the scattering intensity is measured at small incident angles γ (with respect to interfaces), as a function of the wave vector transfer q. For specular reflection (see Figure 5.27)

$$q = \frac{4\pi}{\lambda}\gamma \tag{5.B.12}$$

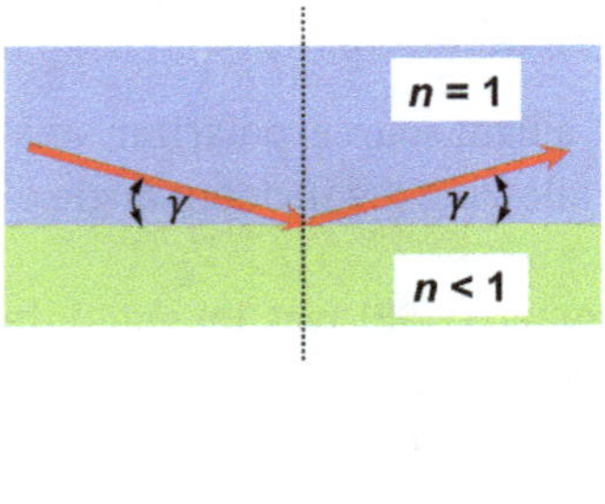

Figure 5.27: Specular X-ray reflection at the boundary between an air ($n = 1$) and a material ($n < 1$).

If an incident angle is less than the critical angle given by eq. (2.37), that is, $\gamma < \gamma_c$, the refection coefficient $\mathcal{R} = 1$ and we observe the total external reflection (see Section 2.3). At $\gamma > \gamma_c$, the X-ray reflectivity is rapidly attenuated with increasing q-values, because of partial X-ray penetration into the material's depth (Figure 5.28). At $q \gg \frac{4\pi}{\lambda}\gamma_c$, for the ideal flat film surface, one obtains the so-called **Porod** law

$$\mathcal{R} \sim \frac{1}{q^4} \tag{5.B.13}$$

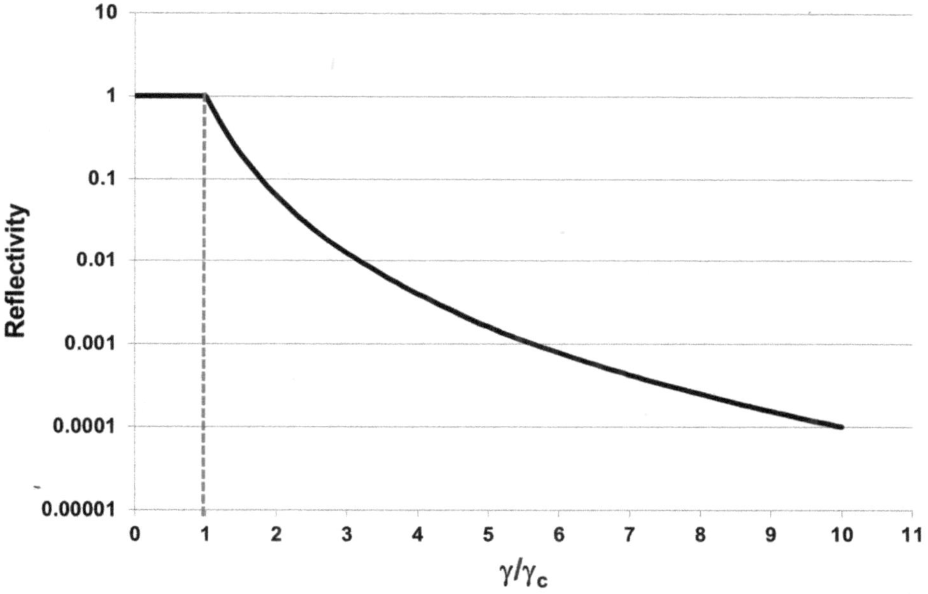

Figure 5.28: X-ray reflectivity (with no absorption), as a function of the normalized glancing angle, γ/γ_c. The end of an angular interval, in which total reflection occurs, is indicated by the dashed vertical line.

For rough surfaces, the reflectivity attenuates slower with increasing q. This is used to estimate the mean root-square roughness parameter integrated over the irradiated film area. Based on the measured critical angle γ_c in the angular region where $\mathcal{R} = 1$ (see Figure 5.28), and using eqs. (2.37) and (3.15), one can extract the electron density ρ_e

$$\rho_e = \frac{\pi}{r_0 \lambda^2} \gamma_c^2 \tag{5.B.14}$$

and atomic density, ρ_a

$$\rho_\text{a} = \frac{A}{ZN_\text{A}}\rho_\text{e} \tag{5.B.15}$$

where N_A is the **Avogadro** number, and A and Z are, respectively, the atomic mass and atomic number of atoms comprising the investigated material.

For multilayers, an iterative approach can be applied, based on **Fresnel** equations, similarly to what we discussed in Appendix 5.A.

Chapter 6
Strong frequency effects in light optics

As we know, even in commonplace glass or water, the refractive index n is wavelength (frequency)-dependent that is used to separate different colors in the light spectrum. These effects, however, are usually not very strong. For example, refractive index of silica glass is changing between $n = 1.45–1.47$ over entire frequency range (430–770 THz) of visible light. In this case, the light dispersion is rather weak because the light energy is far from absorption edges (resonant frequencies) and absorption bands, where dispersion may be much stronger.

To understand this statement deeper, let us calculate (within some simplified model) dielectric polarizability χ of a medium under high-frequency electric field $\boldsymbol{E}$. According to eq. (1.6), in isotropic medium the electric displacement field $\boldsymbol{D}$ is

$$\boldsymbol{D} = \varepsilon \boldsymbol{E} = \varepsilon_m \varepsilon_0 \boldsymbol{E} \tag{6.1}$$

On the other hand, in dielectrics under electric field application, some dielectric polarization $\boldsymbol{P}$ is induced, which equals the dipole moment of unit volume

$$\boldsymbol{P} = \sum_{i=1}^{N} \boldsymbol{p}_i \tag{6.2}$$

where $\boldsymbol{p}_i$ is the induced dipole moment of an individual atom (or molecule) and the summation is performed over unit volume. Therefore,

$$\boldsymbol{D} = \varepsilon_0 \left(\boldsymbol{E} + \frac{\boldsymbol{P}}{\varepsilon_0} \right) = \varepsilon_m \varepsilon_0 \boldsymbol{E} \tag{6.3}$$

In linear dielectrics, polarization $\boldsymbol{P}$ is proportional to the applied field $\boldsymbol{E}$

$$\boldsymbol{P} = \varepsilon_0 \chi \boldsymbol{E} \tag{6.4}$$

with polarizability χ being the proportionality factor. Correspondingly,

$$\boldsymbol{D} = \varepsilon_0 (1 + \chi) \boldsymbol{E} = \varepsilon_m \varepsilon_0 \boldsymbol{E} \tag{6.5}$$

and

$$\varepsilon_m = 1 + \chi \tag{6.6}$$

The next step is calculating the induced polarization of individual atoms (or molecules) in some simple (one-dimensional) model and after that finding polarizability χ using eqs. (6.2) and (6.4). We remind the reader that under an alternating electric field $E = E_0 e^{-i\omega t}$ applied in the x-direction, electrons in atoms experience the forced vibrations which can be described by the aid of the second **Newton** law:

https://doi.org/10.1515/9783111140100-007

$$\frac{d^2x}{dt^2} - \beta\frac{dx}{dt} + \omega_0^2 x = \frac{eE_0}{m}e^{-i\omega t} \tag{6.7}$$

where fundamental constants e and m are, respectively, the electric charge and mass of an electron, parameter β is the damping coefficient, and ω_0 is the resonant frequency. Seeking solution in the form of

$$x = x_0 e^{-i\omega t} \tag{6.8}$$

we find

$$x_0 = \frac{e}{m} \cdot \frac{E_0}{\left(\omega_0^2 + i\beta\omega - \omega^2\right)} \tag{6.9}$$

Respectively, the dipole moment of an individual atom (or molecule) equals

$$p = \xi e x_0 e^{-i\omega t} = \xi\frac{e^2}{m} \cdot \frac{E_0}{\left(\omega_0^2 + i\beta\omega - \omega^2\right)} e^{-i\omega t} \tag{6.10}$$

where ξ is the effective number of electrons participating in the creation of an individual dipole moment. Using definition (6.2) and eq. (6.10), one obtains

$$\boldsymbol{P} = \xi N\frac{e^2}{m} \cdot \frac{\boldsymbol{E}_0}{\left(\omega_0^2 + i\beta\omega - \omega^2\right)} e^{-i\omega t} = \xi N\frac{e^2}{m} \cdot \frac{1}{\left(\omega_0^2 + i\beta\omega - \omega^2\right)}\boldsymbol{E} \tag{6.11}$$

where N is the number of local dipole moments per unit volume. Comparing eqs. (6.11) and (6.4) yields dielectric polarizability

$$\chi = \xi N\frac{e^2}{\varepsilon_0 m} \cdot \frac{1}{\left(\omega_0^2 + i\beta\omega - \omega^2\right)} \tag{6.12}$$

and dielectric constant in the form

$$\varepsilon_{\text{m}} = 1 + \chi = 1 + \xi N\frac{e^2}{\varepsilon_0 m} \cdot \frac{1}{\left(\omega_0^2 + i\beta\omega - \omega^2\right)} \tag{6.13}$$

Recalling eq. (1.27) for $\mu_{\text{m}} = 1$, we finally find

$$n^2 = \varepsilon_{\text{m}} = 1 + \xi N\frac{e^2}{\varepsilon_0 m} \cdot \frac{1}{\left(\omega_0^2 + i\beta\omega - \omega^2\right)} = 1 + C(\omega) \tag{6.14}$$

with $C(\omega) = n^2 - 1$.

Certainly, our model is oversimplified, and real situation may be much more complicated. For example, electric field “felt” by atoms is not external field $\boldsymbol{E}$, but some local field which includes the polarization component. As a result, eq. (6.14) is modified providing the so-called **Clausius–Mossotti** relationship

$$3\,\frac{n^2-1}{n^2+2}=C(\omega) \tag{6.15}$$

with function $C(\omega)$ being somewhat different from that in eq. (6.14) (i.e., $C(\omega)=n^2-1$). Note that both functions practically coincide in case of $n \simeq 1$.

Furthermore, in solid dielectrics atoms strongly interact with each other. For this reason, resonant frequencies and damping constants considerably differ from those being characteristic for individual atoms. Other complications arise for solids having built-in dipole moments as in ferroelectrics.

Considering eq. (6.14), however, still allows us to draw two important conclusions: (i) refractive index is frequency-dependent, and this dependence becomes very strong near resonant frequencies; (ii) refractive index, generally, is a complex number comprising real n_r and imaginary n_{im} parts. These numbers can be expressed via the real $\varepsilon'_{\mathrm{m}}$ and imaginary $\varepsilon''_{\mathrm{m}}$ components of dielectric constant ε_{m}

$$n_{\mathrm{r}}=\sqrt{\frac{\varepsilon'_{\mathrm{m}}+\sqrt{\varepsilon'^2_{\mathrm{m}}+\varepsilon''^2_{\mathrm{m}}}}{2}};\quad n_{\mathrm{im}}=\sqrt{\frac{-\varepsilon'_{\mathrm{m}}+\sqrt{\varepsilon'^2_{\mathrm{m}}+\varepsilon''^2_{\mathrm{m}}}}{2}} \tag{6.16}$$

The presence of the imaginary part of refractive index means energy losses during light propagation. We stress that energy losses can be caused not only by direct light absorption by individual atoms, but what is most interesting for us here, by strong wave scattering (wave expelling), which prevents light propagation into the medium depth (as for evanescent waves at total reflection, described in Sections 2.2 and 2.3).

In this chapter, we will learn more about new and sometimes unexpected phenomena arising in intricate media in which refractive index is complex or even negative number. We start with calculating refractive index for metals by solving **Maxwell** equations in a medium with nonzero electrical conductivity σ. As we saw in Chapter 1, by solving the **Maxwell** equations, we find that, in general, both the magnitude of wave vector $\boldsymbol{k}$ in a material

$$k=\frac{\omega}{c}\sqrt{\mu_{\mathrm{m}}}\left(\varepsilon_{\mathrm{m}}+i\frac{\sigma}{\varepsilon_0\omega}\right)^{1/2} \tag{6.17}$$

and refractive index n

$$n=\frac{k}{k_{\mathrm{v}}}=\sqrt{\mu_{\mathrm{m}}}\left(\varepsilon_{\mathrm{m}}+i\frac{\sigma}{\varepsilon_0\omega}\right)^{1/2} \tag{6.18}$$

are complex quantities (as we mentioned before) comprising the real and imaginary parts:

$$n=n_{\mathrm{r}}+in_{\mathrm{im}} \tag{6.19}$$

$$k = \frac{\omega}{c}(n_r + i n_{im}) \tag{6.20}$$

As we already said, the presence of the imaginary part in eq. (6.20) means that propagation of electromagnetic wave is accompanied by energy losses. In the following section, we show that these losses in metals, in general, grow with increasing frequency.

6.1 Skin effect in metals

Let us consider the propagation of the plane wave

$$\boldsymbol{E} = \boldsymbol{E}_0 \exp[i(\boldsymbol{kr} - \omega t)] \tag{6.21}$$

into a metal depth at normal incidence, that is, along the z-direction (see Figure 6.1).

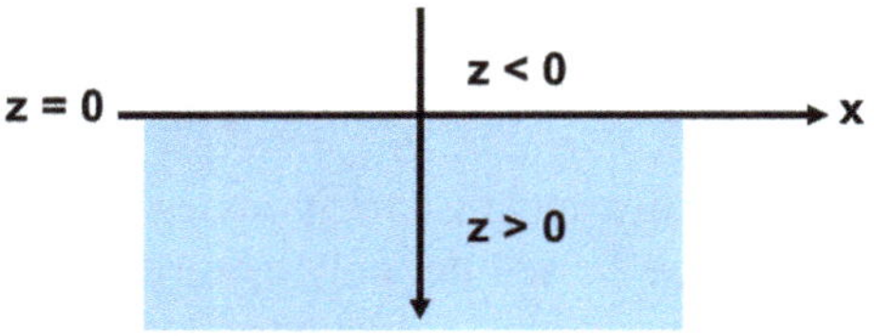

Figure 6.1: Coordinate system used for calculating the skin effect.

Substituting eq. (6.20) into (6.21) yields

$$\boldsymbol{E} = \boldsymbol{E}_0 \exp\left[i\omega\left(\frac{n_r}{c}z - t\right)\right] \exp\left(-\frac{n_{im}\omega}{c}z\right) = \boldsymbol{E}_0 \exp\left[i\omega\left(\frac{z}{V_{pm}} - t\right)\right] \exp\left(-\frac{z}{\delta}\right) \tag{6.22}$$

where

$$V_{pm} = \frac{c}{n_r} \tag{6.23}$$

is the phase velocity V_{pm} for electromagnetic wave propagation within a metal, which is determined by the real part n_r of refractive index (see eq. (6.19)) and differs from that in a vacuum c. Parameter δ is characteristic length

$$\delta = \frac{c}{n_{im}\omega} \tag{6.24}$$

which limits the wave penetration into metal depth. According to eq. (6.24), the penetration depth diminishes with increasing wave frequency. This phenomenon, called skin effect, was discovered by **Horace Lamb** in 1883 and later used by **Nikola Tesla** for impressive demonstrations, which led to the transferring of the electricity system across the globe from dc to ac current. Exact expression for characteristic length δ, considering the frequency dependence of parameter n_{im} is rather complicated. If, however,

$$\varepsilon_m \ll \frac{\sigma}{\varepsilon_0 \omega} \tag{6.25}$$

which is usually valid for metals at optical frequencies, and $\mu_m = 1$, then eq. (6.18) transforms into

$$n \approx \left(i\frac{\sigma}{\varepsilon_0 \omega}\right)^{\frac{1}{2}} = (1+i)\sqrt{\frac{\sigma}{2\varepsilon_0 \omega}} \tag{6.26}$$

with

$$n_r = n_{im} = \sqrt{\frac{\sigma}{2\varepsilon_0 \omega}} \tag{6.27}$$

Substituting eq. (6.27) into eq. (6.24) finally yields

$$\delta = c\sqrt{\frac{2\varepsilon_0}{\sigma\omega}} \tag{6.28}$$

Therefore, in this approximation, the skin depth δ diminishes with increasing frequency as $1/\sqrt{\omega}$ (neglecting possible changes in metal conductivity). For good metals, as copper, $\delta \approx 5 - 10$ mm at 50 Hz, $\delta \approx 0.5$ μm at microwave frequencies of about 10 GHz and is reduced to a few tens of nanometers for optical frequencies of about 500 THz. It implies that most photons are scattered back when light enters a metal. This qualitatively explains the well-known superb light reflectivity by metal surfaces (see next section), which is of great practical importance.

6.2 Light reflection from a metal

To proceed further, let us calculate the light reflection coefficient from a metal in the simple case of normal incidence of electromagnetic wave, that is, along the *z*-axis (see Figure 6.2).

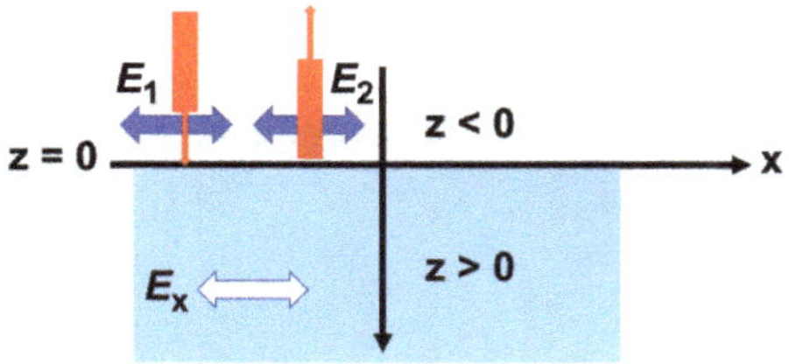

Figure 6.2: Illustration of the light reflection from a metal surface.

Metal surface is located at $z = 0$. Within a metal ($z > 0$), a lone plane wave exists:

$$E_x = E_0 \exp\left[i\omega\left(n\frac{z}{c} - t\right)\right] \tag{6.29}$$

In contrast, in free space ($z < 0$), we have the sum of two waves, the incident and reflected ones:

$$E_x = E_1 \exp\left[i\omega\left(\frac{z}{c} - t\right)\right] + E_2 \exp\left[-i\omega\left(\frac{z}{c} + t\right)\right] \tag{6.30}$$

Boundary condition (at $z = 0$) is

$$E_0 = E_1 + E_2 \tag{6.31}$$

As second equation, the continuity of the magnetic field component H_y at $z = 0$ is used. For this purpose, we again recall the **Maxwell** equation (1.11), which for the plane wave yields the following relationship between the electric $\boldsymbol{E}$ and magnetic field $\boldsymbol{H}$ components (for $\mu_m = 1$):

$$\mu_0 \omega \boldsymbol{H} = \boldsymbol{k} \times \boldsymbol{E} \tag{6.32}$$

Therefore, at $z > 0$

$$H_y = H_0 \exp\left[i\omega\left(n\frac{z}{c} - t\right)\right] = \frac{n}{\mu_0 c} E_0 \exp\left[i\omega\left(n\frac{z}{c} - t\right)\right] \tag{6.33}$$

Substituting the expression for the speed of light c in a vacuum (see eq. (1.17)) into eq. (6.33) yields

$$H_0 = \frac{n}{\mu_0 c} E_0 = nc\varepsilon_0 E_0 \tag{6.34}$$

In free space ($z < 0$), we again have the sum of two waves, the incident and reflected ones:

$$H_y = H_1 \exp\left[i\omega\left(\frac{z}{c} - t\right)\right] + H_2 \exp\left[-i\omega\left(\frac{z}{c} + t\right)\right] \tag{6.35}$$

where

$$H_1 = c\varepsilon_0 E_1 \tag{6.36}$$

$$H_2 = -c\varepsilon_0 E_2 \tag{6.37}$$

Using eqs. (6.33)–(6.37) and the continuity of the magnetic field component at $z = 0$ yields

$$nE_0 = E_1 - E_2 \tag{6.38}$$

Solving eqs. (6.31) and (6.38), we find the ratio between the amplitudes of the incident and reflected waves

$$\frac{E_2}{E_1} = \frac{1-n}{1+n} \tag{6.39}$$

and the reflectivity coefficient

$$\mathcal{R} = \left|\frac{E_2}{E_1}\right|^2 = \left|\frac{1-n}{1+n}\right|^2 \tag{6.40}$$

This result matches eq. (5.A.25) derived in Appendix 5.A. Recalling eq. (6.19), one can express the reflectivity coefficient $\mathcal{R}$ via the real and imaginary parts of the refractive index

$$\mathcal{R} = \frac{(1-n_r)^2 + n_{im}{}^2}{(1+n_r)^2 + n_{im}{}^2} \tag{6.41}$$

At optical frequencies, the light wavelength in metals λ_m is of the order of the penetration depth δ, that is, $\lambda_m \sim \delta \ll \lambda_0$ is much smaller than the wavelength λ_0 in vacuum. Correspondingly, the refraction index $|n| \gg 1$ and, according to eq. (6.40), the reflectivity coefficient indeed is very close to $\mathcal{R} \approx 1$, that is, metal surface reflects almost 100% of incoming photons.

6.3 Plasma frequency

Another key characteristic of free electrons in a metal is the so-called plasma frequency ω_p which is of great importance to optical properties of metals. Plasma frequency is the resonance frequency of free electron vibrations upon an action of high-frequency electromagnetic wave. To calculate the plasma frequency, let us consider again a simple mechanical problem of the one-dimensional electron movement in the time-oscillating electric field of an electromagnetic wave, $E = E_0 e^{-i\omega t}$, like that described by eq. (6.7), but with no spring-like force and no damping:

$$\frac{d^2x}{dt^2} = \frac{e}{m} E_0 e^{-i\omega t} \tag{6.42}$$

Seeking possible solution of eq. (6.42) in the form of

$$x = x_0 e^{-i\omega t} \tag{6.43}$$

yields

$$x_0 = -\frac{e}{m\omega^2} E_0 \tag{6.44}$$

Therefore, an oscillating electric field induces the elementary dipole moment with an amplitude p_0 which is formally calculated via the induced electron displacement

$$p_0 = ex_0 = -\frac{e^2}{m\omega^2}E_0 \tag{6.45}$$

If the concentration of free electrons (per unit volume) is N_0, then the amplitude of the induced dipole moment per unit volume is

$$P_0 = p_0N_0 = -N_0\frac{e^2}{m\omega^2}E_0 \tag{6.46}$$

One can express the electric displacement field $D = D_0e^{-i\omega t}$ in a material via externally applied electric field $E = E_0e^{-i\omega t}$ and the induced temporal polarization $P = P_0e^{-i\omega t}$:

$$D_0 = \varepsilon_m\varepsilon_0E_0 = \varepsilon_0(E_0 + \frac{P_0}{\varepsilon_0}) = \varepsilon_0E_0\left(1 - N_0\frac{e^2}{\varepsilon_0m\omega^2}\right) \tag{6.47}$$

It follows from eq. (6.47) that the dielectric constant ε_m of free electron gas is

$$\varepsilon_m = 1 - \left(\frac{\omega_{pl}}{\omega}\right)^2 \tag{6.48}$$

where ω_{pl} is the plasma frequency defined as

$$\omega_{pl} = \sqrt{\frac{N_0e^2}{\varepsilon_0m}} \tag{6.49}$$

Collective excitations in the electron gas corresponding to electron vibrations at this frequency are called plasmons. The characteristic plasmon energy is

$$E_{pl} = \hbar\omega_{pl} \tag{6.50}$$

For example, in copper $E_{pl} = 10$ eV, while in silver $E_{pl} = 6.8$ eV. Plasmons produce clear peaks in the energy-resolved inelastic electron scattering within an electron microscope. This technique is called the electron energy loss spectroscopy (EELS).

We stress that dielectric constant (6.48) is real and positive at $\omega > \omega_{pl}$, while real and negative at $\omega < \omega_{pl}$. Therefore, applying eq. (6.16) with $\varepsilon''_m = 0$ yields

$$n = \sqrt{\varepsilon_m} \tag{6.51}$$

which fits eq. (1.30) derived in Chapter 1. Therefore, positive value of ε_m means real refractive index, that is, transparent medium. In contrast, negative ε_m corresponds to the imaginary refractive index, indicating opaque medium, having high effective attenuation coefficient for the incoming radiation flux. In the latter case, light will be totally expelled from the medium, as we showed before for light reflection from metal surfaces (see eqs. (6.40) and (6.41)).

In this context, let us consider (as an illustration) practically important phenomenon of the reflection of radio waves from ionosphere and transmission of radio waves through it. The ionosphere layer comprises the gas of positive ions and free electrons produced by ionizing cosmic radiation. Such gas has its own plasma frequency, which is rather low because the electron density is several orders of magnitude smaller than that in metals. Taking typical electron density in ionosphere as 10^{12} electrons per cubic meter, we find, by the aid of eq. (6.49), that $\nu_{\mathrm{pl}} = \frac{\omega_{\mathrm{pl}}}{2\pi} = 9$ MHz is in the radiowave region. Therefore, at $\nu > \nu_{\mathrm{pl}}$ electromagnetic waves will go through the ionosphere, while at $\nu < \nu_{\mathrm{pl}}$ they will be reflected back to the Earth. It means that for radio connection with cosmic objects (high-orbit satellites, cosmic stations, etc.) we must use rather high radio frequencies, $\nu > \nu_{\mathrm{pl}}$. Oppositely, for radio connection at far distances on the surface of the Earth, like trans-Atlantic connection, the lower frequency waves with $\nu < \nu_{\mathrm{pl}}$ are exploited, which are reflected from ionosphere.

6.4 Negative materials and metamaterials

The fact that dielectric constant ε_{m}, under certain circumstances, can be negative and hence refractive index can be an imaginary number is of upmost importance to wave propagation through a medium. In the last decades, a lot of efforts were spent to understand whether the magnetic permeability μ can also be negative together with ε. The interest in this subject was initiated in 1967 by **Victor Veselago**, who published the groundbreaking paper entitled "The electrodynamics of substances with simultaneously negative values of ε and μ." He predicted that novel physical effects would occur in such a medium, being completely different from what we know for regular wave propagation. This seminal paper paved a way to the development of the rapidly growing field of metamaterials with complex structures allowing practical realization of negative ε and μ. Let us briefly follow the **Veselago**'s argumentation leading to new phenomena in the wave propagation and first in light optics.

We remind that the solution of the **Maxwell** eqs. (1.1)–(1.7) for a homogeneous medium with negligible conductivity yields the following dispersion law:

$$k^2 = \frac{\omega^2}{c^2} n^2 \tag{6.52}$$

Here the refractive index n equals

$$n = \sqrt{\varepsilon_{\mathrm{m}} \mu_{\mathrm{m}}} \tag{6.53}$$

which fits eq. (6.51), obtained previously for $\mu_{\mathrm{m}} = 1$.

Let us rewrite again general **Maxwell** equations (1.1)–(1.4) for $\boldsymbol{J} = 0$ and $\rho_f = 0$:

$$\operatorname{div} \boldsymbol{D} = 0 \tag{6.54}$$

$$\operatorname{div} \boldsymbol{B} = 0 \tag{6.55}$$

$$\operatorname{rot} \boldsymbol{E} = -\frac{d\boldsymbol{B}}{dt} \tag{6.56}$$

$$\operatorname{rot} \boldsymbol{H} = \frac{d\boldsymbol{D}}{dt} \tag{6.57}$$

with

$$\boldsymbol{D} = \varepsilon_{\mathrm{m}}\varepsilon_0 \boldsymbol{E} \tag{6.58}$$

$$\boldsymbol{B} = \mu_{\mathrm{m}}\mu_0 \boldsymbol{H} \tag{6.59}$$

For plane waves, $\exp[i(\boldsymbol{kr} - \omega t)]$, eqs. (6.56)–(6.59) yield

$$\boldsymbol{k} \times \boldsymbol{E} = \mu_{\mathrm{m}}\mu_0 \omega \boldsymbol{H} \tag{6.60}$$

$$\boldsymbol{k} \times \boldsymbol{H} = -\varepsilon_{\mathrm{m}}\varepsilon_0 \omega \boldsymbol{E} \tag{6.61}$$

Neglecting possible absorption effects, material constants ε_{m} and μ_{m} are real. If both are positive, that is, $\varepsilon_{\mathrm{m}} > 0$ and $\mu_{\mathrm{m}} > 0$, vectors $\boldsymbol{E}$, $\boldsymbol{H}$, and $\boldsymbol{k}$ form a right-handed triplet, as follows from eqs. (6.60) and (6.61). If so, the direction of the **Poynting** vector $\boldsymbol{S}$:

$$\boldsymbol{S} = \boldsymbol{E} \times \boldsymbol{H} \tag{6.62}$$

that is, the direction of the energy flow is along vector $\boldsymbol{k}$. This implies that the phase velocity (along vector $\boldsymbol{k}$) and group velocity (along energy flow, i.e., along vector $\boldsymbol{S}$) are in the same direction (see Figure 6.3a). In contrast, if simultaneously $\varepsilon_{\mathrm{m}} < 0$ and $\mu_{\mathrm{m}} < 0$, vectors $\boldsymbol{E}$, $\boldsymbol{H}$, and $\boldsymbol{k}$ form a left-handed triplet (as follows again from eqs. (6.60) and (6.61)), which means that the **Poynting** vector $\boldsymbol{S}$ is antiparallel to vector $\boldsymbol{k}$ (see Figure 6.3b). Correspondingly, the phase velocity and group velocity are in opposite directions. This surprising conclusion leads to even more unforeseen physical phenomena.

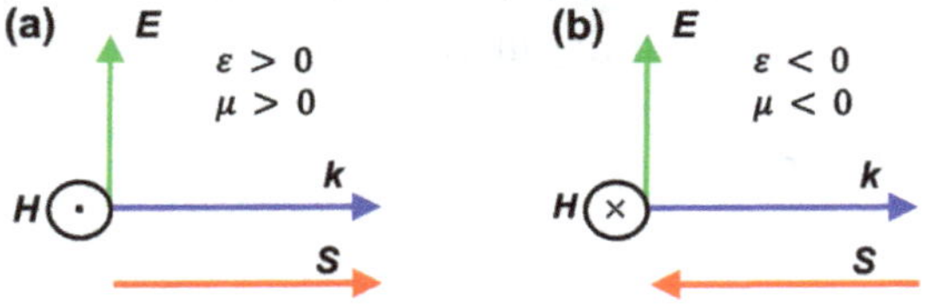

Figure 6.3: Direction of the **Poynting** vector with respect to the light wave vector $\boldsymbol{k}$ in: (a) right-handed (RH) materials and (b) left-handed (LH) materials.

First, we note that in the left-handed materials (LH materials) the **Doppler** frequency shift will be of the opposite sign to the normal **Doppler** effect, that is, the signal frequency will be decreased when the source is moving towards detector with the velocity $u_s \ll c$:

$$\omega = \omega_0\left(1 - \frac{u_s}{c}\right) \tag{6.63}$$

instead of increasing signal frequency

$$\omega = \omega_0\left(1 + \frac{u_s}{c}\right) \tag{6.64}$$

in the conventional right-handed materials (RH materials).

The fact that in LH materials the energy flow is opposite to the wave vector $\boldsymbol{k}$ can be reformulated in terms of negative refractive index $n < 0$. Changing sign of vector $\boldsymbol{k}$ is equivalent to multiplying it by $n = -1$. Formally, we can obtain this result, presenting eq. (6.53) for $\varepsilon_m = -1$ and $\mu_m = -1$, as

$$n = \sqrt{\varepsilon_m \mu_m} = \sqrt{\exp(i\pi)\exp(i\pi)} = \exp\left(i\frac{\pi}{2}\right)\exp\left(i\frac{\pi}{2}\right) = \exp(i\pi) = -1 \tag{6.65}$$

Negative refractive index causes dramatic modifications in a whole geometrical optics and, first, in the light refraction phenomena. The latter proceed very differently within the LH materials (called nowadays as negative materials) as compared to the conventional RH materials. In fact, the first published note about negative refraction and backward propagation of electromagnetic waves was probably written by **Leonid Mandelstam** in the 1940s. **Mandelstam** pointed out that in isotropic medium, the group velocity is directed along wave vector $\boldsymbol{k}$ or in opposite direction depending on the sign of derivative $\frac{\partial\omega}{\partial k}$. In case of negative dispersion, one expects backward-wave propagation in such a medium, and negative refraction should occur at its interface with conventional medium.

Let us consider more carefully the light refraction at the boundary between two homogeneous media, one being a vacuum ($n = 1$) and the second having refractive index n. We remind that kinematics of refraction is governed by the constancy in both media of the wave vector component along the interface, which provides the famous **Snell**'s law (see Figure 6.4). Conservation law of the tangential component of the wave vector yields

$$|\boldsymbol{k}_1|\sin\alpha = |\boldsymbol{k}_2|\sin\beta \tag{6.66}$$

Since in our case

$$|\boldsymbol{k}_2| = n\ |\boldsymbol{k}_1| \tag{6.67}$$

the angle of refraction is

$$\sin\beta = \frac{\sin\alpha}{n} \tag{6.68}$$

in accordance with the **Snell**'s law.

For RH materials $n > 0$, and the incident and refracted waves are situated, respectively, at the left and right sides of the normal to the interface (see Figure 6.4a). In contrast, for the LH materials with $n < 0$

$$\sin\beta = -\frac{\sin\alpha}{|n|} \tag{6.69}$$

and the incident and refracted waves are situated at the same side of the normal to the interface (see Figure 6.4b).

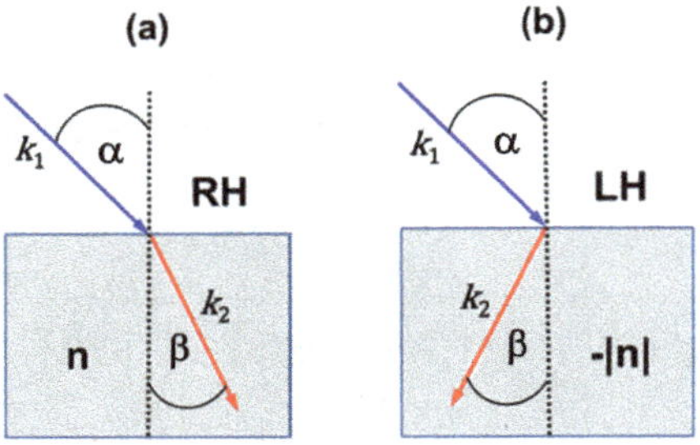

Figure 6.4: Application of **Snell**'s law to light refraction in (a) right-hand (RH) materials and (b) left-hand (LH) materials.

This fact is of great importance to refraction phenomena and leads to further practical consequences. For example, a convex lens made of the LH materials will produce defocusing effect, while concave lens will focus on light, that is, acting oppositely, as compared to conventional lenses made of RH materials. An action of the convex lens built of the RH or LH materials is illustrated, respectively, in Figures 6.5a and 6.5b.

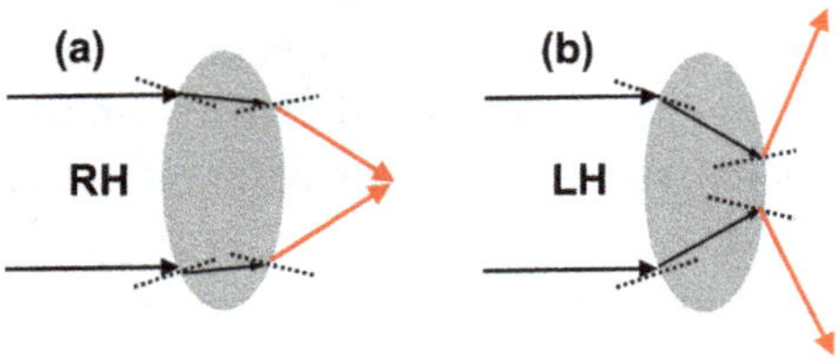

Figure 6.5: The action of a convex lens produced with conventional RH (a) and negative LH (b) materials.

Even flat plate, made of the LH material, will focus light on striking contrast with RH materials. In fact, one can easily prove that by considering the ray optics scheme depicted in Figure 6.6.

The rays emanate from point source A, situated left from the LH plate with thickness d. The source–plate distance is a. Rays entering the LH plate at an incident angle α will be refracted at an angle β, whose absolute value is

$$\sin\beta = \frac{\sin\alpha}{|n|} \tag{6.70}$$

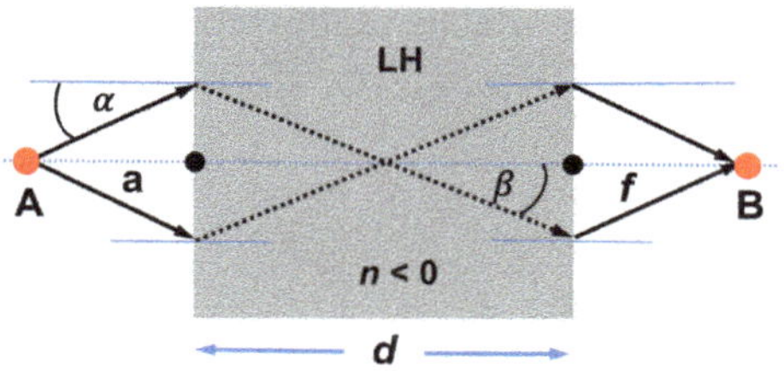

Figure 6.6: Illustration of light focusing by a flat plate built of LH material.

After additional refraction at the right-hand edge of the plate, the rays are focused to point *B*, located at distance *f* from this edge. Simple geometrical calculations yield

$$f = \frac{d\ \tan\beta}{\tan\alpha} - a \tag{6.71}$$

For small incident angles, $\tan\alpha \approx \sin\alpha$, $\tan\beta \approx \sin\beta$, and then using eqs. (6.70) and (6.71), one obtains

$$f = \frac{d}{|n|} - a \tag{6.72}$$

If $n = -1$, then $f = d - a$ for any angle of incidence, that is, in this approximation, such lens have no aberrations.

In 2000, that is, more than 30 years after first **Veselago** publication, **John Pendry** published a seminal paper entitled "Negative refraction makes a perfect lens." He showed that lenses built of negative materials can break the diffraction limit for resolving power of optical instruments. We remind that in classical optics, the latter cannot be better than a fraction of the light wavelength λ (see Sections 3.5 and 10.5). It is also worth mentioning another brilliant proposal of **Pendry** and coworkers in the field of negative materials, which is called "invisible cloak." This term means that it is principally possible using negative materials to make an object invisible under certain conditions. These splendid ideas regarding perfect lenses and invisible cloak are discussed further in Section 11.3.

Other possible optical and microwave applications of negative materials include band-pass filters, microwave couplers, beam steerers, and new types of antennas with improved characteristics and structural protection of a radar antenna (antenna radome). Some applications use nonmagnetic materials with negative dielectric constant only. In several respects, optical metamaterials are overlapped with photonic materials and structures (see Chapter 16). All this facilitated intensive research toward practical realizations using new negative materials and structures.

Note that in the last two decades, a few ideas have been suggested to build a medium with negative magnetic permeability μ_{eff}. For example, the structures containing loops of conducting wires or paired thin metallic strips have resonant properties providing negative magnetic constant in some frequency intervals. A trendy way to achieve negative magnetic permeability is by using the so-called split-ring resonators introduced by **John Pendry** and coworkers (see Figure 6.7).

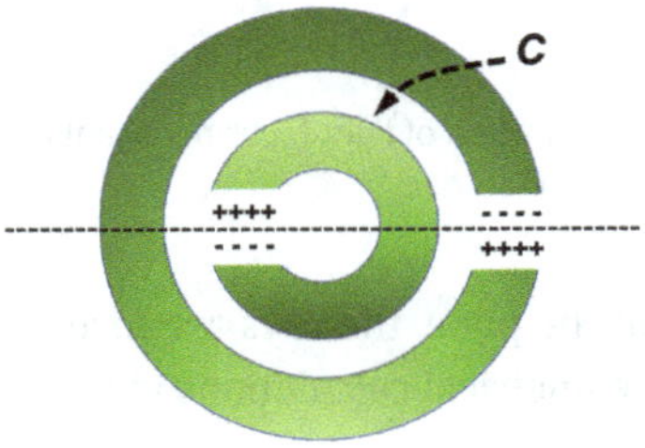

Figure 6.7: Sketch of the split-ring resonator. The letter C indicates the capacitance arising between two adjacent concentric rings.

They consist of metallic circular rings, interrupted by small gaps which produce resonance effect. A time-varying magnetic field applied normally to the rings induces electrical currents along the rings. These currents, depending on the resonant properties of the ring elements, in turn, create magnetic fields being in-phase or in antiphase with the applied field. The splits in the rings facilitate to realize resonances at wavelengths much larger than the diameters of the rings; in other words, there is no half-wavelength requirement for the resonance, as it would be for closed rings. The purpose of the second split ring, whose opening is oriented opposite to that in the first one, is to create a large capacitance in the narrow area between the rings, lowering the resonant frequency ω_0 and concentrating the electric field. Therefore, an individual resonator acts as electrical circuit composed of a capacitor C and effective inductor L. Consequently, the resonance frequency equals $\omega_0 = 1/\sqrt{LC}$. By arranging split-ring resonators into a periodic medium with strong magnetic coupling between the resonators, one obtains unique characteristics of the entire array. Because these resonators respond to the incident magnetic field, the medium can be considered as having an effective permeability $\mu_{\text{eff}}(\omega)$ which for negligible energy losses equals

$$\mu_{\text{eff}}(\omega) = 1 - \frac{g\omega^2}{\omega^2 - \omega_0^2} \tag{6.73}$$

where $g < 1$ is some numerical factor. It follows from eq. (6.73) that $\mu_{\text{eff}}(\omega) < 0$, if $\omega < \frac{\omega_0}{\sqrt{1-g}}$.

We see that the main principle staying behind the designing negative metamaterials is the usage of the negative dispersion of dielectric permittivity and magnetic permeability near resonance frequencies. So, nowadays the projects of negative materials and super-lenses on their basis are well established.

Some additional possibilities of light manipulation offer the so-called plasmonic nanostructures, which are classified as a branch of metamaterials. Their action is based on utilizing the properties of surface plasmons, arising at interfaces between metals and dielectrics (see also Section 7.3).

Chapter 7
Interference phenomena

As we already learned, geometrical optics properly describes light propagation when the light wavelength λ tends to zero. If the wavelength is comparable to the size of characteristic features of the investigated objects, new phenomena, such as interference and diffraction, should be considered. Moreover, they must be understood as playing a leading role in such wave optics. In this chapter, we will describe various interference phenomena, keeping the review of diffraction optics for the following chapters.

The term "interference" means that the intensity of the wave superposition depends critically on the phase difference between the involved individual waves. Let us illustrate this statement by a simple example that considers an interaction of two waves E_1 and E_2 with real amplitudes, respectively A_1 and A_2, and phases φ_1 and φ_2

$$E_1 = A_1 e^{i\varphi_1} \quad E_2 = A_2 e^{i\varphi_2} \tag{7.1}$$

The intensities of the individual waves are

$$I_1 = E_1^* E_1 = A_1^2 \quad I_2 = E_2^* E_2 = A_2^2 \tag{7.2}$$

Using eqs. (7.1) and (7.2), the intensity I of the overall wavefield equals

$$I = \left(E_1^* + E_2^*\right)(E_1 + E_2) = I_1 + I_2 + E_1^* E_2 + E_2^* E_1 = I_1 + I_2 + 2\sqrt{I_1}\sqrt{I_2}\cos(\Delta\varphi) \tag{7.3}$$

where the phase difference is $\Delta\varphi = \varphi_2 - \varphi_1$. We see that in addition to the individual intensities I_1 and I_2, eq. (7.3) contains the third term, the so-called interference term, which depends directly on the phase difference $\Delta\varphi$.

Here we introduce the term "wave coherence," which means that there are no stochastic changes in the phase difference $\Delta\varphi$, during intensity measurements. Otherwise, averaging the phase difference over the measurement period leads to eliminating the interference term in eq. (7.3). If the incoming light is characterized by the spectral interval $\Delta\lambda$ (or energy interval ΔE) around the central wavelength $\lambda = \frac{2\pi}{k}$ (or central energy E_l), one can introduce the longitudinal coherence length L_c

$$L_c = \frac{2\pi}{\Delta k} = \frac{\lambda^2}{\Delta\lambda} = \frac{E_l}{\Delta E}\lambda \tag{7.4}$$

The longitudinal coherence length defines the characteristic size of the interaction region within which interference can indeed be considered. For characteristic X-ray lines, $E_l \approx 12\cdot 10^3$ eV, $\lambda \approx 1$ Å, and $\Delta E \approx 0.5$ eV; therefore, $L_c \simeq 2.5\cdot 10^4$ Å ≈ 2.5 µm. For synchrotron radiation, monochromatization of $\frac{\Delta E}{E_l} \simeq 10^{-5}$ can be achieved easily, which gives $L_c \simeq 10^5$ Å ≈ 10 µm. For the unique (recoil-free) **Mössbauer** radiation, issuing from Co^{57} with $E_l = 14.4$ keV, $\lambda \approx 0.86$ Å, the energy width is $\Delta E = 5\cdot 10^{-8}$ eV and then $L_c \simeq 24\cdot 10^{10}$

https://doi.org/10.1515/9783111140100-008

Å ≃ 24 m. For visible light $\lambda \simeq 500$ nm; therefore, for $\frac{\Delta\lambda}{\lambda} \simeq 10^{-5}$, $L_c \simeq 5$ cm, which is enough for most practical applications. If $\frac{\Delta\lambda}{\lambda} \simeq 10^{-6}$, the coherence length is increased up to $L_c \simeq 0.5$ m.

In the case of divergent beams, the so-called transverse coherence is also considered: $L_t = \frac{2\pi}{\Delta k} = \frac{2\pi}{k\alpha} = \frac{\lambda}{\alpha}$, where α is the angular divergence.

For coherent waves, the resultant intensity depends strongly on the $\Delta\varphi$-value and, according to eq. (7.3), on the sign and magnitude of $\cos(\Delta\varphi)$. Let us consider a few illustrative cases (Figure 7.1).

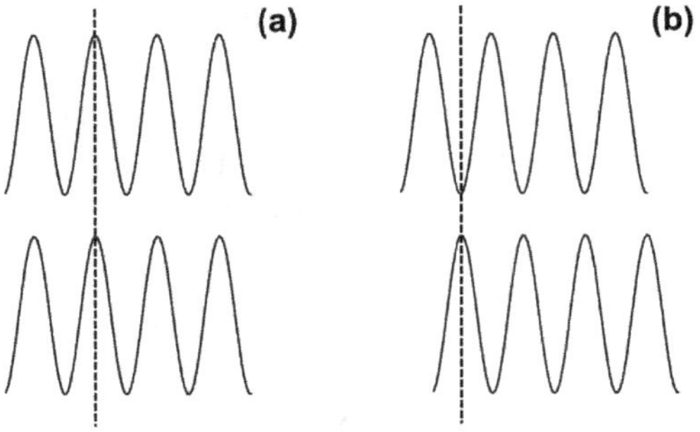

Figure 7.1: Illustration of the constructive (a) and destructive and (b) wave interference.

For example, if $\Delta\varphi = 2\pi m$, where $m = 0, \pm1, \pm2, \ldots$, that is, $\Delta\varphi$ is an even number of π, the individual waves are in phase (constructive interference, see Figure 7.1a), and the overall intensity equals

$$I = I_1 + I_2 + 2\sqrt{I_1}\sqrt{I_2} = (\sqrt{I_1} + \sqrt{I_2})^2 = (A_1 + A_2)^2 \tag{7.5}$$

In other words, in the case of constructive interference, we must summate the wave amplitudes and then calculate the wavefield intensity by squaring the summation result. This is a drastic distinction compared to the result of the incoherent wave summation (zeroing the interference term in eq. (7.3)), that is, as compared with

$$I = I_1 + I_2 \tag{7.6}$$

In this case, according to eq. (7.6), we must summate the individual wave intensities. If we have N waves with equal amplitudes A, incoherent summation gives us the total intensity

$$I_{\text{incoh}} = NA^2 = N \cdot I \tag{7.7}$$

in striking contrast to much higher (N times higher) overall intensity in case of coherent summation

$$I_{\text{coh}} = (NA)^2 = N^2 \cdot I = NI_{\text{incoh}} \tag{7.8}$$

We will consider the interference of multiple waves in more detail in the chapters devoted to light and X-ray diffraction.

Coming back to eq. (7.3) and setting $\Delta\varphi = \pi + 2\pi m$, where $m = 0, \pm1, \pm2, \ldots$, that is, $\Delta\varphi$ being an odd number of π, we find

$$I = I_1 + I_2 - 2\sqrt{I_1}\sqrt{I_2} = \left(\sqrt{I_1} - \sqrt{I_2}\right)^2 = (A_1 - A_2)^2 \tag{7.9}$$

In this situation, individual waves are in antiphase (destructive interference, see Figure 7.1b) and cancel (or weaken) each other. If $A_1 = A_2$, the overall intensity is zero.

All intermediate situations between constructive and destructive interference are also possible, depending on the value of $\Delta\varphi$. For example, setting $\Delta\varphi = \frac{\pi}{2} \pm \pi m$, we find that $I = I_1 + I_2$, as for incoherent summation.

Probably the most renowned practical example of destructive interference is the stealth aircraft technology, which reduces the visibility of military jets on radar screens dramatically. This is achieved by fabricating optimal shapes for aircraft parts that facilitate the desirable (destructive) interference conditions for the reflected electromagnetic waves.

Below we further develop the interference aspects focusing on their importance in different optical technologies.

7.1 Newton's rings

As a classical example of interference phenomena in optics we point to the so-called **Newton**'s rings. Despite the first report regarding these rings having been published by **Robert Hooke** in his 1665 book *Micrographia*, **Newton**'s rings are named after Sir **Isaac Newton** who carried out experiments in 1666 while being quarantined at home during the Great Plague in London. **Newton**'s rings (or ellipses, in the general case) arise when looking from the top of a convex glass lens with a large radius of curvature R, which is placed on a flat microscope glass slide. Apart of the contact point (which will be the center of the concentric rings or ellipses), there is a narrow air gap between the two glass surfaces. The gap magnitude T_g increases with lateral distance r from the central point (see Figure 7.2).

Let us assume that monochromatic light with fixed wavelength λ_0 is passing through the lens at normal incidence and is partially reflected from the bottom surface of the lens (ray 1 in Figure 7.2). The transmitted light passes through the glass lens and the air gap and is reflected from the surface of the glass slide (ray 2 in Figure 7.2). The two reflected rays, 1 and 2, will interfere and produce a characteristic interference pattern on the top surface of the lens, which is completely defined by the difference ΔS in optical passes equal to the double air gap: $\Delta S = 2T_g$. To quantify the interference pattern, we recall the already established fact (see **Appendix 5.A**) that the wave phase changes by π when light is coming from a less optically dense medium and is reflected from an

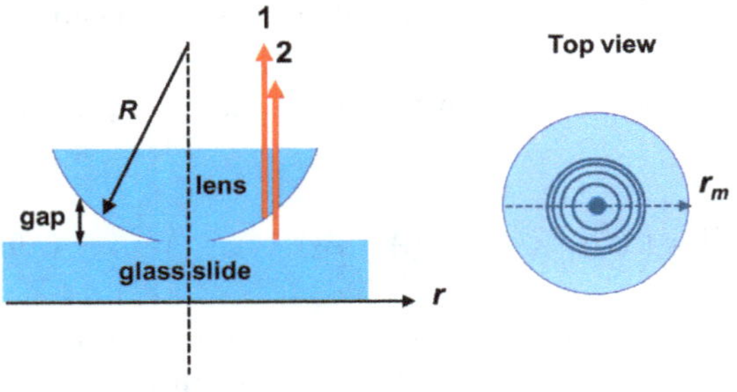

Figure 7.2: Illustration of the formation of **Newton**'s rings.

optically denser medium (i.e., for ray 2). Changing the phase by π means adding half a wavelength to the optical path. In other words, the effective optical path difference is

$$\Delta = \Delta S + \frac{\lambda_0}{2} = 2T_g + \frac{\lambda_0}{2} \tag{7.10}$$

The next step is to relate the gap T_g to the lateral distance r at which point the incoming light is striking the bottom surface of the lens (see Figure 7.2):

$$R^2 - (R - T_g)^2 = r^2 \tag{7.11}$$

At $T_g \ll R$,

$$r = \sqrt{2RT_g} \tag{7.12}$$

$$T_g = \frac{r^2}{2R} \tag{7.13}$$

Using eqs. (7.10) and (7.13), the condition of constructive interference (intensity maxima) is expressed as

$$\Delta = 2T_g + \frac{\lambda_0}{2} = 2m\frac{\lambda_0}{2} \tag{7.14}$$

or

$$\frac{r^2}{R} = (2m - 1)\frac{\lambda_0}{2} \tag{7.15}$$

where $m = 1, 2, 3, \ldots$. Respectively, for intensity minima $\Delta = 2T_g + \frac{\lambda_0}{2} = (2m + 1)\frac{\lambda_0}{2}$ or

$$\frac{r^2}{R} = 2m\frac{\lambda_0}{2} \tag{7.16}$$

where $m = 0, 1, 2, 3, \ldots$. Note that $m = 0$ in eq. (7.16) corresponds to the central point of the interference pattern which, therefore, is dark. One can unite eqs. (7.15) and (7.16) and obtain the following expression for the rings' radii:

$$r = \sqrt{\frac{R\lambda_0}{2}(m-1)} \tag{7.17}$$

with even and odd m-values ($m = 1, 2, 3, \ldots$) for the bright and dark rings, respectively. Note that at oblique angles of incidence, circular rings become elliptic.

The “rainbows” (sometimes called iridescence) observed in thin films of oil on water or in soap bubbles have similar origin. In this case, the varying film thickness plays a role of air gap in **Newton**’s experiment. Under white light illumination, the interference maxima for distinct wavelengths are seen in film sections of different thickness (see eq. (7.14)), which explains the visible “rainbow” effect. The iridescence effect is also observed on the surface of some metals (e.g., steels) covered by oxide films which arise during thermomechanical processing. We will elaborate on the interference in thin films in more detail in the following section.

7.2 Interference in thin films

Let us consider light reflection from two sides of a thin film with refractive index n, placed in the air and having thickness T (see Figure 7.3).

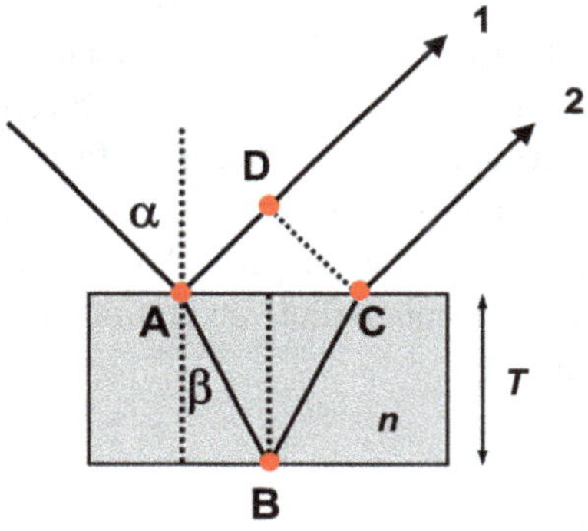

Figure 7.3: Calculating the interference effect in thin films.

Incoming light experiences scattering at points A and B, and, as a result, two scattered waves, marked 1 and 2, interfere close to the exit surface of the film. The interference outcome depends on the optical path difference ΔS between these two trajectories ABC and AD, that is, $\Delta S = 2nAB - AD$. From the geometrical construction in Figure 7.3, one obtains

$$\Delta S = 2n\frac{T}{\cos\beta} - 2T \cdot \tan\beta \cdot \sin\alpha \tag{7.18}$$

Using **Snell's** law (2.26) yields

$$\Delta S = 2Tn\cos\beta \tag{7.19}$$

Applying **Snell's** law again, we finally find

$$\Delta S = 2T\sqrt{n^2 - \sin^2\alpha} \tag{7.20}$$

There is an additional phase change by π for ray 1 after reflection at the air/film interface. Correspondingly, the effective optical path difference equals $\Delta = \Delta S + \frac{\lambda_0}{2}$. Therefore, for constructive interference, the magnitude of Δ should be

$$2T\sqrt{n^2 - \sin^2\alpha} = (m + 1/2)\lambda_0 \tag{7.21}$$

where $m = 0, 1, 2, 3, \ldots$. At normal incidence ($\alpha = \beta = 0$, and for $m = 0$), eq. (7.21) transforms into

$$T = \frac{\lambda_0}{4n} = \frac{\lambda}{4} \tag{7.22}$$

which defines the respective film thickness, being a quarter of the wavelength in a medium, $\lambda = \frac{\lambda_0}{n}$. Such film (or plate) is called quarter-wave film (or plate).

Recently, this result was employed by researchers from Technion-Israel Institute of Technology to improve the yield of light-assisted water splitting into hydrogen and oxygen. In this method, the key issue is the optimal design of thin-film photoanodes toward more effective light absorption and generation of charge carriers, which can reach the anode surface and oxidize water before recombination occurs. Specifically, in the discussed application, a quarter-wave α-Fe_2O_3 photoanode film was deposited on the back-reflector substrate, the latter being coated with a silver-gold alloy. In this way, the resonant light trapping is realized within a photoanode film, facilitating high quantum efficiency of the light-to-charge conversion.

In turn, destructive interference takes place when

$$2T\sqrt{n^2 - \sin^2\alpha} = m\lambda_0 \tag{7.23}$$

The first minimum ($m = 1$) at normal incidence $(\alpha = \beta = 0)$ occurs at the film thickness

$$T = \frac{\lambda_0}{2n} = \frac{\lambda}{2} \tag{7.24}$$

that is used for producing special optical coatings (half-wave films or plates) exploited in different optical devices. One example is antireflective-coated optics. In **Appendix 5.A** we learned that each reflection at the glass/air interface leads to about 4% of reflected light. The presence of several reflecting surfaces within an optical system leads to significant light losses. To surmount this problem, the reflecting surface is covered by a film with a refractive index differing from that of glass. The film thickness is chosen to provide destructive interference between waves reflected from the two film's surfaces.

What does constructive interference mean in the case of X-rays? Since X-rays experience scattering by atomic planes, the rays 1 and 2 in Figure 7.3 originate from two parallel atomic planes separated by interplanar spacing d. This means that thickness T in eq. (7.21) should be replaced by the d-value. Furthermore, in the X-ray frequency range $n \simeq 1$, that is, there is no additional phase jump of π accompanying the scattering process. Correspondingly, eq. (7.21) transforms into

$$2d\sqrt{1-\sin^2\alpha} = 2d\cos\alpha = m\lambda_0 \tag{7.25}$$

with $m = 1, 2, 3, \ldots$. In the field of X-rays, the incident angle $\theta = 90° - \alpha$ is not counted from the normal to an atomic plane, but from the plane itself, which yields the famous **Braggs'** condition for diffraction angles, $2d\sin\theta_B = m\lambda_0$. Note that in the X-ray literature, one can often find a slightly different equation

$$2d\sin\theta_B = \lambda_0 \tag{7.26}$$

in which an integer m is hidden within the d-spacing, the latter being proportionally reduced for multiple reflections. In the X-ray domain, intentionally grown multilayers are widely used as elements of X-ray optics. They are produced by successive depositing of individual sublayers with different scattering powers (e.g., W and Si). In this case, the multilayer period T equals the sum of the thicknesses of two adjacent sublayers. Correspondingly, eq. (7.26) transforms into

$$2T\sin\theta_B = \lambda_0 \tag{7.27}$$

In this way, the **Bragg** angle can be substantially reduced, which is important for some applications. We will discuss these issues further in Sections 10.2 and 10.3.

It is possible to expand the light optics approach to multilayers, composed of several materials, which are arranged in layers with different thicknesses and refractive indices. In the case of a heterostructure (two layers with thicknesses T_1 and T_2 and refractive indices n_1 and n_2), when calculating the optical path difference one obtains expression such as eq. (7.19):

$$\Delta S = 2T_1n_1\cos\beta_1 + 2T_2n_2\cos\beta_2 \tag{7.28}$$

where β_1 and β_2 are refraction angles in materials 1 and 2, respectively. Correspondingly, constructive interference occurs when

$$\Delta S = 2t_1n_1\cos\beta_1 + 2t_2n_2\cos\beta_2 = (m + 1/2)\lambda_0 \tag{7.29}$$

while destructive interference when

$$\Delta S = 2t_1n_1\cos\beta_1 + 2t_2n_2\cos\beta_2 = m\lambda_0 \tag{7.30}$$

7.3 Structural colors

Light interference in thin films governed by eqs. (7.29) and (7.30) is responsible for specific optical effects in living organisms (and beyond), which are called structural colors. Note that colors in nature are mostly produced by chemical dyes and pigments, in which certain molecules selectively absorb and emit light in different spectral intervals. In some cases, however, nature also uses light interference within multilayered structural components, in which interference conditions for different wavelengths are determined by refractive indices and thicknesses of respective constituent sublayers (see eqs. (7.29) and (7.30)). Evidently, for this purpose the thicknesses of individual sublayers (or crystal blocks) should be on the light wavelength scale.

If multiple scattering at interfaces is substantial, we must account for diffraction phenomena (see Chapters 10 and 12). The latter, as for electron waves in regular crystals or light propagation in photonic structures (see Chapter 16), leads to the formation of the forbidden frequency (wavelength) bands. This means that photons having these wavelengths will be expelled from a multilayer. In summary, we can say that structural colors result from the combined action of interference and diffraction phenomena.

The most famous examples in nature are the colors of butterfly wings (Figure 7.4) and bird feathers (Figure 7.5), which originate in the periodic arrangement of submicron thick chitin plates.

Figure 7.4: Blue structural color of butterfly wings (credit: Wikimedia Commons).

Alpha-chitin is a long-chain polymer with a general formula $(C_8H_{13}O_5N)_n$. It has an orthorhombic structure (space group $P2_12_12_1$) with lattice parameters $a = 0.474$ nm, $b = 1.886$ nm, and $c = 1.032$ nm. The adjacent polysaccharide chains, running in antiparallel directions along the c-axis, form the main motif of the chitin structure. Chitin has a high refractive index $n = 1.55$, which facilitates effective control of natural colors. It is worth mentioning that diffraction patterns from a bird feather were observed already

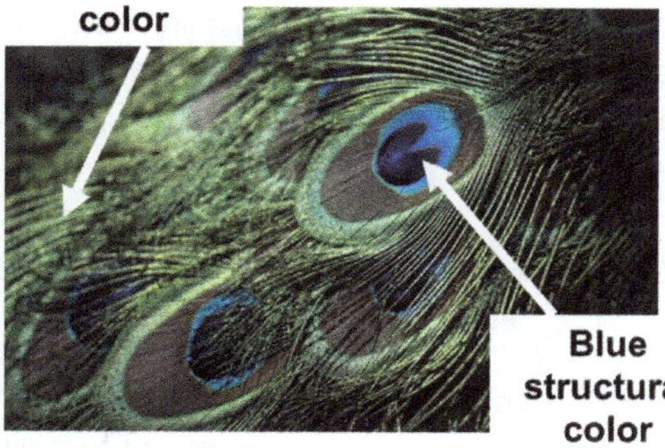

Figure 7.5: Structural colors in peacock feathers (credit: Wikimedia Commons).

in the seventeenth century by **James Gregory**, about a year after **Newton**'s prism experiments. In the same way, natural diffraction gratings were discovered far earlier than their industrial production in the nineteenth century. Diffraction gratings and their functioning toward color splitting of white light will be discussed in more detail in Section 10.1.

Light manipulation in various organisms is also achieved by the usage of ordered arrays of plate-shaped guanine crystals. Anhydrous guanine crystals with the chemical formula $C_5H_5N_5O$ have a monoclinic structure (space group $P2_1/c1$) with H-bonded layers stacked along the shortest a-axis ($a = 0.355$ nm). We point out that the refractive index along this axis is extremely high ($n = 1.83$), which explains why guanine nanocrystals are widely used in nature for producing structural colors. The most well-known examples include fish scales, planktonic crustaceans, reptiles, and arthropods as well as some animal eyes. Furthermore, several animals can intentionally alter the mutual orientation of guanine crystals and in that way achieve very important biological functions such as camouflage. The most striking example of the latter is the skin color change (very fast and reversible) of chameleons. They can regulate the distance between guanine nanocrystals within quasi-periodic arrays, for example, by changing the osmotic pressure.

New possibilities to produce structural colors are offered by so-called plasmonic nanostructures, which are classified as a branch of metamaterials introduced in Chapter 6. Their light manipulation is based on utilizing the properties of surface plasmons (collective oscillations of free electrons), arising at interfaces between metals and dielectrics. Surface plasmons have lower frequencies than bulk plasmons; moreover, the resonant frequencies, which govern the resonant absorption and scattering of light, strongly depend on the geometry and sizes of characteristic surface nanofeatures. For example, when the light strikes a metallic nanoparticle, the electric field of the incoming light wave forces free electrons to oscillate inside the particle, leading to enhanced optical absorption at certain resonant wavelengths. Such localized surface plasmon resonance modes are capable of confining light within nano-size volumes that is used to design color filters with extremely high spatial resolution. For this purpose, various metal meta-surfaces are employed, including sub-wavelength gratings and nano-patches as well as nano-disk and nano-hole arrays.

7.4 Lippmann's color photography

In 1908, **Gabriel Lippmann** was awarded the Nobel Prize in Physics "for his method of reproducing colors photographically based on the phenomenon of interference." Why did **Lippmann**'s discovery draw such great attention and deserve the highest scientific appreciation? To answer this question, we need to delve into the history of photography.

For many years following its invention in the 1820–1830s by **Nicephore Nièpce** and **Louis Daguerre**, photography remained black and white, despite numerous efforts to reproduce colors. Among these attempts, it is worth mentioning experiments by **Edmond Becquerel**, who in 1848 demonstrated photochromatic images produced on a silver plate coated with a thin layer of silver chloride. The colored images were stable in the dark but disappeared upon being exposed to daylight. Another approach was suggested in the 1850s by **James Clerk Maxwell** but developed systematically only at the end of the nineteenth century and the beginning of the twentieth century by **Sergey Prokudin-Gorsky**.

This method is based on capturing three black-and-white photographs, sequentially taken on glass photographic plates through the red, green, and blue filters. These photographs are then projected through filters of the same colors and precisely superimposed on a screen, synthesizing the original color range, as the human eye does routinely. Another way is individually viewing an additive image through an optical device known as a chromoscope or photochromoscope. The latter contains colored filters and transparent reflectors that visually combine the red, yellow, and blue components into one full-color image. Even after 100 years since the invention of this method, the quality of colored photographs is still remarkable (see Figure 7.6). The method, however, is time-consuming and rather difficult for processing, especially regarding the care that must be taken to precisely superimpose the filtered images. Therefore, it is not surprising that the scientific community was greatly impressed by the brand-new technique of color photography invented by **Gabriel Lippmann**, who used the interference phenomenon, that is, wave optics.

Lippmann's idea is depicted schematically in Figure 7.7. Light from a colored object is passing through a photographic plate, placed on the cuvette filled with liquid mercury, and then is reflected from the surface of the latter.

Taking the positive z-direction outward from the entrance surface of the liquid mercury, one can express the incoming wave as

Figure 7.6: Self-portrait of **Prokudin-Gorsky** made using his method of color photography (credit: Wikimedia Commons).

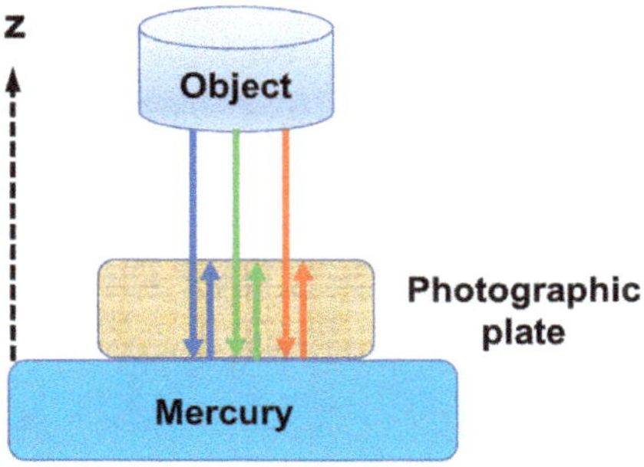

Figure 7.7: Principal scheme of **Lippmann**'s experiment.

$$E_1 = E_0 \exp[i(-k_z z - \omega t)] \tag{7.31}$$

and the reflected wave as

$$E_r = E_0 \exp[i(k_z z - \omega t)] \tag{7.32}$$

Interference of these two waves within the photographic plate produces a standing wave across the plate thickness

$$E_s = E_1 + E_r = 2E_0 \exp(-i\omega t)\cos(k_z z) \tag{7.33}$$

with the intensity equal to

$$I_s = E_s^* E_s = 4E_0^2 \cos^2(k_z z) = 4E_0^2 \cos^2\left(\frac{2\pi}{\lambda} z\right) \tag{7.34}$$

Correspondingly, intensity maxima are situated at depth z_m, calculated from the surface of the liquid mercury

$$z_m = \frac{\lambda}{2} m \tag{7.35}$$

where $m = 0, 1, 2, \ldots$.

At this point, we emphasize two aspects explaining the choice of liquid mercury: (i) at room temperature liquid mercury is a metal with practically 100% light reflectance, which justifies using equal amplitudes E_0 in eqs. (7.31) and (7.32); (ii) liquid mercury has a perfectly flat surface, which is crucial for the lateral homogeneity of the standing wave condition (7.34) across the photographic plate.

Coming back to eq. (7.35), we find for each wavelength from the color spectrum, its own system of intensity maxima. After the development of the photographic plate, these intensity maxima form the wavelength-mediated systems of sublayers parallel to the plate surface. Upon illumination of the developed photographic plate by white light, each wavelength will produce constructive interference patterns originating in the respective system of sublayers. Superposition of such multiple patterns will restore the entire wavefront coming from an object, creating its colored image. The quality of thus produced images is amazing (see Figure 7.8) and the photographic plates have not deteriorated even after being stored for more than 100 years. Nowadays, we can consider

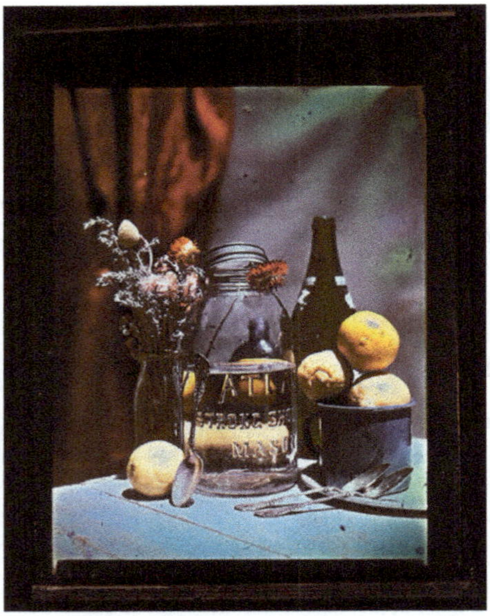

Figure 7.8: Nature morte (still life) picture taken using **Lippmann**'s method (reproduced with permission from Nick Brandreth's Lippmann plates).

this method as the first important step toward holography invented about 50 years later. The latter imaging technique is discussed below in Section 7.5.

7.5 Gabor holography

In 1971, **Dennis Gabor** was awarded the Nobel Prize in Physics "for his invention and development of the holographic method." What kind of experience and knowledge were behind this groundbreaking discovery? As **Gabor** himself related, in 1940s he began thinking about the potential ways for improving the resolving power of electron microscopy toward seeing individual atoms. At some stage, he assumed that it could be done in principle by including the phases of the scattered waves in the "optical" image formation in addition to the scattering intensity. Later, it turned out that it is much easier to implement these ideas in the field of light optics rather than electron optics. Nevertheless, it was the revolutionary change of the existing optical paradigm, which historically considered an optical image as the spatial distribution of light intensity only (as in the human eye). We remind the reader that conventionally, light intensity is defined as the product of the complex conjugate ($E^* = A^* e^{-i\varphi}$) and direct ($E = Ae^{i\varphi}$) wavefields, that is, $E^*E = |A|^2$, within which the phase information (about function φ) is lost. Certainly, **Dennis Gabor** stood on "the shoulders of giants." As **Gabor** mentioned in his Nobel Prize lecture, he was greatly inspired by previous phase manipulations in optics realized by **Gabriel Lippmann** (see previous Section 7.4) and **Frits Zernike** – the father of the phase-contrast microscopy (see Section 9.1).

The principal scheme of the holography experiment is depicted in Figure 7.9. Light issuing from the source is split by a semitransparent mirror (beam splitter), producing two coherent beams. One (illumination beam) is further reflected from an object and contains complete information on the amplitudes $A_{ob}(x,y)$ and phases $\varphi(x,y)$ of waves scattered by different object points (x,y).

This wavefield is analytically described as

$$E_{ob} = A_{ob}(x,y)e^{i\varphi(x,y)} \tag{7.36}$$

The second (reference) beam is directed straightforward to the photographic emulsion, in which it is coherently mixed with the scattered one. The reference wavefield is described as

$$E_{ref} = A_{ref}(x,y)e^{i\psi(x,y)} \tag{7.37}$$

while the mixed wavefield is the sum of eqs. (7.36) and (7.37)

$$E_{mix} = A_{ob}(x,y)e^{i\varphi(x,y)} + A_{ref}(x,y)e^{i\psi(x,y)} \tag{7.38}$$

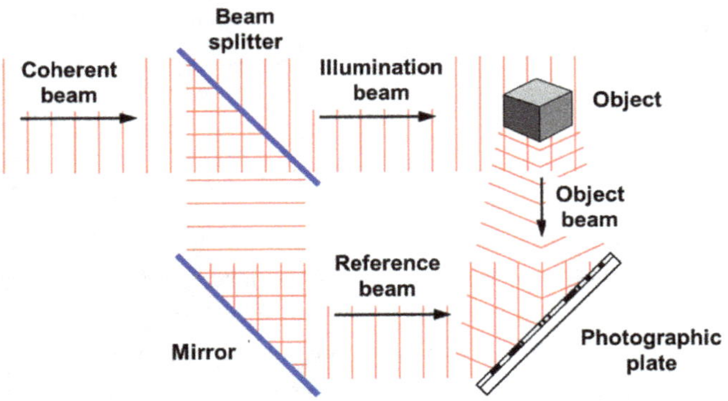

Figure 7.9: Principal scheme of a hologram recording.

Correspondingly, the intensity of the arising interference pattern, which is recorded within the photographic emulsion, is

$$I_{\text{mix}} = E^*{}_{\text{mix}} E_{\text{mix}} = |A_{\text{ob}}|^2 + |A_{\text{ref}}|^2 + A^*{}_{\text{ob}} A_{\text{ref}} e^{i(\psi - \varphi)} + A^*{}_{\text{ref}} A_{\text{ob}} e^{i(\varphi - \psi)} \tag{7.39}$$

The intensity distribution contains complete information on the wavefront scattered by the object, which is recorded as a pattern of dark and bright elements, originating in the minima and maxima of the interference pattern (7.39). Therefore, in contrast with conventional photography, in this case the photographic plate does not contain the direct image of the object. Rather it holds the holographic information on the related wavefront, information that looks like a collection of dark and bright fragments.

To visualize the object, we must reconstruct the wavefront information written on the photographic plate. This is done by illuminating the latter by the same reference wave that was used for the hologram recording. As a result, the reference wave passing through the plate is modulated by the written (hidden) intensity distribution (7.39)

$$I_{\text{out}} = I_{\text{mix}} E_{\text{ref}} = \left[|A_{\text{ob}}|^2 + |A_{\text{ref}}|^2 + A^*{}_{\text{ob}} A_{\text{ref}} e^{i(\psi - \varphi)} + A^*{}_{\text{ref}} A_{\text{ob}} e^{i(\varphi - \psi)} \right] A_{\text{ref}} e^{i\psi} = I_1 + I_2 + I_3 + I_4 \tag{7.40}$$

For us, the most interesting item in the above equation is the fourth term

$$I_4 = |A_{\text{ref}}|^2 A_{\text{ob}} e^{i\varphi} \tag{7.41}$$

which carries information on the object wavefield (7.36), that is, $E_{\text{ob}} = A_{\text{ob}}(x, y) e^{i\varphi(x,y)}$. If illumination of the hologram is done by the conjugate reference wave $E^*{}_{\text{ref}} = A^*{}_{\text{ref}}(x, y) e^{-i\psi(x,y)}$, the modulated signal has the following form:

$$I_{out} = I_{mix}E^*{}_{ref} = \left[|A_{ob}|^2 + |A_{ref}|^2 + A^*{}_{ob}A_{ref}e^{i(\psi-\varphi)} + A^*{}_{ref}A_{ob}e^{i(\varphi-\psi)}\right]A^*{}_{ref}e^{-i\psi} = I_1 + I_2 + I_3 + I_4 \tag{7.42}$$

In this case, the third term

$$I_3 = |A_{ref}|^2 A^*{}_{ob}e^{-i\varphi} \tag{7.43}$$

represents the object wavefield. To correctly reconstruct the object's wavefront, we must use the terms I_3 or I_4 and eliminate the contributions I_1 and I_2. The well-accepted method to do this is by using the inclined reference wave, the propagation direction of which makes some angle with respect to the optical axis connecting an object and the photographic plate. This technique was invented in 1962 by **Emmet Leith** and **Juris Upatnieks**. The images (the real or virtual ones, as in Figure 7.10) obtained in this way give us a three-dimensional representation of the object.

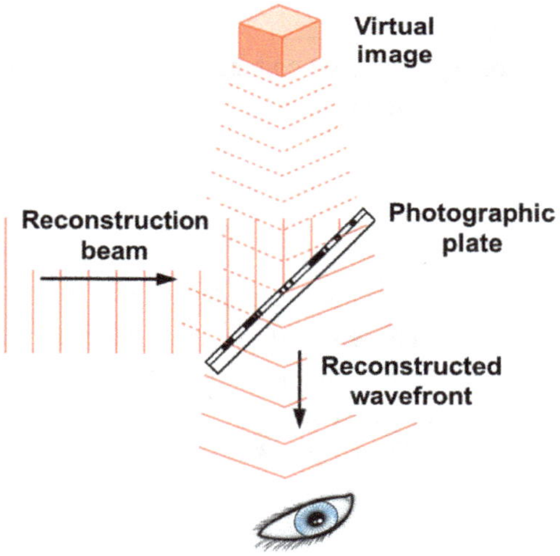

Figure 7.10: Reconstruction of the holographic image.

In holography, the coherence of the light source is of paramount importance. For this reason, the first holograms recorded using conventional light sources, as high-pressure mercury lamps with only $L_c = 0.1$ mm coherence length, were of low quality. In fact, at such short coherence length, only a small number of spatial fringes, about $\frac{2L_c}{\lambda} \simeq 200$, could be detected. Hologram quality was greatly improved (see Figure 7.11) after the invention of lasers, which provided a high photon flux with coherence length on a meter scale. For example, multimode helium–neon lasers have a typical coherence length of 0.2 m, while the coherence length of single-mode lasers can exceed 100 m. Similar numbers are achieved with semiconductor lasers.

Figure 7.11: Example of a 3D holographic image reconstruction: virtual image of "flying" elephant (reproduced with permission from Voxon Photonics).

Nowadays, holography is a well-developed technique that is used for data storage, coding and decoding, security (e.g., for face recognition), art collections, visualization of tiny static and dynamic object displacements, and in many other fields.

7.6 Cherenkov radiation

Another key phenomenon, which can be connected to light interference, is the so-called **Cherenkov** radiation. It was discovered by **Pavel Cherenkov** in 1934 and explained later by **Ilya Frank** and **Igor Tamm** (the 1958 Nobel Prize in Physics). **Cherenkov** radiation emerges when a charged particle is moving within a medium with the velocity V_{cp}, overcoming the phase velocity of light in that medium V_p, that is,

$$V_{cp} > V_p = \frac{c}{n} \tag{7.44}$$

where n is the refractive index. Before this discovery, it was commonly accepted that only accelerated charged particles emit radiation, but not those moving with constant speed. Incidentally, **Cherenkov** radiation provides the bright blue color of water in the water-protected nuclear reactors.

Let us derive the kinematic condition (7.44) using wave interference. For this purpose, we divide the particle path by segments of length a and assume that light emission occurs at the end of each segment (see Figure 7.12). This is done for the sake of convenience only; the length a will be eliminated from calculations at the end. For constructive interference, the phase difference between rays A and B in Figure 7.12 should be $\Delta\varphi = \varphi_B - \varphi_A = 2\pi m$.

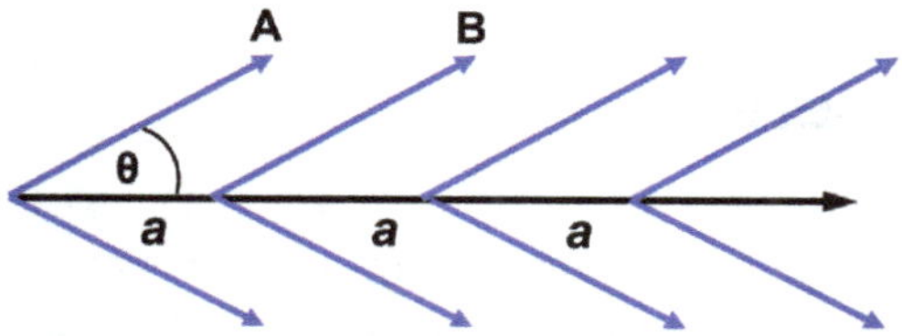

Figure 7.12: Illustration of **Cherenkov** radiation emission.

Taking $m = 0$ and attributing $\varphi_A = 0$, we find

$$\Delta\varphi = ka\cos\Theta - \omega t = nk_0 a\cos\Theta - k_0 c\frac{a}{V_{\text{cp}}} = 0 \tag{7.45}$$

where k_0 is the wave vector of light in a vacuum. It follows from eq. (7.45) that

$$\cos\Theta = \frac{c}{nV_{\text{cp}}} \tag{7.46}$$

Since $\cos\Theta < 1$, one obtains the **Cherenkov** radiation condition (7.44), that is, $V_{\text{cp}} > \frac{c}{n}$.

Chapter 8
Light and X-ray interferometers

Remarkable applications of wave interference are realized with the aid of optical instruments called interferometers. These devices are widely used for careful measurements of different physical parameters, especially for precise measurements of phase shift, path length, and related quantities. All interferometers exploit the same principle: splitting of the incident light into two coherent beams in one spatial location, propagation of these beams along different trajectories, and merging the beams in another place by optical means, followed by subsequent detection of the arising interference pattern. The latter is especially sensitive to subtle changes in phase difference accumulated during beams' propagation. In this chapter we will discuss the construction and applications of the **Michelson**, **Fabry–Pêrot**, and **Mach–Zehnder** light interferometers, as well as the **Bonse–Hart** X-ray interferometer.

8.1 Michelson interferometer

Perhaps the most famous interferometer is the one invented by **Albert Michelson** at the end of the nineteenth century and used for numerous applications, some of which are mentioned below. For these activities, **Michelson** was awarded the 1907 Nobel Prize in Physics "for his optical precision instruments and the spectroscopic and metrological investigations carried out with their aid."

The schematic design of the **Michelson** interferometer is depicted in Figure 8.1.

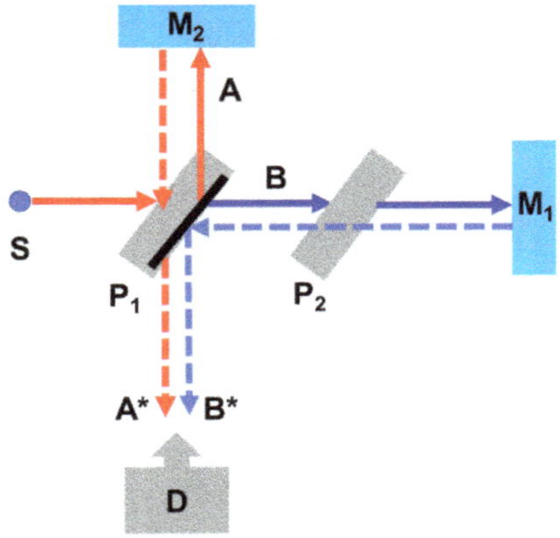

Figure 8.1: Scheme of the **Michelson** interferometer.

https://doi.org/10.1515/9783111140100-009

A light source (S) is placed in the focal point of a lens, which provides a parallel light beam. The latter passes through a semitransparent glass plate (P_1) covered by a reflecting thin silver layer. Here the light splits into two beams, a reflected (A) one and a transmitted (B) one, having practically equal intensities. Beam A is then reflected by the mirror M_2 and returns to plate P_1, where it again splits into two components: the component, which goes toward the source (not our focus here), and the beam A*, which participates in the formation of the interference pattern in the detector (D). Beam B reflects from the mirror M_1 and splits again into another two components when it passes through plate P_1. One component is not important for us since it propagates toward the source, in contrast to the second one (B*), which contributes to the interference pattern. Note that the beam B/B* passes through plate P_1 only once, whereas the beam A/A* does so three times. To compensate for the arising (nondesirable) phase difference, a plate identical to plate P_1, i.e., plate P_2, is placed between plate P_1 and the mirror M_1. Now, the phase difference, accumulated between the beams A/A* and B/B*, is determined by the difference between the respective optical paths only. If the mirrors M_1 and M_2 are parallel to each other, the interference pattern resembles a sequence of concentric bright and dark rings, similar to **Newton**'s rings, which we discussed in Section 7.1. As we know, the shift of the interference pattern between the dark and bright rings corresponds to the optical path change by $\lambda/2$ (see eqs. (7.15) and (7.16)).

In the 1890s, such an interferometer was used by **Michelson** to develop his etalon and define the length standard (1 m) via the wavelength ($\lambda = 643.8469$ nm) of the red light emitted by the thermally excited Cd atoms. He established a rather complicated experimental procedure, which included the comparison of the interference patterns sequentially observed with several etalons of different lengths. Each etalon represented the rigid construction of two parallel mirrors separated by a certain empty interval of length l (see Figure 8.2).

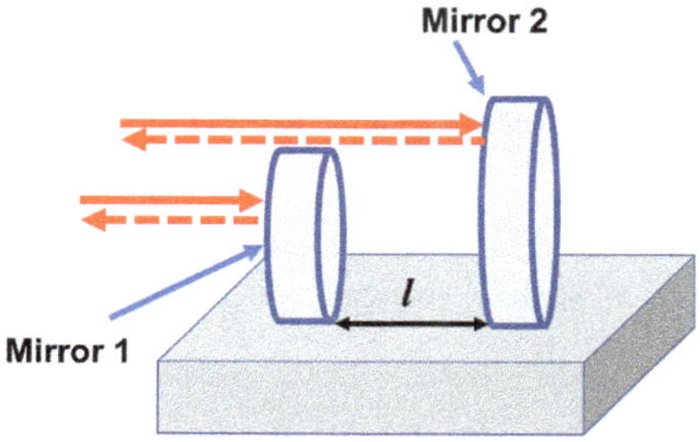

Figure 8.2: Design of the **Michelson** etalon.

At this point, a sequence of nine etalons, each being twice as long as the preceding one, was fabricated. The first etalon was $l_1 = 0.39$ mm long, whereas the length of the last one was $l_9 = 2^9 \cdot l_1 \approx 100$ mm. In the scheme depicted in Figure 8.1, these etalons were placed (one by one) instead of the mirror M_2. The last etalon was compared with standard meter in the International Bureau of Weights and Measures (Sêvres, France). Statistical treatment of the obtained results gave the following result: 1 m corresponds to 1,553,163.5 wavelengths of the Cd-114 red light.

Work on length standardization continued until 1960, when 1 m was defined in the International System of Units (SI) as equal to 1,650,763.73 wavelengths of the orange-red emission line ($\lambda = 605.78$ nm) of the Kr-86 atom in a vacuum. With time, it turned out that the speed of light can be measured more precisely than the wavelength (see Section 8.2), and since 1983, 1 m has been defined as the length of the path travelled by light in vacuum during a time interval of 1/299,792,458 of a second.

8.2 Fabry–Pêrot interferometer

Nowadays, the wavelengths and frequencies of many spectral lines are known with high precision. For these measurements, the so-called **Fabry–Pêrot** interferometer (etalon) is often used. This interferometer was invented by **Charles Fabry** and **Alfred Pêrot** in 1899. In its simplest form, the **Fabry–Pêrot** interferometer comprises two semitransparent glass plates with internal surfaces covered by reflecting silver films, which are placed face-to-face with respect to each other. These mirrors are separated by a certain distance l and fixed to be strictly parallel (Figure 8.3).

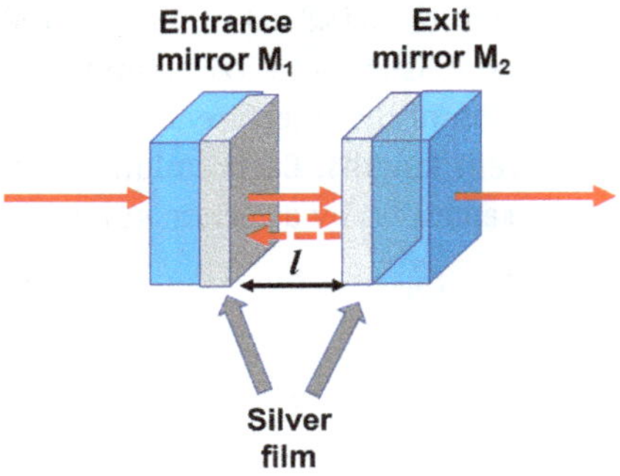

Figure 8.3: Design of the **Fabry–Pêrot** etalon.

Monochromatic rays, being reflected from the mirrors, produce a ring-like interference pattern. If, for two different wavelengths λ_1 and λ_2, the central spot is bright (constructive interference), then according to eq. (7.14)

$$2l = (m_1 + 1/2)\lambda_1; \quad 2l = (m_2 + 1/2)\lambda_2 \tag{8.1}$$

where l is the length of the etalon and m_1 and m_2 are integers. If the length l is approximately known, the integers m_1 and m_2 can unquestionably be determined and the etalon length can be expressed via wavelengths λ_1 and λ_2. Knowing the light frequency $\omega = 2\pi\nu$ and wavelength λ, the speed of light c is then calculated using the relationship

$$c = \lambda\nu \tag{8.2}$$

The **Fabry–Pêrot** interferometer, however, has much wider applications based on the complex character of the wavefield within it. In fact, if the light reflection coefficient ζ from each mirror is close to 1, the multiple waves will propagate back and forth between the mirrors with rather large amplitudes and different phases (see Figure 8.4).

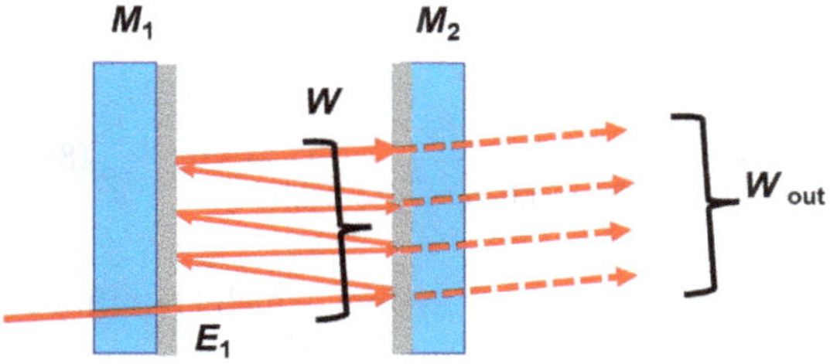

Figure 8.4: Wave propagation within interferometer (etalon) **Fabry–Pêrot**.

The entire wavefield amplitude W within such a "cavity" close to the exit mirror M_2 is the sum of N waves, including the incident wave E_1 just behind the entrance mirror M_1 and $N-1$ waves, which experienced double reflection (i.e., weakened by factor ζ^2) from the mirrors M_2 and M_1:

$$W = E_1 + E_2 e^{i\varphi} + E_3 e^{2i\varphi} + \cdots + E_N e^{i(N-1)\varphi} \tag{8.3}$$

With no physical absorption of light within an interferometer,

$$E_n = \zeta^2 E_{n-1} \tag{8.4}$$

Furthermore, the phase φ accumulated when light propagates the double distance $2l$ between the mirrors (i.e., back and forth) is

$$\varphi = \frac{2\pi}{\lambda} 2l \tag{8.5}$$

Combining eqs. (8.3) and (8.4) yields

$$W = E_1 \left[1 + \zeta^2 e^{i\varphi} + \zeta^4 e^{2i\varphi} + \cdots + \zeta^{2(N-1)} e^{i(N-1)\varphi} \right] \tag{8.6}$$

Equation (8.6) represents the geometric progression with the common ratio q:

$$q = \zeta^2 e^{i\varphi} \tag{8.7}$$

Its sum is

$$W = E_1 \frac{1-q^N}{1-q} = E_1 \frac{1-\zeta^{2N} e^{iN\varphi}}{1-\zeta^2 e^{i\varphi}} \tag{8.8}$$

Since $\zeta < 1$, for $N \gg 1$, the amplitude ratio $\frac{W}{E_1}$ becomes

$$\frac{W}{E_1} \simeq \frac{1}{1-\zeta^2 e^{i\varphi}} \tag{8.9}$$

Respectively, the intensity ratio is

$$\mathcal{R}_{\text{in}} = \left|\frac{W}{E_1}\right|^2 = \frac{1}{\left(1-\zeta^2 e^{i\varphi}\right)} \cdot \frac{1}{\left(1-\zeta^2 e^{-i\varphi}\right)} = \frac{1}{(1-\zeta^2)^2 + 4\zeta^2 \sin^2\frac{\varphi}{2}} \tag{8.10}$$

We see that the **Fabry–Pêrot** interferometer is a resonant system, which amplifies an incident intensity many times, up to

$$\mathcal{R}_{\text{in}} = \frac{1}{(1-\zeta^2)^2} \tag{8.11}$$

at $\sin\frac{\varphi}{2} = 0$, that is, when all waves are in-phase, $\varphi = 2m\pi$. With the aid of eq. (8.5), the resonant condition has the following form:

$$2l = m\lambda \tag{8.12}$$

If individual waves are in antiphase,

$$2l = (2m+1)\frac{\lambda}{2} \tag{8.13}$$

$\frac{\varphi}{2} = \frac{\pi}{2}(2m+1)$, $\sin^2\frac{\varphi}{2} = 1$, and the intensity ratio drops to

$$\mathcal{R}_{\text{in}} = \frac{1}{(1+\zeta^2)^2} \tag{8.14}$$

which for $\zeta \simeq 1$, equals $\mathcal{R}_{\text{in}} \simeq \frac{1}{4}$.

Near the resonances, the phase equals $\varphi = (2\pi m - y)$ with $y \ll 1$, and the intensity ratio is described by the **Lorentzian** function

$$\mathcal{R}_{\text{in}} = \frac{1}{(1-\zeta^2)^2 + \zeta^2 y^2} = \frac{(1-\zeta^2)^{-2}}{1+\left(\frac{\zeta}{1-\zeta^2}\right)^2 y^2} = \frac{(1-\zeta^2)^{-2}}{1+\left(\frac{y}{\Gamma}\right)^2} \tag{8.15}$$

having half width at half maximum on the phase scale, Γ_y, equal to

$$\Gamma_y = \frac{1-\zeta^2}{\zeta} \tag{8.16}$$

When $\zeta \simeq 1$, the **Lorentzian** width becomes very narrow (Figure 8.5), which means that our resonator is of high quality. We see in Figure 8.5 that even a modest increase in parameter ζ from $\zeta = 0.92$ up to $\zeta = 0.98$ leads to drastic resonance sharpening. The sharpest resonances allow us to carefully measure the distances, wavelengths, and related physical parameters with the highest precision.

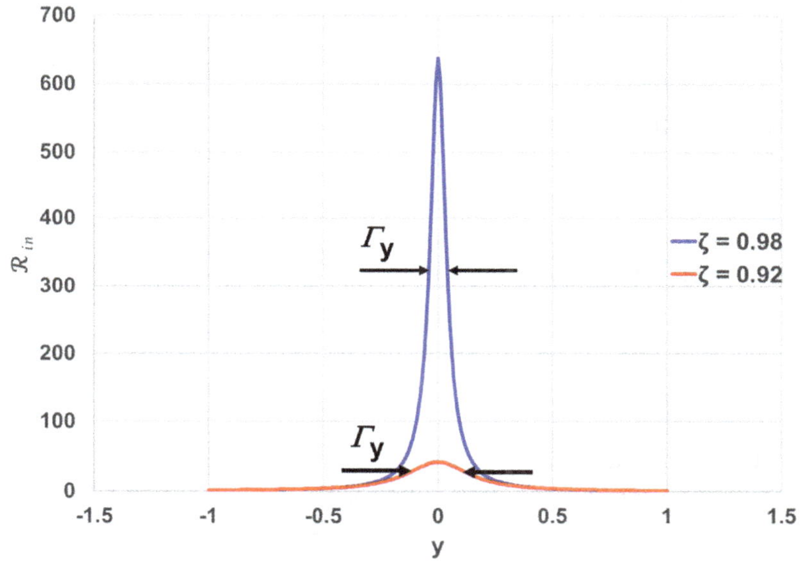

Figure 8.5: Drastic narrowing of the resonance signal in line with the increasing reflection coefficient of the mirror system.

For example, a cavity resonator with adjustable length was used to measure the speed of light with 0.14 ppm precision. In this experiment, to detect successive resonances separated by half wavelength intervals, the electromagnetic wave frequencies were tuned to around 9 GHz, while the resonator length was systematically varied (see eq. (8.12)). Due to the sharpness of the resonances, the wavelength was precisely determined, which, together with the tuned frequency, enabled the velocity calculation using eq. (8.2).

The fields of applications for cavity resonators are very wide and include spectroscopy, frequency filtering in telecommunications, and laser technology. For us, the latter is extremely interesting since most common lasers have, as a necessary component, the

Fabry–Pêrot resonator (or cavity) filled by active (gain) medium, which triggers stimulated light emission (see Section 8.2.1).

In this context, we stress that all expressions derived earlier in this section are related to the wavefield inside the resonator. To find the output intensity, we first have to multiply each term in eq. (8.3) by factor $(1-\zeta)$. Correspondingly, the output amplitude W_{out} is

$$W_{\text{out}} = E_1(1-\zeta)\left(\frac{1-\zeta^{2N}e^{iN\varphi}}{1-\zeta^2 e^{i\varphi}}\right) \tag{8.17}$$

and the intensity ratio $\left|\frac{W_{\text{out}}}{E_1}\right|^2$ for $N \gg 1$ becomes

$$\mathcal{R}_{\text{out}} = \left|\frac{W_{\text{out}}}{E_1}\right|^2 = \frac{(1-\zeta)^2}{(1-\zeta^2)^2 + 4\zeta^2 \sin^2\frac{\varphi}{2}} \tag{8.18}$$

Even in the resonance, when $\sin\frac{\varphi}{2} = 0$, the ratio equals

$$\mathcal{R}_{\text{out}} = \frac{(1-\zeta)^2}{(1-\zeta^2)^2} = \frac{1}{(1+\zeta)^2} \tag{8.19}$$

that is, it is rather small, $\mathcal{R}_{\text{out}} \simeq \frac{1}{4}$. If individual waves are in antiphase, then $\sin^2\frac{\varphi}{2} = 1$ and

$$\mathcal{R}_{\text{out}} = \frac{(1-\zeta)^2}{(1-\zeta^2)^2 + 4\zeta^2} = \frac{(1-\zeta)^2}{(1+\zeta^2)^2} \tag{8.20}$$

that is, the output radiation at $\zeta \simeq 1$ is practically negligible.

The situation is changed drastically when in the active medium within resonator, the stimulated emission (lasing) conditions are realized. To proceed, we must now consider previously ignored light absorption and related reemission processes. This is done below in the following section.

8.2.1 Stimulated light emission

The abbreviation "**LASER**" stands for "Light Amplification by Stimulated Emission of Radiation." Stimulated emission was theoretically predicted by **Albert Einstein** in 1916, when he tried to develop the theory of light interaction with matter being in accordance with **Planck's** quantum theory of radiation emission. In fact, it is commonly accepted that the photon flux I is attenuated when propagating through matter, mainly due to conventional absorption by atoms. The corresponding differential equation in the one-dimensional case (light propagation along the x-axis) has the following simple form:

$$\frac{dI}{dx} = -I\sigma_a n_0 \tag{8.21}$$

where σ_a is the absorption cross section and n_0 is the concentration of atoms (per unit volume) located in the so-called ground state (i.e., ready for excitation). Stopping for the moment here, we find the simple exponential law for attenuation of the initial light intensity I_0 during its propagation through a medium

$$I = I_0 \exp(-\sigma_a n_0 x) = I_0 \exp(-\beta x) \tag{8.22}$$

with $\beta = \sigma_a n_0 > 0$. Just after absorbing the photons, however, the atoms become excited, that is, energetically unstable, trying to return to their ground state. The energy excess is released in the form of photons, electrons, and phonons. Some of these photons will have the same energy $\hbar\omega$, as in the incident beam and will be emitted in the direction of the latter, thus increasing the transmitted beam intensity. Correspondingly, eq. (8.21) should be corrected to include this so-called spontaneous emission

$$\frac{dI}{dx} = -I\sigma_a n_0 + \chi n_1 \tag{8.23}$$

where n_1 is the concentration of the excited atoms and χ is some proportionality factor. The absorption and emission terms have different signs, indicating that spontaneous emission increases the transmitted light intensity in reverse to the absorption trend. In the thermal equilibrium, $\frac{dI}{dx} = 0$, and we obtain the following equilibrium flux, I_{eq}

$$I_{eq} = \frac{\chi n_1}{\sigma_a n_0} \tag{8.24}$$

Since the excited atoms have extra energy $\hbar\omega$ compared to their ground state, then at equilibrium, according to the **Boltzmann** statistics, we obtain

$$\frac{n_1}{n_0} = \exp\left(-\frac{\hbar\omega}{k_B T}\right) \tag{8.25}$$

and

$$I_{eq} = \frac{\chi}{\sigma_a} \exp\left(-\frac{\hbar\omega}{k_B T}\right) \tag{8.26}$$

where T is the absolute temperature and k_B is the **Boltzmann** constant. This result is in strict contradiction to **Planck's** theory, according to which the magnitude of the equilibrium photon flux I_{eq} should be

$$I_{eq} \sim \frac{1}{\left[\exp\left(\frac{\hbar\omega}{k_B T}\right)\right] - 1} \tag{8.27}$$

Einstein noticed that the equilibrium photon flux fits **Planck**'s theory, after correcting eq. (8.24) in the following way

$$I_{eq} = \frac{\chi n_1}{\sigma_a(n_0 - n_1)} \tag{8.28}$$

Indeed, substituting eq. (8.25) into eq. (8.28) yields

$$I_{eq} \sim \frac{\exp\left(-\frac{\hbar\omega}{k_B T}\right)}{1 - \exp\left(-\frac{\hbar\omega}{k_B T}\right)} = \frac{1}{\left[\exp\left(\frac{\hbar\omega}{k_B T}\right)\right] - 1} \tag{8.29}$$

Subtracting n_1 from n_0 in the denominator of eq. (8.28) is very reasonable, since if there are n_1 excited atoms of the total amount of atoms n_0, then the number of atoms available for further excitation, that is, atoms that are ready to absorb photons, is $n_0 - n_1$. If so, we also have to correct eq. (8.23), as follows

$$\frac{dI}{dx} = -I\sigma_a n_0 + \chi n_1 + I\sigma_a n_1 \tag{8.30}$$

Equation (8.30) yields eq. (8.28) at $\frac{dI}{dx} = 0$. An additional (third) term in eq. (8.30) has deep physical meaning. It implies that besides the spontaneous emission, there is another emission process, whose strength is proportional to the light flux I. For this reason, it is called *light-stimulated emission*. It proceeds in the direction of the transmitted beam and is coherent with the latter. This coherence is the stimulated emission's "visiting card." If the stimulated emission is more significant than the spontaneous emission, one obtains

$$\frac{dI}{dx} \approx I\sigma_a(n_1 - n_0) \tag{8.31}$$

According to eq. (8.31), amplification of the transmitted beam during its propagation through the medium occurs if the medium is characterized by the inverse population of the respective energy states

$$n_1 > n_0 \tag{8.32}$$

Under this condition, light intensity is exponentially increasing along the propagation path x

$$I = I_0 \exp[\sigma_a(n_1 - n_0)x] = I_0 \exp(\beta_s x) \tag{8.33}$$

with $\beta_s = \sigma_a(n_1 - n_0) > 0$.

Accounting for light absorption modifies eq. (8.17) as follows

$$W_{out} = E_1(1-\zeta)\left(\frac{1-\gamma^N\zeta^{2N}e^{iN\varphi}}{1-\gamma\zeta^2 e^{i\varphi}}\right) \tag{8.34}$$

where

$$\gamma = \exp(-2\beta l) \tag{8.35}$$

for a conventional (passive) medium (see eq. (8.22)) or

$$\gamma = \exp(2\beta_s l) \tag{8.36}$$

for an active medium under stimulated emission (see eq. (8.33)).

In a passive medium (eq. (8.35)), absorption only weakly modifies the previously obtained results. For example, eq. (8.18) transforms into a structurally similar expression

$$\mathcal{R}_{out} = \left|\frac{W_{out}}{E_1}\right|^2 = \frac{(1-\zeta)^2}{(1-\gamma\zeta^2)^2 + 4\gamma\zeta^2\sin^2\frac{\varphi}{2}} \tag{8.37}$$

In contrast, in the case of an active medium, using eq. (8.34) with the γ-value given by eq. (8.36), yields the output intensity exponentially growing with the increasing light path $2Nl$

$$\mathcal{R}_{out} = \left|\frac{W_{out}}{E_1}\right|^2 = \frac{(1-\zeta)^2}{(1-\gamma\zeta^2)^2 + 4\gamma\zeta^2\sin^2\frac{\varphi}{2}}\left[\left(1-\gamma^N\zeta^{2N}\right)^2 + 4\gamma^N\zeta^{2N}\sin^2\frac{N\varphi}{2}\right] \tag{8.38}$$

In resonance, $\sin^2\frac{\varphi}{2} = \sin^2\frac{N\varphi}{2} = 0$, and

$$\mathcal{R}_{out} = \left|\frac{W_{out}}{E_1}\right|^2 = \frac{(1-\zeta)^2}{(1-\gamma\zeta^2)^2}(1-\gamma^N\zeta^{2N})^2 \sim \gamma^{2N} = e^{4\beta_s Nl} \tag{8.39}$$

There exist different ways to arrange inverse occupation of the atomic or molecular energy levels. One way is the so-called three-level scheme, in which atoms in ground state 1 absorb photons and are transferred to excited level 3 (see Figure 8.6). The latter is short-living, which results in the fast nonradiative transition down to the lower excited level 2. If the relaxation time of level 2 is much longer than that of level 3, the inverse population of level 2, with respect to ground level 1, will exist for the prolonged time required to produce stimulated light emission. This scheme, for the first time, was realized with the Cr impurity ions in ruby (Al_2O_3) crystals. Following this work, more effective pumping schemes (e.g., the four-level scheme) were developed.

To reach high laser intensities, a lengthy light path within an active medium must be organized (see eqs. (8.33) and (8.39)). As we know, in laboratory conditions, this is achieved by forcing light to propagate back and forth along a reasonable distance by

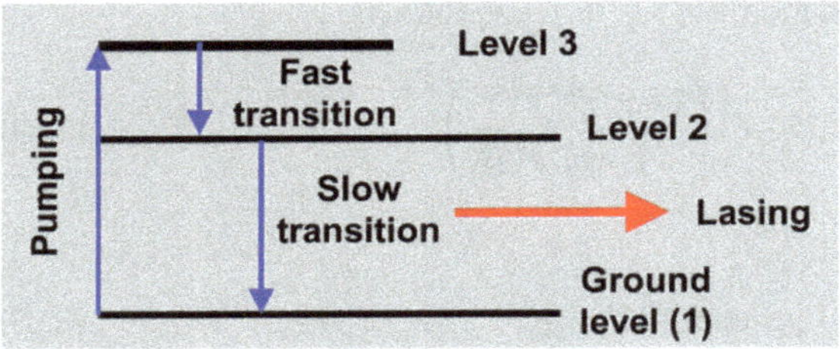

Figure 8.6: Classical three-level pumping system illustrating the laser principle.

using reflecting mirrors of the **Fabry–Pêrot** resonator. In 1964, **Alexander Prokhorov** and **Nicolay Basov**, together with **Charles Townes**, were awarded the Nobel Prize in Physics "for fundamental work in the field of quantum electronics, which has led to the construction of oscillators and amplifiers based on the maser–laser principle." As well **Arthur Schawlow** is commonly credited for laser invention. He was awarded the 1981 Nobel Prize in Physics (shared with **Nicolaas Bloemberger**) "for their contribution to the development of laser spectroscopy."

In the beginning of the 1960s, **William Dumke** analyzed the perspectives of observing the stimulated light emission in semiconductors. He theoretically showed that it is principally possible to organize a lasing regime in direct-gap materials (specifically, in GaAs). As a necessary condition, he formulated that a significant part of the electrons should be pumped through the semiconductor bandgap E_g from the valence band into the conduction band, producing a metal-like gas of free electrons in there, with effective "**Fermi** energy" E_{Fc}. Correspondingly, the miniband of empty states (holes) appears in the valence band, which is characterized by the effective hole "**Fermi** energy" E_{Fv} (see Figure 8.7).

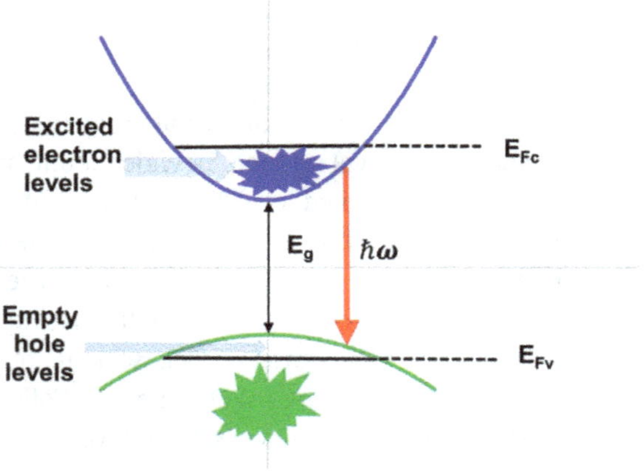

Figure 8.7: Principle of the semiconductor laser.

Lasing principally occurs under the condition

$$E_g < \hbar\omega < E_{Fc} - E_{Fv} \tag{8.40}$$

We stress that the energy of the emitted radiation is below the threshold energy for light absorption, which is $E_{Fc} - E_{Fv}$.

In 1962, based on these ideas, the coherent-stimulated light emission from a GaAs semiconductor diode was demonstrated in the General Electric Research Center (Schenectady, NY, USA) by **Robert Hall**, who is regarded as an inventor of the semiconductor laser.

Semiconductor lasers operate under forward bias. Carriers of different signs injected into the active region recombine here, causing photon emission due to radiative recombination. If the electric current is larger than some threshold value, then more and more carriers participate in the light emission, and the gain can exceed the light losses. The gain is amplified using a resonant cavity, which is formed by two-end mirrors that are fabricated by cleaving the front and back faces of a semiconductor crystal. Typical cavity sizes are a few hundreds of microns long. In fact, such cavities are a kind of **Fabry–Pêrot** resonator. This design enables us to generate stimulated emission, that is, coherent light with a narrow line width.

An innovative step toward the development of reliable semiconductor laser technology was made by the implementation of heterostructures. In these devices, a layer of narrow-bandgap semiconductor is sandwiched between two wide bandgap semiconductor layers, which allow us to obtain a lasting effect at room temperature. The commonly used pair of materials is GaAs combined with $Al_xGa_{1-x}As$. The major advantages of semiconductor lasers are their compact size and rather low power consumption, which facilitated their usage, for example, in laser printers. Semiconductor lasers find various applications in many different areas including optical communication systems, optical data storage, laser material processing, and medical treatments.

8.3 Mach–Zehnder interferometer

The basic principle of the **Mach–Zehnder** interferometer is graphically illustrated in Figure 8.8. In gross mode, its design is similar to what we described in previous sections of this chapter. The incoming light is split into two coherent beams that, after propagating along equal optical paths, are merged to initiate the interference pattern. The investigated object is placed in one of the beam paths, causing the phase difference of interest, which is then extracted from the arising interference pattern. This device was invented by **Ludwig Mach** and **Ludwig Zehnder** at the beginning of 1890s and comprises two mirrors and two semitransparent plates (splitters, see Figure 8.8), separated by rather large distances. Such a construction offers a lot of free space, which can be used to place large-scale objects. Interference patterns, affected by the object, are recorded by photographic plates (or screens). The interference fringes can be adjusted so they are in the same plane as the investigated object, allowing the fringes and test object to be photographed together.

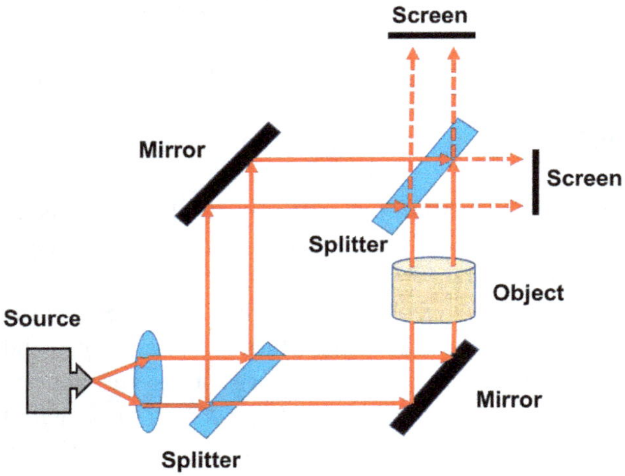

Figure 8.8: Principle scheme of the **Mach–Zehnder** interferometer.

Mach–Zehnder interferometers are widely used for visualizing gas and liquid flows in aerodynamics, plasma physics, and studies of heat transfer, aimed at measuring pressure, density, and temperature changes. Another well-established application field is electro-optics, where **Mach–Zehnder** interferometers serve as an integrated part of light modulators in optical communication systems (see also Section 16.4).

Mach–Zehnder interferometers contributed a lot to the contemporary understanding of quantum mechanics, especially in clarifying such fundamental phenomenon as quantum entanglement. The latter term is used to describe the situation when particles (in our case, photons) interact in such a way that the quantum state of an individual particle (photon) cannot be described independently of the states of the other particles within an ensemble, even if the particles are separated by large distances. Therefore, we speak about a cooperative phenomenon, in which the change in the quantum state of one particle modifies to some extent the quantum states of all other particles in the ensemble.

How can interferometry help in better understanding this phenomenon? For this purpose, the **Mach–Zehnder** interferometer, comprising two splitters, two mirrors, and two detectors, is indeed well suited (see Figure 8.9).

In this application, photons issued from a light source pass through the first splitter (S1) and continue to propagate along two different paths, marked A and B in Figure 8.9. Photons A and B experience reflections by mirrors M1 and M2, respectively, and then pass through the second splitter (S2) toward mixing and producing interference effects

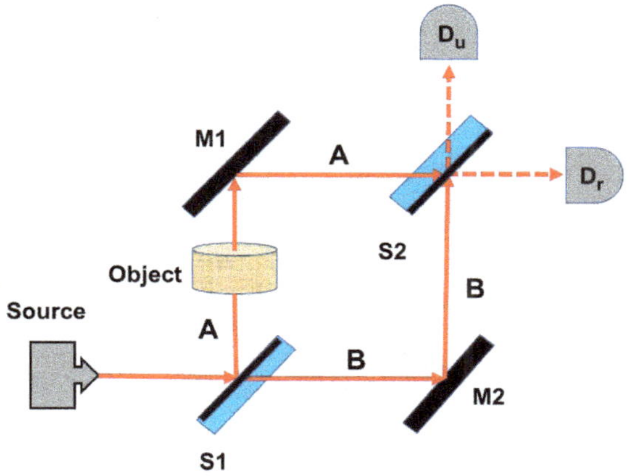

Figure 8.9: Application of the **Mach–Zehnder** interferometer to the quantum entanglement problem.

in the right-hand (D_r) and upper (D_u) detectors. If there is no quantum entanglement, the optical paths, A and B, should be considered completely independent, that is, photons are not interacting and should be counted separately.

Suppose that we have no objects placed in the light trajectories A and B, and their geometrical paths are equal. After passing the first splitter (S1), the wave functions of photons A and B are

$$\psi_{1A} = \frac{1}{\sqrt{2}} e^{i\Phi_{1A}}; \psi_{1B} = \frac{1}{\sqrt{2}} e^{i\Phi_{1B}} \tag{8.41}$$

that is,

$$|\psi_{1A}|^2 = |\psi_{1B}|^2 = \frac{1}{2} \tag{8.42}$$

Phase Φ_{1A} originates in the light reflection from the metal coating of glass plate S1, that is, from the optically denser (than an air) medium and equals

$$\Phi_{1A} = \pi \tag{8.43}$$

as is shown before in **Appendix 5.A**. Phase Φ_{1B} arises when photon B straightforwardly propagates through glass plate S1

$$\Phi_{1B} = \Phi \tag{8.44}$$

Furthermore, each specular reflection in mirrors M1 and M2 produces an additional phase

$$\Phi_m = \pi \tag{8.45}$$

The second splitter (S2) works in a bit more complicated manner. Considering detector D_r, for photon B we already have the usual phase jump

$$\Phi_{2\mathrm{Br}} = \pi \tag{8.46}$$

while for photon A

$$\Phi_{2\mathrm{Ar}} = \Phi \tag{8.47}$$

Considering detector D_u, we find for photon B

$$\Phi_{2\mathrm{Bu}} = \Phi \tag{8.48}$$

whereas for photon A

$$\Phi_{2\mathrm{Au}} = 2\Phi \tag{8.49}$$

The last result signifies that photon A passes the glass plate of splitter S2 twice and is reflected from its back surface, that is, from the optically less dense medium that does not lead to the additional phase jump by π.

Summarizing all the mentioned above, we can calculate the phase changes that are accumulated until photons A and B enter the respective detector, D_r or D_u. Considering detector D_r and photon A, we must summate

$$\Phi_{\mathrm{Ar}} = \Phi_{1\mathrm{A}} + \Phi_m + \Phi_{2\mathrm{Ar}} = \pi + \pi + \Phi = 2\pi + \Phi \tag{8.50}$$

whereas for photon B

$$\Phi_{\mathrm{Br}} = \Phi_{1\mathrm{B}} + \Phi_m + \Phi_{2\mathrm{Br}} = \Phi + \pi + \pi = 2\pi + \Phi \tag{8.51}$$

We find that here the wavefields A and B are in phase (constructive interference), which means the maximum intensity

$$I_\mathrm{r} = \frac{1}{4}\left|\left(e^{i\Phi_{\mathrm{Ar}}} + e^{i\Phi_{\mathrm{Br}}}\right)\right|^2 = \frac{1}{4}\left|\left[e^{i(2\pi+\Phi)} + e^{i(2\pi+\Phi)}\right]\right|^2 = 1 \tag{8.52}$$

Note that factor $\frac{1}{4}$ appears in eq. (8.52) because of two splitters, each acting according to eq. (8.42).

Considering detector D_u and photon A, we find

$$\Phi_{\mathrm{Au}} = \Phi_{1\mathrm{A}} + \Phi_\mathrm{m} + \Phi_{2\mathrm{Au}} = \pi + \pi + 2\Phi = 2\pi + 2\Phi \tag{8.53}$$

whereas for photon B, the accumulated phase is

$$\Phi_{\mathrm{Bu}} = \Phi_{1\mathrm{B}} + \Phi_\mathrm{m} + \Phi_{2\mathrm{Bu}} = \Phi + \pi + \Phi = \pi + 2\Phi \tag{8.54}$$

Therefore, we see that here the wavefields A and B are in antiphase (destructive interference; phase difference equals π), which means zero intensity in the upper detector D_u

$$I_\mathrm{u} = \frac{1}{4}\left|\left(e^{i\Phi_{\mathrm{Au}}} + e^{i\Phi_{\mathrm{Bu}}}\right)\right|^2 = \frac{1}{4}\left|\left[e^{i(2\pi+2\Phi)} + e^{i(\pi+2\Phi)}\right]\right|^2 = 0 \tag{8.55}$$

The situation changes drastically if an object is placed into the propagation trajectory A between the first splitter (S1) and mirror M1. Suppose that the object introduces an additional phase φ into the total phase balance. In this case, phase equations (8.51) and (8.54) remain unchanged, whereas eqs. (8.50) and (8.53) are modified as follows:

$$\Phi'_{\mathrm{Ar}} = \Phi_{1\mathrm{A}} + \varphi + \Phi_{\mathrm{m}} + \Phi_{2\mathrm{Ar}} = 2\pi + \Phi + \varphi \tag{8.56}$$

$$\Phi'_{\mathrm{Au}} = \Phi_{1\mathrm{A}} + \varphi + \Phi_{\mathrm{m}} + \Phi_{2\mathrm{Au}} = 2\pi + 2\Phi + \varphi \tag{8.57}$$

Using eqs. (8.51) and (8.56), we find the light intensity I_{r} coming to the right-hand detector D_{r}

$$I_{\mathrm{r}} = \frac{1}{4}\left|\left(e^{\Phi'_{\mathrm{Ar}}} + e^{i\Phi_{\mathrm{Br}}}\right)\right|^2 = \frac{1}{4}\left|\left[e^{i(2\pi+\Phi+\varphi)} + e^{i(2\pi+\Phi)}\right]\right|^2 = \cos^2\left(\frac{\varphi}{2}\right) \tag{8.58}$$

Similarly, eqs. (8.54) and (8.57) provide the light intensity entering the upper detector D_{u}

$$I_{\mathrm{u}} = \frac{1}{4}\left|\left(e^{i\Phi'_{\mathrm{Au}}} + e^{i\Phi_{\mathrm{Bu}}}\right)\right|^2 = \frac{1}{4}\left|\left[e^{i(2\pi+2\Phi+\varphi)} + e^{i(\pi+2\Phi)}\right]\right|^2 = \sin^2\left(\frac{\varphi}{2}\right) \tag{8.59}$$

In the case of quantum entanglement, we are not able to indicate which path (A or B) the photons are taking, and therefore, the detectors' counts will differ from those predicted by eqs. (8.58) and (8.59). In fact, in this case the first splitter (S1) is practically inactive and, for the purpose of estimating the entanglement effect, can simply be removed from the scheme. Accordingly, the detected intensity is determined by splitter S2 and is the same for both detectors, D_{r} and D_{u}

$$I_{\mathrm{r}} = I_{\mathrm{u}} = \frac{1}{2} \tag{8.60}$$

that is, as that given by eq. (8.42).

The research works, carried out by **Alain Aspect**, **John Clauser**, and **Anton Zeilinger**, fundamentally contributed to deep comprehension of the quantum entanglement problem and its potential technological applications. They were awarded the 2022 Nobel Prize in Physics "for experiments with entangled photons, establishing the violation of **Bell** inequalities and pioneering quantum information science."

8.4 Bonse–Hart X-ray interferometer

As already mentioned, the refractive index for X-ray is very close to one. This implies that in contrast to the optics of visible light, the refraction phenomenon cannot be used to significantly change X-ray trajectories, as is done in the case of light splitting and subsequent merging. The same conclusion can be drawn regarding X-ray reflection from the sample surface, which has considerable intensity under total external reflection only. The latter, however, causes once again just tiny angular changes in the X-ray propagation.

Fortunately, a proper solution can be realized using X-ray diffraction. We discuss this topic in detail in Sections 10.2 and 13.3, but the main principle is clear from the **Braggs** diffraction condition obtained in Section 7.2 (see eq. (7.26)). In fact, if the interplanar spacing (d-spacing) in a crystal is comparable with X-ray wavelength λ, then the **Bragg** angle θ_B defined via $\sin\theta_B = \frac{\lambda}{2d}$ can be large enough up to $\theta_B = 90°$ at $d = \frac{\lambda}{2}$. Another important aspect is that the X-ray diffraction intensity must be comparable with the transmitted intensity to provide a high-contrast interference pattern when interacting with the transmitted beam. In **Chapter 15**, we show that achieving a necessarily high diffraction intensity is possible in the framework of so-called dynamical diffraction, which considers multiple X-ray scattering and continuous dynamic interaction between the diffracted and transmitted X-ray beams during their propagation within perfect single crystals. As a result, the diffracted and transmitted beams can have nearly equal intensities, but only within a very narrow angular interval (a few seconds of arc or about 10 μrad) around the **Bragg** angle θ_B.

Note that to produce the interference pattern we need at least two crystals, one for splitting the X-ray beams, and the second to merge them back and bring to certain spatial region where the interference pattern is created (see Figure 8.10). In practical terms, mutual alignment of individual single crystals with a second of arc accuracy and, moreover, maintaining the alignment during the long-period measurements, requires time-consuming hard work. The most serious problem, however, is to maintain the optical paths on the X-ray wavelength scale unchanging during measurements influenced by mechanical vibrations, thermal fluctuations, and so on.

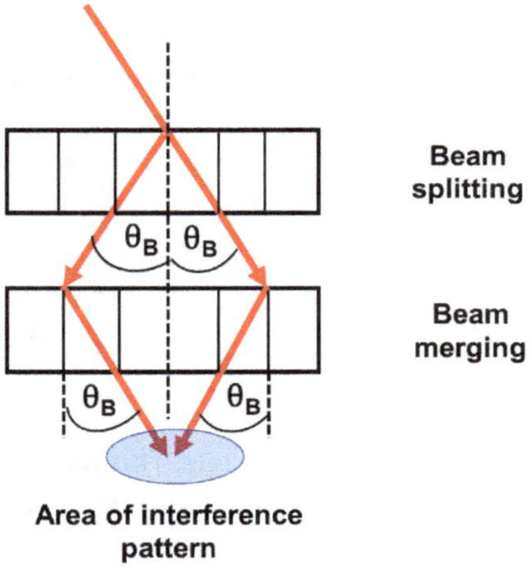

Figure 8.10: Producing an interference pattern using subsequent X-ray diffraction from two perfect crystals.

In 1965, **Ulrich Bonse** and **Michael Hart** found a very elegant solution to these problems, by fabricating an X-ray interferometer as a monolithic unit. The latter is appropriately cut from an ingot of a perfect silicon single crystal and comprises three crystalline blocks (Figure 8.11). The first crystal in this design acts as a beam splitter (S), the second one serves as a mirror (M), which facilitates beam merging, and the third one acts as analyzer (A) of the arising interference pattern.

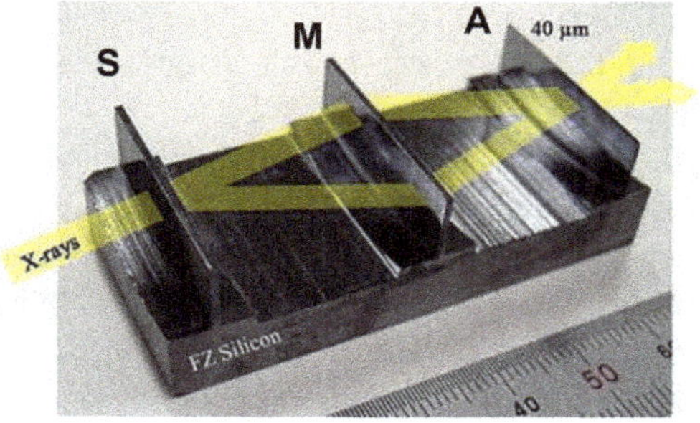

Figure 8.11: Original design of the **Bonse–Hart** interferometer. Letters S, M, and A stand for the splitter, merger (mirror), and analyzer, respectively.

The first application of the **Bonse–Hart** interferometer was intended for measuring tiny strains in crystals induced by heating or ion implantation. This is done by comparison of lattice parameters in distorted and perfect standard samples. Note that high-resolution X-ray diffraction also allows careful lattice parameter measurements with relative precision of 10^{-5}. With the aid of X-ray interferometry, this precision was greatly improved up to 10^{-7}–10^{-8}. Based on the ability to measure the lattice parameter a of single silicon crystals with $\frac{\Delta a}{a} \simeq 10^{-8}$ precision, X-ray interferometry was included in the large-scale project aimed at precise measurement of the **Avogadro** number N_A. The latter is related to the silicon lattice parameter a via the following relationship

$$N_A = 8\frac{V_m}{a^3} = 8\frac{A}{\rho a^3} \tag{8.61}$$

where V_m is the silicon mole volume, A is atomic mass of Si, and ρ is the silicon mass density. In this manner, the **Avogadro** number was determined with higher than 10^{-7} precision. The significance of this result is greatly increased by the fact that the **Avogadro** number is directly associated with the inverse mass of a proton.

It is worth mentioning the successful usage of **Bonse–Hart** interferometry in the field of phase-contrast X-ray tomography (see also Sections 9.4 and 12.5.1). This technique is particularly important for imaging of biological and other materials composed of light atoms, in which photoelectric absorption is too weak to produce substantial image contrast. When conducting such measurements, an object is placed in the path of one of the rays between the mirror and analyzer (Figure 8.12).

In fact, the phase difference $\Delta\varphi$, accumulated when an X-ray wave propagates a distance z through a material, compared to the propagation of the same distance in a vacuum, is

$$\Delta\varphi = 2\pi\frac{1-n}{\lambda}z = 2\pi\frac{\delta}{\lambda}z \tag{8.62}$$

where the refractive index $n = 1 - \delta$ (see eq. (2.33)). For reasonable estimations, we can take typical values of $\lambda = 1$ Å and $\delta = 5 \times 10^{-6}$, which yield $\Delta\varphi = 2\pi \cdot (5 \times 10^{-6}) \frac{z}{\lambda}$. In other words, the phase difference, $\Delta\varphi = 2\pi$, is accumulated through the short distance of $z = 20$ µm. Certainly, an interference pattern is sensitive to much smaller phase changes, which can be recorded after X-ray propagation over much thinner specimen slices. Today, using advanced computer software, it is possible to retrieve phase shifts from experimental interference patterns and convert this information into the spatial distribution of the electron density. The achieved spatial resolution is on a micrometer scale, which has accelerated the successful application of this method in medicine. One prominent example of the latter is X-ray phase-contrast mammography.

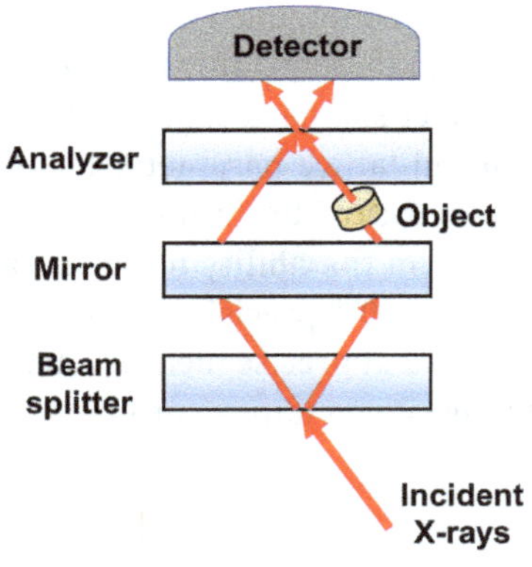

Figure 8.12: Application of **Bonse–Hart** interferometry to phase-contrast X-ray imaging.

Chapter 9
Phase-contrast microscopy

The term "phase-contrast microscopy" is generally applied to the imaging of transparent samples, in which the direct amplitude changes and associated intensity variations during light propagation are too subtle; correspondingly, the spatially resolved phase differences are recorded and converted into visible contrast modulation. These techniques are especially important for thin biological specimens, which represent the so-called phase objects.

9.1 Zernike microscope

The most influential discovery for the imaging of phase objects was done in the 1930s by **Frits Zernike**. To properly describe the invention of his phase-contrast microscope, we must go back by nearly 100 years. In 1930, **Zernike** had studied some auxiliary intensity effects arising when using not completely perfect diffraction gratings (the latter are described in Section 10.1). He found that in addition to regular (strong) scattering intensity in the expected spatial directions, dictated by diffraction conditions, there are weak intensity satellites (ghosts), which are phase shifted by 90° with respect to the former. The genius conjecture of **Zernike** was to "translate" this finding into the language of microscopy, aimed at the imaging of phase objects.

In fact, the wavefield after passing through the phase object can be described roughly as

$$F = Ae^{i\Phi} \tag{9.1}$$

where wave amplitude A remains practically unchanged, whereas the accumulated phase Φ is the result of wave interaction with an object. Correspondingly, the intensity,

$$I = |F|^2 = |A|^2 \tag{9.2}$$

is constant, reflecting the unchanged amplitude A. To further proceed, let us expand the exponent in eq. (9.1) into a series:

$$F \simeq A(1 + i\Phi) = A\left(1 + \Phi e^{i\frac{\pi}{2}}\right) \tag{9.3}$$

The first term in eq. (9.3) represents the directly transmitted light, whereas the second term describes the weakly scattered light (a "ghost" in the **Zernike**'s terminology), which is a 90°-phase shifted with respect to the transmitted one. **Zernike** understood that this 90°-phase shift is the reason that the intensity is not sensitive to phase changes. Actually, neglecting second-order terms over Φ, we find:

https://doi.org/10.1515/9783111140100-010

$$I = |F|^2 = |A|^2 \ (1 + i\Phi)(1 - i\Phi) \simeq |A|^2 \tag{9.4}$$

The situation could be changed drastically if the way to change the relative phase between the first and the second term by an additional 90° (clockwise or counterclockwise) could be found. After doing this, the intensity becomes equal:

$$I = |F|^2 \simeq |A|^2 \ (1 \pm \Phi)^2 \simeq |A|^2(1 \pm 2\Phi) \tag{9.5}$$

In other words, the phase change induced by the object, is indeed converted into intensity variation, which is called phase contrast, positive or negative, depending on the sign [+] ($e^{i\pi/2}e^{-i\pi/2} = 1$) or [–] ($e^{i\pi/2}e^{i\pi/2} = -1$) in eq. (9.5). We stress that in both cases, the intensity change is proportional to the accumulated phase Φ. This is a "visiting card" of the **Zernike** method, which allows us to characterize the object's thickness variations or the variations in its refractive index, since the accumulated phase is proportional to both the thickness and refractive index.

To implement these ideas into practical microscopy, a transparent dielectric ring with refractive index n is placed in the path of the transmitted beam. The thickness T of the ring is

$$T = \frac{\lambda}{4} = \frac{\lambda_0}{4n} \tag{9.6}$$

or

$$T = \frac{3}{4}\lambda = \frac{3}{4n}\lambda_0 \tag{9.7}$$

where λ and λ_0 are the light wavelengths in a dielectric material and in vacuum, respectively. As we learned in Section 5.4, such dielectric plates shift the wave phase by 90° (a quarter-wave plate, see eq. (9.6)) or by 270°, equivalent to (–90°) (three-quarters-wave plate, see eq. (9.7)).

In more detail, the **Zernike** microscope scheme is shown in Figure 9.1. Light issuing from the source is formed by the condenser system and focused on the specimen region of interest. After passing through the object, the directly transmitted light (colored in red in Figure 9.1) is focused by objective lens onto some point in the image plane situated on the optical axis. Before coming to the image plane, the transmitted beam passes through the phase shift ring (colored in yellow in Figure 9.1), where its phase is changed by 90°. Conversely, the slightly scattered light (colored in green in Figure 9.1) propagates to the image plane with no additional phase shift. Finally, the transmitted light (often called the background) and scattered light merge with the aid of the objective lens and interfere in the image plane of the microscope. As a result, the contrast modulation is observed, as described by eq. (9.5).

In 1953, **Frits Zernike** was awarded the Nobel Prize in Physics for "for his demonstration of the phase contrast method, especially for his invention of the phase contrast microscope."

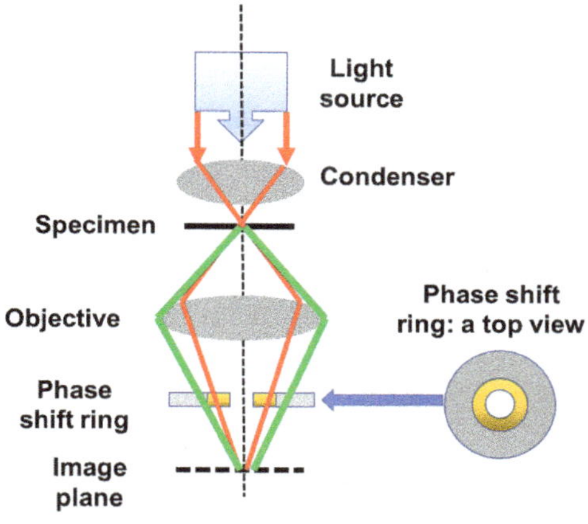

Figure 9.1: Principal scheme of the **Zernike** microscope.

9.2 Nomarski microscope

An alternative technique, which is called differential interference contrast (DIC) microscopy, **Nomarski** interference contrast (NIC), or **Nomarski** microscopy, was developed by **Georges Nomarski** in 1952. It uses polarized light and relies on the changes in the polarization state arising during light propagation through different points of the investigated phase object. As key polarizing elements, the two **Nomarski** prisms are used (see Section 5.3 and Figure 5.17 in Chapter 5).

The principal scheme of the **Nomarski** microscope is presented in Figure 9.2. The incoming light is polarized by the primary polarizer to have a polarization direction at 45° with respect to the main optic axis in the entrance block of the first **Nomarski** prism. Here, the light splits into two coherent rays orthogonally polarized with respect to each other. When passing through the second block of the first **Nomarski** prism, these rays merge at some point outside the prism, which serves as a source of a condenser system illuminating the investigated specimen. The condenser system focuses the orthogonally polarized rays onto two adjacent points in the specimen, separated by a tiny distance of about 200 nm (i.e., close to the limiting resolution of the microscope). When passing through the specimen, the additional phase shift between the rays is accumulated, due to the differences in the refractive index or thickness along the trajectories of the rays. The rays further travel through the objective lens and are focused onto the second **Nomarski** prism.

Numerous pairs of rays originating in the adjacent points in the specimen form the bright field images for both orthogonal polarizations. Information on the accumulated phase shift, however, cannot be used since orthogonal polarizations prevent interference between these two images. To overcome this problem, the second **Nomarski** prism is used, which recombines the two rays into one polarized at 135° with respect to the main optic axis here. Ray unification leads to the light interference, that is, brightening

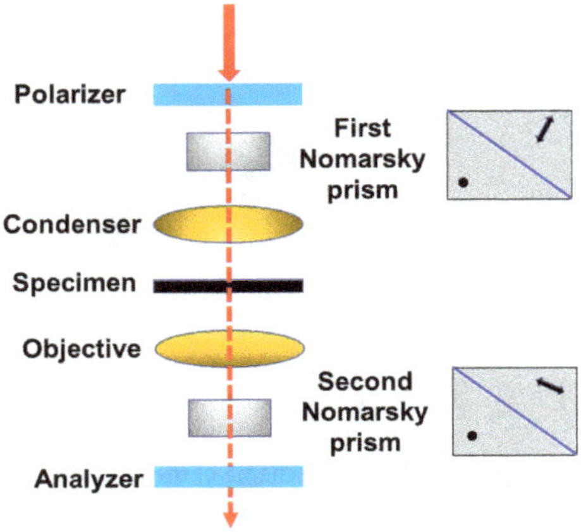

Figure 9.2: Principal scheme of the **Nomarski** microscope.

or darkening of local points in the image, according to the local phase difference. As we see, the method is based on the DIC, which arises when the interfering beams pass through the object at a tiny distance from each other. In this way, gradients of phase differences in the object are visualized by converting them into the intensity variations. The image quality is excellent, revealing sharp edges (no halo) of the features of interest (see the illustrative example in Figure 9.3).

Figure 9.3: *Micrasterias furcata* (green alga about 100 μm in diameter) imaged by transmitted DIC (**Nomarski**) microscopy (licensed under the Creative Commons Attribution-Share Alike 2.5 Generic license).

9.3 Modulation-contrast microscopy

Another important imaging technique is modulation-contrast microscopy invented in 1975 by **Robert Hoffman**. The principal scheme of his microscope is shown in Figure 9.4.

In addition to the standard bright field setup, two extra elements are installed. These are a special narrow slit aperture below the condenser and a modulator that is placed within the objective system in the **Fourier** plane, conjugate to the slit. The latter is needed for proper illumination of the investigated specimen. The heart of the

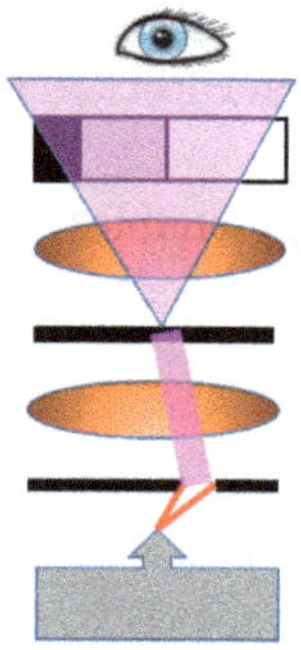

Figure 9.4: Principal scheme of the **Hoffman** method.

system, however, is the amplitude modulator, which consists of a little glass disc with three areas having different optical absorption: (A) a "bright" region with maximal (about 100%) transmission, (B) a grey region with about 15% transmission, and (C) a dark region with close to zero (less than 1%) transmission (see Figure 9.5). It is essential that the modulator itself does not introduce any additional phase shifts into the propagating light.

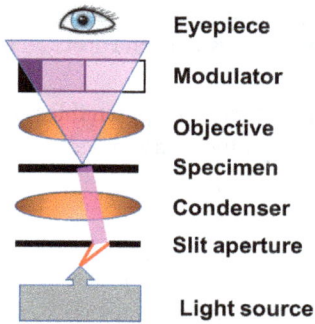

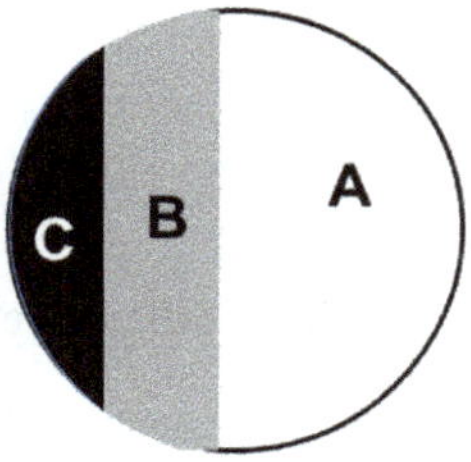

Figure 9.5: Design of **Hoffman**'s amplitude modulator.

Let us consider a phase object (specimen) with gradients of optical density (phase gradients) induced, for example, by the gradients of the refractive index. As we learned in Chapter 1 (see eqs. (1.39) and (1.41)), these gradients cause gentle bending of the light trajectories. Such trajectory changes do not influence the image formation in bright field microscope and the phase object remains invisible. The situation changed drastically when using the amplitude modulator, which converts the phase gradients into the intensity variations in the image.

In fact, when passing through the bright region of the modulator, slightly scattered rays will form a bright image of the respective optical gradient, while the rays passing through the grey region of the modulator form the grey background of the image. Consequently, the respective gradient image is brighter than the background and becomes visible. An opposite gradient causes the scattered light to be highly attenuated by the dark region of the modulator, thus creating a dark gradient image. In other words, opposite gradients result in deflection of the slit image to the very dark part or the bright section of the modulator. The local image intensity is proportional to the first derivative of the optical density. Alternating bright-grey-dark contrasts provide a kind of optical shadowing, which the human visual system interprets as a three-dimensional image.

Providing high contrast images, the modulation-contrast microscopy enables us to visualize transparent, unstained, and living cells, as well as other phase objects.

9.4 Phase-contrast X-ray imaging

There exists a rich compendium of different techniques allowing phase-contrast X-ray imaging. These techniques are especially effective when applied to thin biological samples and samples composed of atoms with low atomic number Z, in which the absorption contrast is very weak. Most techniques are based on tight interrelation between small (or ultra-small) angular scattering and induced phase changes. In fact, for the directly transmitted wave, the phase is

$$\varphi_{\mathrm{T}} = \boldsymbol{kr} - \omega t \tag{9.8}$$

while for the scattered waves, it equals:

$$\varphi_{\mathrm{S}} = (\boldsymbol{k} + \Delta\boldsymbol{k})\boldsymbol{r} - \omega t \tag{9.9}$$

with

$$|\Delta\boldsymbol{k}| = |\boldsymbol{k}|\alpha \tag{9.10}$$

where α is the small scattering angle (see Figure 9.6). Therefore, the phase difference equals

$$\Delta\varphi = \varphi_{\mathrm{S}} - \varphi_{\mathrm{T}} = \Delta\boldsymbol{kr} = \frac{1}{2}|\boldsymbol{k}| \cdot |\boldsymbol{r}|\ \alpha^2 \tag{9.11}$$

and, indeed, is directly related to the scattering angle α.

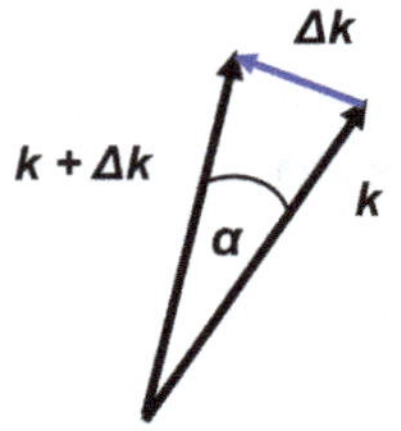

Figure 9.6: Wave vector's change at small-angle X-ray scattering.

To convert the angle-mediated phase shifts into phase contrast imaging, several approaches can be applied. In early stages, X-ray interferometers such as those described in Section 8.4 have been employed for this purpose. In such devices, superposition of the phase-shifted wave (in one of the interferometer's arms) with a reference wave (in another arm) takes place. The reference wave is generated by coherent splitting of the amplitude of the incident wave by the beam divider and, then, is superimposed with the phase-shifted wave in the analyzer. The phase information is extracted from the obtained interference pattern using specially developed computer algorithms. The **Bonse–Hart** interferometer offers extraordinary angular resolution for phase-contrast imaging, but, due to the intensity issue, can mostly be used on synchrotron beamlines rather than in the laboratory environment.

An alternative technique, using simpler double-crystal diffractometers, was proposed in 1995 by **Viktor Ingal** and **Elena Beliaevskaya**. Conventional double-crystal diffractometers comprise two identical perfect crystals (e.g., Si or Ge) in a parallel setting, with the ability to slightly change the angle α between the diffractive planes in a crystal couple (see Figure 9.7). Placing an object between the crystals and gently rotating the second crystal with respect to the first one, one can choose the rays that are scattered by the object at certain angle α. If the crystals are parallel, only non-scattered X-rays (red arrows in Figure 9.7) go through the system. Rotating the second crystal by angle α allows us to obtain, at the exit, the X-rays deflected by the object at the same angle α (green arrows in Figure 9.7).

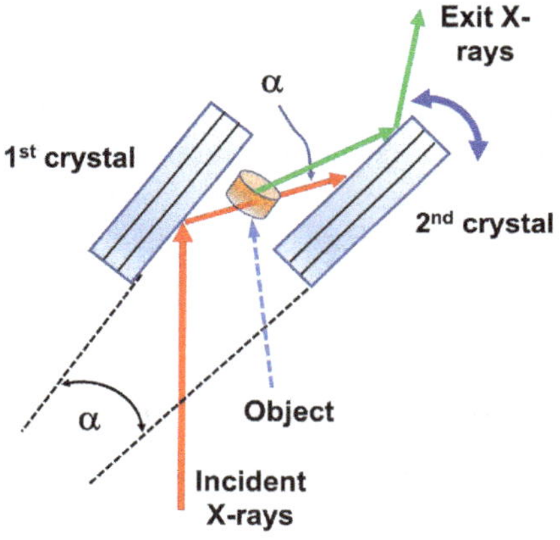

Figure 9.7: Scheme of a double-crystal diffractometer used for phase-contrast X-ray imaging.

Double-crystal diffractometers still provide angular resolution (few seconds of arc) and sufficient intensity that are good enough even when working with sealed X-ray tubes. The image contrast is achieved because of the small-angle X-ray scattering at the boundaries between the regions with tiny differences in the electron density values and, hence, in the refractive index. This technique is explained in more detail in Section 15.4.

As an important step forward made at the end of the 1990s, one can mention the use of diffraction gratings (described in Section 10.1) instead of single crystals. Several research groups contributed to the development of this technique, but the impact made by **Atsushi Momose** and **Christian David** should be especially noted. One of their workable experimental setups looks like the **Bonse–Hart** interferometer, but here, three perfect Si crystals are replaced by three diffraction gratings (Figure 9.8). As usual, the first grating (G1) is used to diffract incident X-rays; the second one (G2) works as a beam merger, while the third one (G3) serves as an analyzer of diffraction pattern, being phase shifted by the investigated object. The usage of diffraction gratings instead of perfect crystals allows us to considerably increase the wavelength bandwidth of incoming X-rays suitable for phase-contrast imaging (for details, see Section 12.5.1).

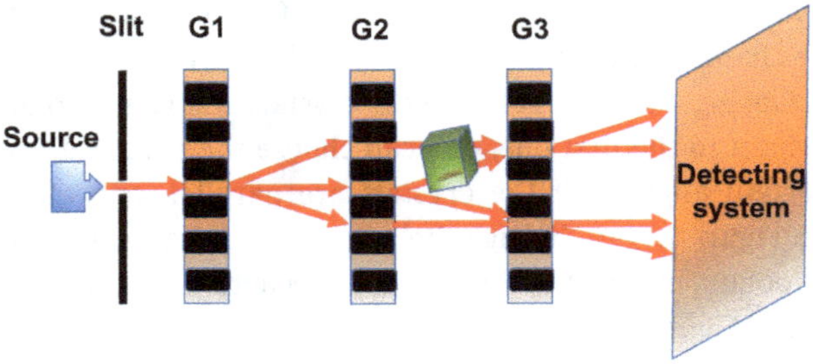

Figure 9.8: Scheme of the grating-based X-ray interferometer for phase-contrast X-ray imaging.

Another option is the grating-based phase-contrast imaging technique that uses the so-called **Talbot** effect (see Section 12.5). This effect creates an interference pattern, which, at a certain distance from the grating, repeats the structure of the latter exactly. By placing a phase object in the path of the beam, the position of the interference pattern is altered with respect to that taken with no object. The phase shift between these two interference patterns is detected with the help of a second grating, and in this way, information about the real part of the refractive index can be retrieved using appropriate reconstruction methods (for details, see Section 12.5.1).

At the end of the 1990s, a completely different approach, called the propagation-based imaging technique, was developed at the European Synchrotron Radiation Facility (ESRF, Grenoble, France), mainly due to the efforts of two research groups headed by **Anatoly Snigirev** and **Jose Baruchel**. Their key idea can be understood by once again considering the phase shift in eq. (9.11), which depends not only on the scattering angle α, but also on the distance $|r|$ from the object to the detector. Correspondingly, the method is based on the detection of interference fringes, which arise during free-space propagation.

The basic experimental setup consists of an X-ray source, a sample, a detector, and, optionally, a beam deflector. By recording interference patterns at multiple angular positions α and at a few distances $|\boldsymbol{r}|$, it is possible to retrieve the phase-shift distribution and, hence, the distribution of the refractive index.

Chapter 10
Fraunhofer diffraction

The term "diffraction" in optics, initially, was used to explain the deviations of light propagation from the trajectories dictated by geometrical (ray) optics. Later, due to the ideas of **Christiaan Huygens** and fundamental discoveries by **Thomas Young** and **Augustin-Jean Fresnel**, wave optics was established. In the framework of the latter, the diffraction phenomenon is treated as the interference of all waves scattered by an object, with the focus on the proper wave summation, taking account of the phases of individual waves. As a result, in addition to the directly transmitted beam, the propagation of diffracted beams is allowed along the directions differing from the incident one. As will be shown later in this **Chapter**, diffraction angles are mostly defined by the ratio $\frac{\lambda}{S_c}$ of the wavelength λ to the characteristic size S_c of the diffractive object. Correspondingly, to obtain significant diffraction angles, the size S_c should be comparable with λ. We will elaborate further on this key issue below.

The diffraction evolving under parallel beam illumination of an object, that is, using plane waves is generally termed **Fraunhofer** diffraction. Another important feature of **Fraunhofer** diffraction is its formation rather far from the object, in the so-called far-field regime. Even though **Joseph Fraunhofer** did not make a great contribution to diffraction theory, he is considered a pioneer in using diffraction gratings for separating spectral colors (by wire grating in 1821) and was the first to measure the wavelengths of spectral lines using diffraction gratings (see next Section 10.1).

10.1 Diffraction gratings

Diffraction gratings are dispersive optical devices that convert incident light into several diffracted beams propagating along different directions defined by the constructive interference conditions. This is achieved by periodic arrangement of elements or regions with distinct scattering amplitudes (or phases) or different absorption (transmission). Diffraction angles of these beams depend on the angle of incidence, the spacings between periodically arranged elements, and the light wavelength. Because of this, diffraction gratings are commonly used in monochromators and spectrometers, as well as for various linked applications.

For a majority of practical functions, two kinds of gratings are employed, namely, the reflective and transmissive ones. Reflective gratings are produced by fabricating the ridges or grooves on their surfaces, while transmissive gratings comprise periodically arranged hollow slits. These are amplitude-modulated gratings. The phase-modulated gratings are commonly produced by holography-like methods, which use two-beam interference patterns with equally spaced interference fringes. The latter are projected onto a photoresist and are fixed by the photolithography treatments.

https://doi.org/10.1515/9783111140100-011

Diffraction gratings can create rainbow colors when illuminated by a white light. Rainbow-like colors from closely spaced narrow imprints on the surface of the optical data storage disks, such as CDs or DVDs, are nice examples of light diffraction produced by this type of diffraction gratings.

Let us consider two key examples of a grating's functioning under illumination by monochromatic light having wavelength λ. The first example relates to light passing at normal incidence through a transmissive grating with spatial periodicity d (see Figure 10.1).

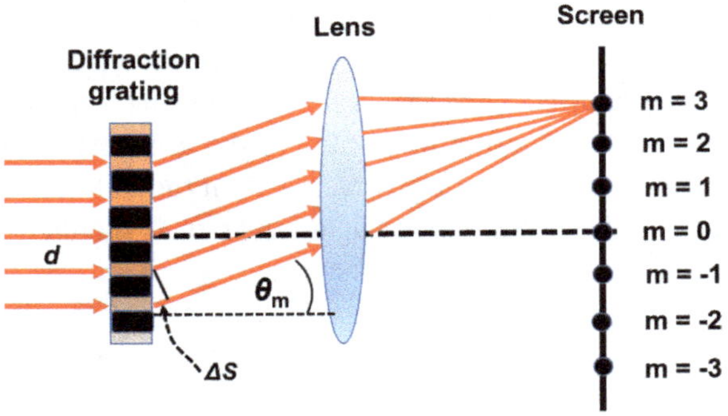

Figure 10.1: Action of a transmissive diffraction grating.

The diffraction angles θ_m are defined by the constructive interference, that is, by the condition that the path difference ΔS between the adjacent diffracted rays equals an integer number of wavelength λ. Using Figure 10.1, this condition can be rewritten as:

$$\Delta S = d \sin \theta_m = m\lambda \qquad (10.1)$$

where the diffraction orders are enumerated by integer numbers $m = 0, \pm 1, \pm 2 \ldots$ Placing a lens behind the diffraction grating, one can focus diffraction beams of different orders to bright spots on the screen or onto another detecting system (see Figure 10.1). Zero-order diffraction ($m = 0$) corresponds to light transmission with no angular deviation. As we already mentioned, diffraction angles,

$$\theta_m = \arcsin\left(m\frac{\lambda}{d}\right) \qquad (10.2)$$

depend critically on the ratio $\frac{\lambda}{d}$; moreover, there is no diffraction if $d < \lambda$. To achieve significant diffraction angles in the case of visible light ($\lambda \simeq 500$ nm), the d-values should be on a micrometer scale, while for hard X-rays ($\lambda \simeq 1$ Å = 0.1 nm) – on a nanometer scale.

Regarding reflective gratings, let us calculate the diffraction angles θ_m when the plane wave enters at an arbitrary angle θ_i to the grating normal (see Figure 10.2).

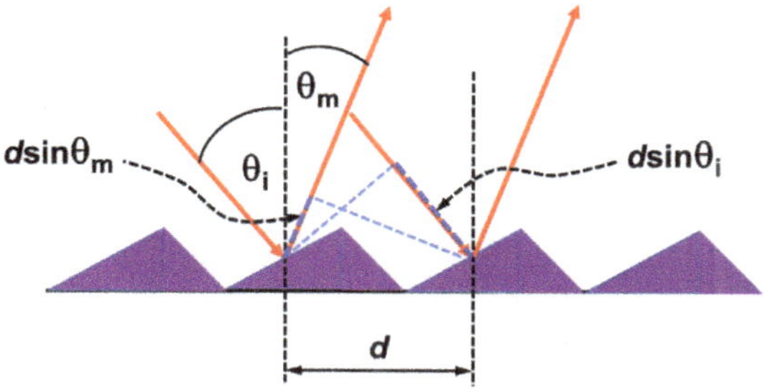

Figure 10.2: Action of a reflective diffraction grating.

Also, in this case, to be in phase, the adjacent scattered waves should encounter path difference $\Delta S = m\lambda$. With the aid of Figure 10.2, one finds:

$$\Delta S = d(\sin\theta_i - \sin\theta_m) = m\lambda \tag{10.3}$$

where $m = 0, \pm 1, \pm 2, \ldots$. For $m = 0$, we have specular reflection, $\theta_i = \theta_m$, as expected. Finally, the permitted diffraction angles are defined as:

$$\theta_m = \arcsin\left(\sin\theta_i - \frac{m\lambda}{d}\right) \tag{10.4}$$

10.2 Kinematic diffraction of X-rays in the Bragg scattering geometry

Immediately after the epoch-making discovery of X-rays by **Wilhelm Konrad Röntgen** in 1895 (awarded the first Nobel Prize in Physics, 1901), prolonged discussions about the nature (corpuscular or wave one) of this novel kind of radiation began. These discussions ended in 1912 when **Max von Laue** proposed searching for X-ray diffraction in crystals (the 1914 Nobel Prize in Physics), while **Walter Friedrich,** with the aid of **Paul Knipping,** performed successful experiments and indeed detected X-ray diffraction from nearly perfect crystals of copper sulfate.

To properly assess the brilliant idea of **Max von Laue**, one must understand that in those days, no one knew either the exact values of interatomic and interplanar distances in crystals or the X-ray wavelengths. Certainly, based on known X-ray energies, on one hand, and atomic densities, on the other hand, it was possible to assume that both interatomic distances and X-ray wavelengths could probably be comparable, being on a sub-nm scale. Believing in these rather feeble assumptions, but keeping in mind translational symmetry in crystals, M**ax von Laue** brought about a real revolution by proving that X-rays are short-wavelength electromagnetic waves, which can diffract by atomic networks in crystals. This discovery paved the way to structural analysis of crystals, a field of enormous importance to physics, chemistry, biology, and medicine.

Let us describe X-ray diffraction in crystals in more detail using the **Bragg** approach, in which the general three-dimensional diffraction problem is reduced to the one-dimensional problem, that is, to the scattered wave summation along the normal to the set of parallel atomic planes separated by spacing d (see Figure 10.3).

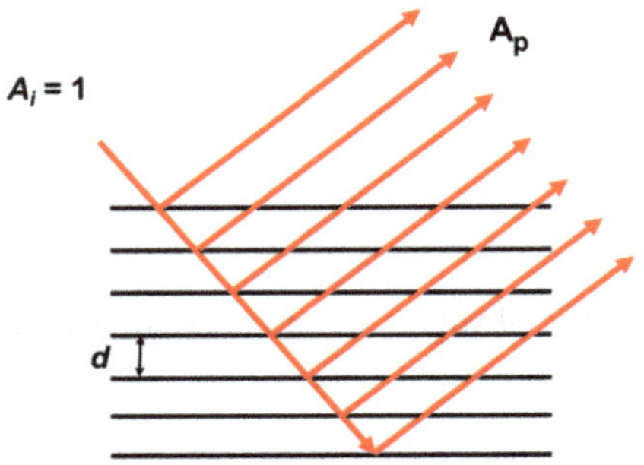

Figure 10.3: X-ray **Bragg** diffraction from a set of parallel atomic planes separated by spacing d.

For simplicity, we will treat the X-ray scattering using kinematic approximation, that is, neglecting an interaction between the transmitted and diffracted waves and, especially, ignoring the attenuation of the transmitted wave due to repumping of scattered quanta into the diffraction direction. A more general theory of dynamical X-ray diffraction, which deals with these issues, will be treated in Chapter 15.

Staying within kinematic approximation, let us take the amplitude of the incident wave as $A_i = 1$, and the even amplitudes of the scattered waves from each individual plane as A_p. The phase difference between two adjacent scattered waves is $\Delta\varphi$. Summating individual scattering amplitudes with their phases, we find that the total scattering amplitude A_{tot} from N planes is determined by geometrical progression:

$$A_{\text{tot}} = A_p \left[1 + e^{i\Delta\varphi} + e^{2i\Delta\varphi} + e^{3i\Delta\varphi} + \cdots + e^{i(N-1)\Delta\varphi}\right] \tag{10.5}$$

with the common ratio $e^{i\Delta\varphi}$. Therefore,

$$A_{\text{tot}} = A_p \frac{1 - \exp(iN\Delta\varphi)}{1 - \exp(i\Delta\varphi)} \tag{10.6}$$

The relative scattering intensity (reflectivity) is:

$$I = \left|A_{\text{tot}}/A_i\right|^2 = \left|A_p\right|^2 \frac{\sin^2\left(\frac{N\Delta\varphi}{2}\right)}{\sin^2\left(\frac{\Delta\varphi}{2}\right)} \tag{10.7}$$

It has the main maxima:

$$I_{\max} = |A_p|^2 N^2 \tag{10.8}$$

at $\Delta\varphi = 2\pi m$ (where m is an integer number), and rapidly decreases when moving away from it. Aside from main maxima, the scattering intensity exhibits an oscillating behavior (the so-called thickness fringes) with lateral minima and maxima being shifted along the phase axis (the $\Delta\varphi$-axis) from the main maximum by $2j\pi/N$ and $(2j+1)\pi/N$ ($j = \pm1, \pm2, \pm3, \ldots$), respectively. Such intensity distribution is really obtained from single-crystalline thin films by using nearly parallel and monochromatic X-rays (see Figure 10.4). The distance between successive fringes is inversely proportional to N, that is, inversely proportional to the film thickness, $T = Nd$. The latter relationship is used to measure the film thickness with sub-nm precision.

The interval between the first-order intensity minima around the main peak (i.e., between the minima defined by $j = \pm 1$) is also inversely proportional to N; the same conclusion is valid for the width Γ of the main intensity peak:

$$\Gamma \sim \frac{1}{N} \sim \frac{1}{T} \tag{10.9}$$

Intensities of secondary maxima rapidly diminish with increasing index j, as:

$$I_j = \frac{4}{(2j+1)^2\pi^2}|A_p|^2 N^2 \tag{10.10}$$

As we see, the ratio of $\frac{I_1}{I_{max}} = \frac{4}{9\pi^2}$ is already 4.5% only.

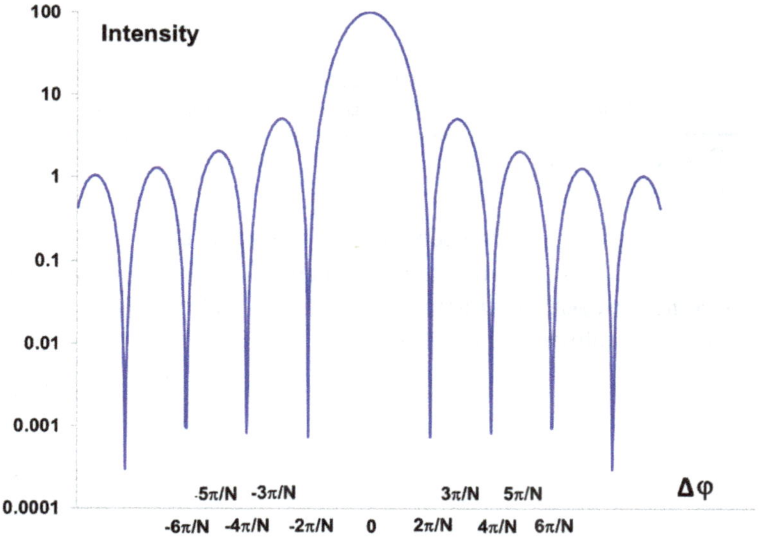

Figure 10.4: Thickness fringes in X-ray diffraction profile (on the phase $\Delta\varphi$-scale) taken from a thin film.

Therefore, we can estimate the integrated scattering intensity I_{int} by considering the area under the main diffraction peak. Using eqs. (10.8) and (10.9) yields:

$$I_{\text{int}} \sim I_{\max} \cdot \Gamma \sim \frac{|A_p|^2 N^2}{N} \sim N \sim T \tag{10.11}$$

Thus, we have obtained one of the key results of kinematic diffraction theory, which states that the integrated diffraction intensity from a non-absorbing crystalline plate is proportional to its thickness T. In more general terms, we can say that in this approximation, the diffraction intensity is proportional to the amount of material participating in the diffraction process. The latter statement is the basis of quantitative analysis of the crystalline phases comprising the investigated sample.

On the other hand, the fact that integrated diffraction intensity I_{int} in the kinematic approximation is linearly proportional to the crystal thickness elucidates serious intrinsic problem of kinematic diffraction theory. In fact, it implies that above some critical thickness T_{cr}, diffraction intensity may exceed the intensity of the incident beam I_0 – which is clearly nonsense (see Figure 10.5). This problem is overcome in dynamical diffraction theory (see Chapters 14 and 15), which takes proper account of the interaction between transmitted and diffracted waves.

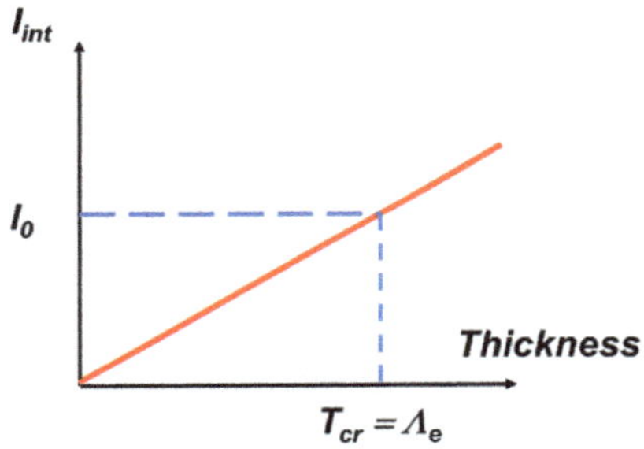

Figure 10.5: Proportionality of the integrated diffraction intensity I_{int} to crystal thickness T in kinematic approximation, which is valid for thin crystals with $T < T_{cr} = \Lambda_e$.

This interaction becomes more and more essential as the crystals become thicker, and the total diffraction amplitude grows. The critical thickness T_{cr} can be estimated assuming that the accumulated scattering amplitude from the number N_{cr} of atomic planes comes close to the amplitude of the incident X-ray wave, that is,

$$|A_p| N_{\text{cr}} = A_i = 1 \tag{10.12}$$

or

$$T_{\text{cr}} = N_{\text{cr}} d = \frac{d}{|A_p|} = \Lambda_e \tag{10.13}$$

Note that the exact expression for scattering amplitude A_p is derived in Section 12.1. Nevertheless, even in the form (10.13), this expression for T_{cr} is the same as for the characteristic penetration depth Λ_e, which appears in the dynamical diffraction theory and describes the in-depth exponential decay of transmitted X-ray waves in **Bragg** scattering geometry, due to the intensity "repumping" into the diffracted wave towards its total reflection (see Sections 14.4 and 15.1.3).

As the last point in this section, let us elaborate more about the phase difference ($\Delta\varphi$) issue. We understand that for constructive interference, $\Delta\varphi$ should be an integer number of 2π. In fact, in Section 7.2, the **Bragg** law (eq. (7.26)) has been already derived:

$$2d \sin \theta_{\mathrm{B}} = \lambda \tag{10.14}$$

as a consequence of the constructive interference condition. We repeat that the **Bragg** law defines the diffraction (**Bragg**) angles θ_{B} depending on the X-ray wavelength λ and d-spacings between parallel atomic planes in the crystal used for X-ray diffraction. As mentioned in Section 7.2, an integer m is hidden in eq. (10.14) within the d-spacing, the latter being proportionally reduced for multiple reflections.

Multiplying both parts of eq. (10.14) by 2π and expressing wavelength λ via the magnitude of wave vector $k = \frac{2\pi}{\lambda}$ yields:

$$2dk \sin \theta_{\mathrm{B}} = 2\pi \tag{10.15}$$

This means that the phase difference equals $\Delta\varphi = 2dk \sin \theta_{\mathrm{B}}$. With the aid of Figure 10.6, we also find that:

$$\boldsymbol{k}_{\mathrm{d}} - \boldsymbol{k}_{\mathrm{i}} = \boldsymbol{Q} \tag{10.16}$$

where $\boldsymbol{Q}$ is the so-called diffraction vector, that is, the wave vector transferred to the crystal in the diffraction process. It equals the difference between the wave vectors of the diffracted ($\boldsymbol{k}_{\mathrm{d}}$) and incident ($\boldsymbol{k}_{\mathrm{i}}$) X-rays. At the exact **Bragg** position, diffraction vector $\boldsymbol{Q}$ is perpendicular to a certain set of parallel atomic planes.

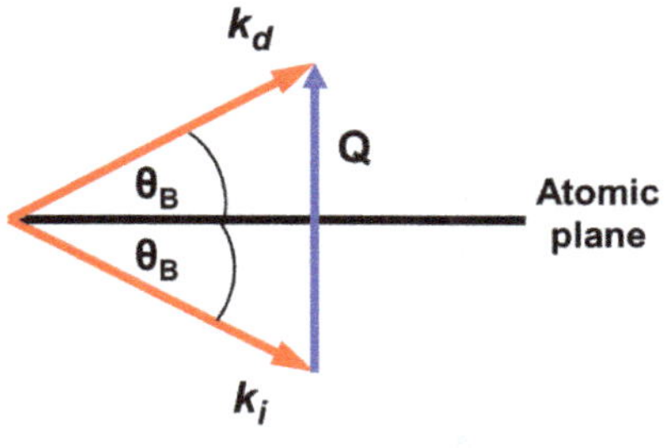

Figure 10.6: Wave vector scheme of X-ray diffraction: $\boldsymbol{k}_{\mathrm{i}}$ and $\boldsymbol{k}_{\mathrm{d}}$ are wave vectors of the incident and diffracted waves, respectively; $\boldsymbol{Q} = \boldsymbol{k}_{\mathrm{d}} - \boldsymbol{k}_{\mathrm{i}}$ is the wave vector transferred to the crystal during scattering; θ_B is the **Bragg** angle.

Furthermore, for elastic scattering:

$$|\boldsymbol{k}_{\mathrm{i}}| = |\boldsymbol{k}_{\mathrm{d}}| = k = \frac{2\pi}{\lambda} \tag{10.17}$$

one obtains

$$2k\sin\theta_B = |\boldsymbol{Q}| = Q \tag{10.18}$$

Combining eqs. (10.15) and (10.18) yields an alternative expression for the phase difference:

$$\Delta\varphi = Qd \tag{10.19}$$

which provides the interrelation between momentum (wave vector) space, where $\boldsymbol{Q}$-vectors exist and real crystalline space where d-spacings "live." Experimentally, employing X-ray diffraction, we change the magnitude and direction of the $\boldsymbol{Q}$-vectors until a strong diffraction signal is obtained ($\Delta\varphi = 2\pi$). Using eq. (10.19), the respective d-spacings, $d = \frac{2\pi}{Q}$, are then extracted. This is the basis of the X-ray analysis of crystal structures (X-ray crystallography).

10.3 Producing parallel X-ray beams with a Göbel mirror

Another interesting and important real-life application of X-ray diffraction is in producing parallel X-ray beams by parabolic mirrors. As already discussed in Section 3.3.1, a parabolic profile,

$$y = px^2 \tag{10.20}$$

is ideally suited for focusing and defocusing of visible light and X-rays. The focal distance F (segment OF in Figure 10.7) according to eq. (3.19) equals:

$$F = \frac{1}{4p} \tag{10.21}$$

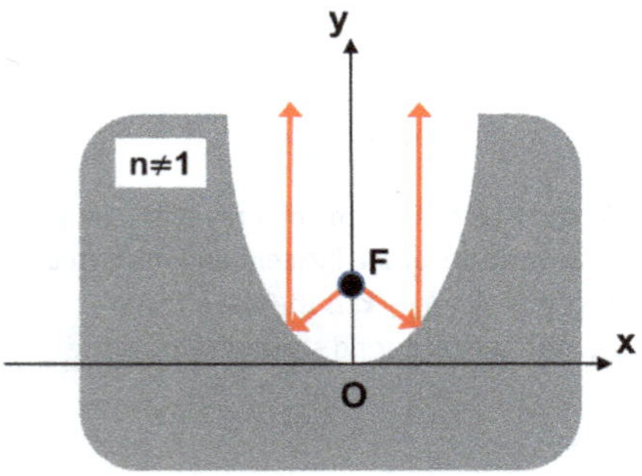

Figure 10.7: Action of a parabolic mirror, which converts the rays issuing from the point focus F to the parallel beam along the *y*-axis.

As is seen in Figure 10.7, a large part of rays, issuing from focal point F, will strike the parabolic surface at rather large angles. This is not a problem for visible light since its reflectivity, for example, by metal surfaces, remains very high for any angle of incidence (see Section 6.2). This is not the case with X-rays, for which a 100% reflectivity is realized only if the incident angle is below the critical angle of total external reflection (see Section 2.3). Since critical angles are very small (about 0.25°), the usage of total external reflection for producing parallel X-ray beams becomes impractical in many cases.

An alternative methodology towards obtaining high X-ray intensity is X-ray diffraction. The problem with the latter is that the **Bragg** angle θ must vary along the parabolic surface. In fact, as is clearly seen in Figure 10.8,

$$\theta + \delta = 90^\circ \tag{10.22}$$

where angle δ is the tangent slope to the local parabolic point $M(x,y)$. The slope δ is evidently a function of (x,y); according to the relationship (10.22), the same is valid for angle θ.

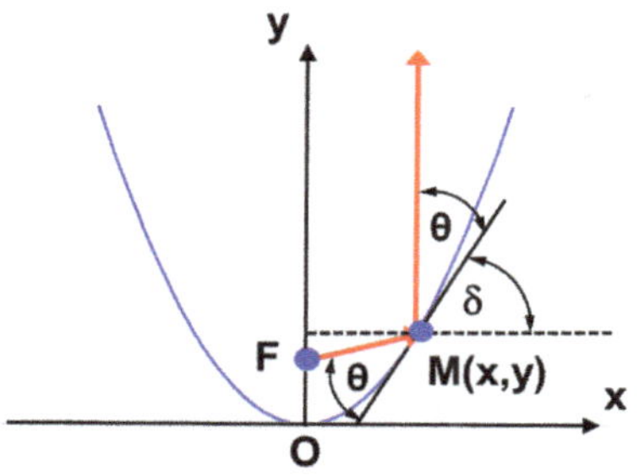

Figure 10.8: Scheme showing the relationship between local diffraction angle θ and local tangent slope δ, that is, $\theta + \delta = 90°$.

The problem was solved in the early 1990s by **Herbert Göbel**, who suggested depositing a multilayered coating on the parabolically shaped substrate, the multilayer with variable period T_v to fit local **Bragg** angles applying the relationship given by eq. (7.27):

$$2T_v \sin\theta = \lambda \tag{10.23}$$

Furthermore, using eqs. (10.20)–(10.22), one obtains

$$\cot\theta = \tan\delta = \frac{dy}{dx} = 2px = \frac{x}{2F} = \frac{\sqrt{4Fy}}{2F} = \sqrt{\frac{y}{F}} \tag{10.24}$$

The key parameter here is the distance FM between focal point F and local point $M(x,y)$ on the parabolic surface:

$$FM = f = \sqrt{x^2 + (y - F)^2} \tag{10.25}$$

Squaring both parts of eq. (10.25) yields:

$$f^2 = x^2 + (y - F)^2 = 4Fy + y^2 - 2Fy + F^2 = (y + F)^2 \tag{10.26}$$

and

$$f = y + F; \quad y = f - F \tag{10.27}$$

Combining eqs. (10.25) and (10.27), we find:

$$\cot\theta = \sqrt{\frac{f-F}{F}} = \sqrt{\frac{f}{F} - 1} \tag{10.28}$$

Therefore, the multilayer period T_v (considering eqs. (10.23) and (10.28)) should vary with distance $FM = f$, as

$$T_v = \frac{\lambda}{2\sin\left[\operatorname{arccot}\left(\sqrt{\frac{f}{F} - 1}\right)\right]} \tag{10.29}$$

Thus, the variable period provides a right entrance angle (**Bragg** angle θ) at any point $M(x,y)$ on the surface of the mirror at which X-rays strike the multilayered coating (colored in green in Figure 10.9).

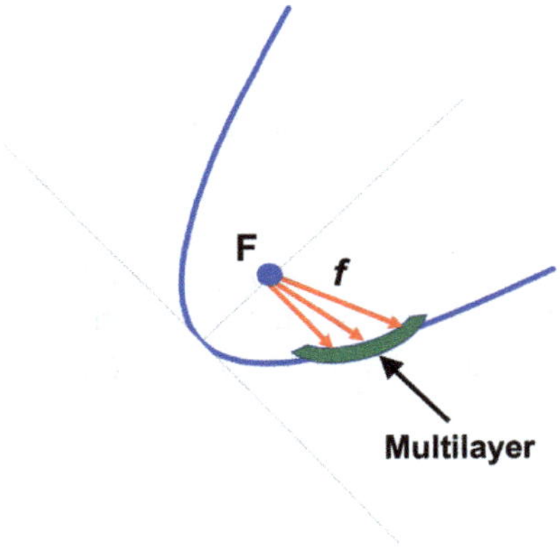

Figure 10.9: Design of a **Göbel** mirror.

The **Göbel** invention allowed the brilliance of laboratory X-ray diffraction instruments to be greatly increased. The combination of a parabolic mirror with a double-crystal monochromator (see Section 15.4) is called parallel beam optics and is a good compromise solution for diffraction measurements in thin films, benefiting from the highly enhanced X-ray flux, while keeping the instrumental angular width of the setup small enough, about 0.01°.

10.4 Fraunhofer diffraction of light by a circular hole

Coming back to light optics, one of the classical examples of **Fraunhofer** diffraction is the transmission of initially parallel beams of light through a circular hole of radius D in a non-transparent screen. Within the framework of geometrical optics, behind the screen, the non-zero transmitted intensity will be detected just in front of the hole (see Figure 10.10).

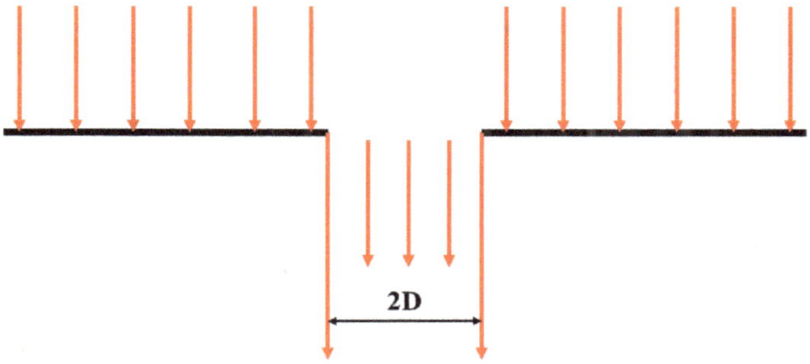

Figure 10.10: Light transmission through a circular hole of radius D in geometrical optics.

This means that after passing through the screen, the direction of light propagation does not change; the only effect is the total light intensity reduction in proportion, dictated by the area of the hole $S_h = \pi D^2$ with respect to the cross section of the incident beam. The light scattering by the border of the hole and the interference of the scattered and transmitted waves, however, can substantially modify this result and provide additional intensity in spatial directions, which differ by angle θ from the initial direction of light propagation before the screen (see Figure 10.11, upper panel).

In other words, after passing through the screen, light is propagating not only in one direction, which is defined by the initial wave vector $\boldsymbol{k}_\text{i}$, but also in many other directions defined by vectors $\boldsymbol{k}_\text{s} = \boldsymbol{k}_\text{i} + \boldsymbol{q}$. Here, $\boldsymbol{q}$ is a variable wave vector transfer to the screen during scattering events (see Figure 10.12).

As we already noted, for elastic scattering processes,

$$|\boldsymbol{k}_\text{s}| = |\boldsymbol{k}_\text{i}| = 2\pi/\lambda \tag{10.30}$$

where λ is the wavelength of light. Considering eq. (10.30) and the axial symmetry of the scattering problem (at fixed scattering angle θ, (see Figure 10.12)), we find that:

$$|\boldsymbol{q}| = q \approx \frac{2\pi}{\lambda}\theta \tag{10.31}$$

For each $\boldsymbol{q}$ value, the light scattering amplitude is given by the **Fourier**-component $u_{\boldsymbol{q}}$ of the wavefield $u(\boldsymbol{r})$ just after the screen:

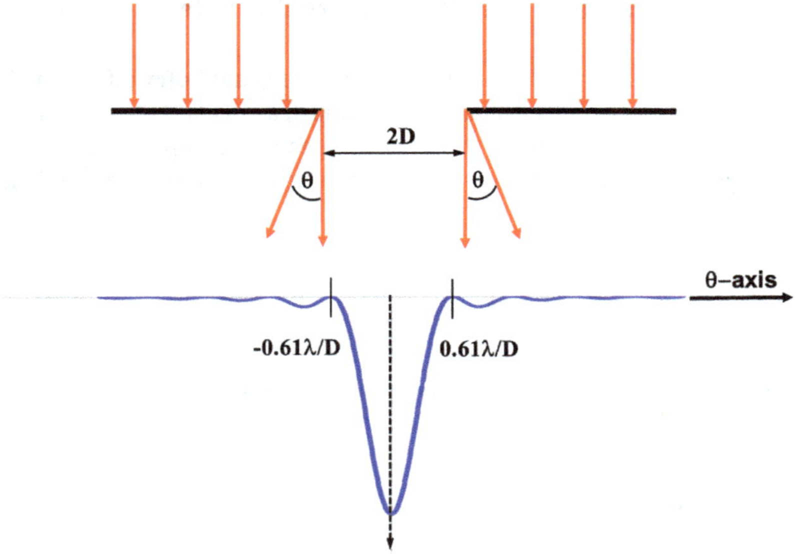

Figure 10.11: Upper panel: Light transmission through a circular hole of radius *D* considering the diffraction phenomenon (**Fraunhofer** diffraction). Bottom panel: Transmitted intensity as a function of angular deviation θ.

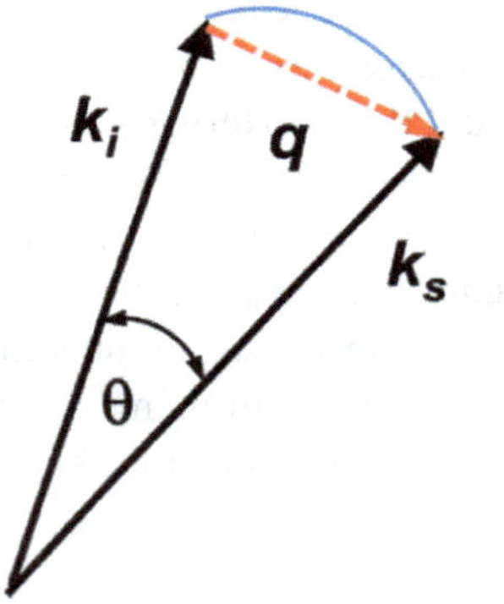

Figure 10.12: Wave vector change $\boldsymbol{q}$ during elastic scattering of propagating light. Wave vectors of the incident and scattered waves are designated $\boldsymbol{k}_\mathrm{i}$ and $\boldsymbol{k}_\mathrm{s}$, respectively.

$$u_q = \iint u(\boldsymbol{r})e^{-i\boldsymbol{qr}}dxdy \tag{10.32}$$

In the first approximation, however, we can set $u = u_0$, that is, u equals the amplitude of the homogeneous wavefield before the screen, and then express the scattering amplitude u_q as:

$$u_q = \iint u_0 e^{-iqr} dxdy \tag{10.33}$$

where the integration proceeds over the entire area S_h of the hole. The diffraction intensity (relative to that in the incident beam) within the element $d\Omega$ of solid angle Ω for a given $\boldsymbol{q}$ value is expressed as follows:

$$dI_{\text{rel}} = \lambda^{-2} \left| \frac{u_q}{u_0} \right|^2 d\Omega \tag{10.34}$$

To find u_q, let us introduce polar coordinates r and ϕ within the circular hole. In this coordinate system, eq. (10.33) transforms into:

$$u_q = u_0 \int_0^D \int_0^{2\pi} e^{-iqr\cos\phi}\ r\ d\phi\ dr = 2\pi u_0 \int_0^D J_0(qr) r\ dr \tag{10.35}$$

where J_0 is a **Bessel** function of zero order. Note that deriving eq. (10.35), we use the fact that for small scattering angles θ, vector $\boldsymbol{q}$ is nearly situated in the plane of the hole. One can express the integral in (10.35) via the **Bessel** function of first order J_1:

$$\int_0^D J_0\ (qr) r\ dr = \frac{D}{q}\ J_1(Dq) \tag{10.36}$$

and, finally,

$$u_q = \frac{2\pi u_0 D}{q}\ J_1(Dq) \tag{10.37}$$

Substituting eq. (10.37) into eq. (10.34) and using eq. (10.31), one obtains:

$$dI_{\text{rel}} = \frac{D^2}{\Theta^2}\ J_1^2\left(\frac{2\pi D}{\lambda}\theta\right) d\Omega \tag{10.38}$$

The distribution of the transmitted light intensity (10.38), as a function of the scattering angle θ, is displayed in Figure 10.11 (bottom panel). When increasing the absolute value of the angle θ, the light intensity shows fast overall reduction, on which the pronounced oscillating behavior is superimposed. Intensity oscillations are revealed as lateral maxima of diminishing height, separated by zero intensity points. The latter are determined by the zeroes of the J_1-function. Most of the diffraction intensity (about 84%) is confined within the angular interval, $-\theta_0 \le \theta \le \theta_0$, which is defined by the first zero of the **Bessel** function J_1:

$$\frac{2\pi}{\lambda}\ D\ \theta_0 = 3.832 \tag{10.39}$$

that is,

$$\theta_0 = 0.61 \frac{\lambda}{D} \tag{10.40}$$

It follows from eq. (10.40) that diffraction is important, when wavelength λ is a significant part of the D-value. If $\frac{\lambda}{D} \ll 1$, angular deviations are subtle, which implies that diffraction effects (deviations from geometrical optics) are weak. For visible light with $\lambda \approx 0.5$ µm, the diffraction phenomena are regularly observed for objects having characteristic size D ranging from a few microns and up to about 10^3 microns.

The diffraction of light imposes a strong limitation on the resolving power of optical instruments. For a telescope, the resolution is defined on an angular scale and is given by the **Rayleigh** criterion. The latter states that two objects (stars) can be separately resolved, if an angular distance $\Delta\theta_c$ between the maxima of their intensity distributions (eq. (10.38)), exceeds the θ_0-value defined by eq. (10.40). This implies that angular resolution of a telescope is given by eq. (10.40).

For a microscope, length limitations are most essential, helping us evaluate the size of the smallest objects still visible using a particular optical device. To "translate" the **Rayleigh** criterion into the length-scale language, let us consider the simplified equivalent scheme of a microscope. The latter is represented by a circular lens of radius D and focal length F and transforms an object of size Y into its image of size Y' (see Figure 10.13). We learned in Sections 3.4 and 3.6 that to achieve high magnification, an object (Y) should be placed close to the focus (left side of the lens in Figure 10.13).

Figure 10.13: Illustration of the diffraction-limited spatial resolution of a microscope.

If so,

$$\theta \approx \frac{Y}{F} \tag{10.41}$$

Applying the **Rayleigh** criterion means that $\theta > \theta_0$ and hence,

$$Y > \Delta = F\theta_0 = 0.61\frac{\lambda}{D}F \tag{10.42}$$

For the focusing effect (see Figure 10.14), we illuminate our lens by a wide parallel beam and obtain small spot Y' in the focal plane (the right side of the lens in Figure 10.14).

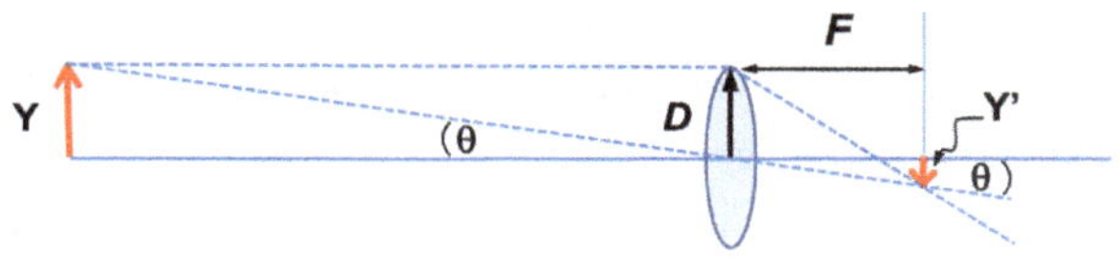

Figure 10.14: Illustration of the diffraction-limited focusing spot size, which is achievable by using a lens.

Now,

$$\theta = \frac{Y'}{F} \tag{10.43}$$

Again applying the **Rayleigh** criterion and eq. (10.40), we find that spot size Y' cannot be smaller than parameter Δ given by eq. (10.42), that is,

$$Y' > \Delta = F\theta_0 = 0.61\frac{\lambda}{D}F \tag{10.44}$$

Therefore, the spatial resolution Δ when using the circular focusing element is completely defined by its size D, focal length F, and radiation wavelength λ. We will use these results in Section 12.3, when describing the focusing effect of **Fresnel** zone plates for X-ray optics.

10.5 Rayleigh criterion and diffraction limit of optical instruments

We see that to determine the resolving power of a microscope, the crucial parameter is some characteristic size D of the optical system, which is called its entrance pupil. To calculate the magnitude of D, let us follow the **Abbe** ideas and consider the generalized scheme of a microscope under parallel beam illumination (Figure 10.15).

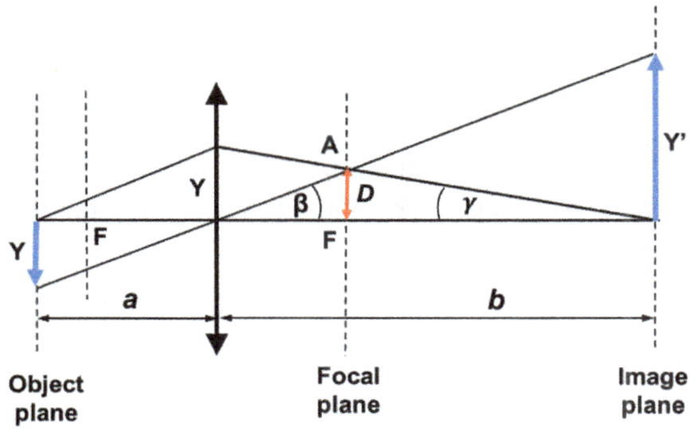

Figure 10.15: Constructing the entrance pupil D of a whole optical system.

We see that the image size Y' of an object Y is defined by the intersection point A of refracted rays with the trace of the focal plane, namely, the segment $FA = D$, which is the entrance pupil. The larger the D-value, the greater is the image size Y' and, correspondingly, the object magnification and resolving power. Certainly, this is true if the physical aperture is open wide.

Let us prove that the trajectories of all the refracted rays are intercepted at the same height $FA = D$ in the right-hand focal plane. Considering small angles β and γ in Figure 10.15, we find

$$\beta = \frac{Y}{a}; \quad \gamma = \frac{Y}{b} \tag{10.45}$$

Rays passing through the center of the optical system (with no refraction) traverse the trace of the focal plane at height D:

$$D = \beta \cdot F = \frac{F}{a} Y \tag{10.46}$$

The refracted rays, shown in Figure 10.15, are crossing the trace of the focal plane potentially at another height D^*:

$$D^* = \gamma \cdot (b - F) \tag{10.47}$$

Recalling **Newton**'s formula (eq. (3.33)), one obtains

$$\frac{1}{b} = \frac{1}{F} - \frac{1}{a} = \frac{a - F}{aF} \tag{10.48}$$

and then

$$D^* = \gamma \cdot (b - F) = \gamma \left(\frac{aF}{a - F} - F \right) = \gamma \frac{F^2}{a - F} \tag{10.49}$$

Finally, substituting eqs. (10.45) and (10.48) into eq. (10.49) yields

$$D^* = \frac{Y}{b} \frac{F^2}{(a - F)} = Y \frac{(a - F)}{aF} \cdot \frac{F^2}{(a - F)} = Y \frac{F}{a} = D \tag{10.50}$$

that is, indeed $D^* = D$, and, namely, this size defines the entrance pupil of a whole optical system, which should be used when applying the **Rayleigh** criterion for the generalized microscope scheme.

Therefore, one can say that $\beta > 0.61 \frac{\lambda}{D}$ or if an immersion medium with refractive index n is used

$$\beta > 0.61 \frac{\lambda_0}{nD} \tag{10.51}$$

where λ_0 is the light wavelength in vacuum. Recalling that $\beta = \frac{Y}{a}$ (see eq. (10.45)) and substituting it into eq. (10.51) yields

$$Y > 0.61\frac{\lambda_0}{nD}a \tag{10.52}$$

Since in a microscope $a \approx F$,

$$Y > 0.61\frac{\lambda_0}{nD}F \tag{10.53}$$

In this form, the resolving power, that is, the minimal size of an object still visible with the aid of a microscope is very similar to that given before by eq. (10.42). Note that the ratio F/D, known in photography as the *f*-factor, is used to specify the angular aperture of a lens.

The widely accepted expression for resolving power is obtained when the entrance pupil D is expressed as $D = \beta \cdot F$ (see eq. (10.46)). If so,

$$Y > 0.61\frac{\lambda_0}{n\beta} \tag{10.54}$$

or more precisely,

$$Y > 0.61\frac{\lambda_0}{n\sin\beta} \tag{10.55}$$

In this form, the resolving power (except arithmetic factor of 0.61) looks like our earlier estimation (3.49).

As already mentioned in Section 3.5, the product $NA = n\sin\beta$ is called the numerical aperture and cannot be much larger than 1. The acceptable maximal value of NA for optical microscopes is $NA \approx 1.4$. Therefore, the best diffraction-limited spatial resolution is about $\frac{\lambda_0}{2}$, that is, a few hundred nanometers. This fundamental result from the nineteenth century is engraved on the tombstone of **Ernst Abbe** in Jena, Germany (see Figure 10.16).

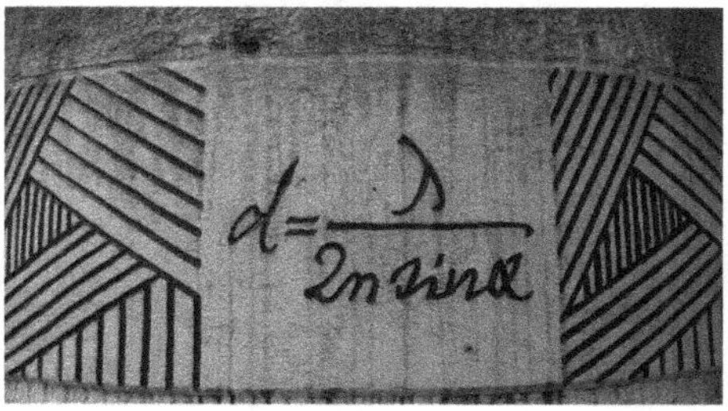

Figure 10.16: The tombstone of **Ernst Abbe** in Jena (Germany) showing the engraved record of the diffraction limit (credit: Wikimedia Commons).

Most conventional light microscopes are not able to image the sample features on a sub-micron length scale. Drastic improvement in resolving power was achieved following the invention of electron microscopy by **Ernst Ruska** (1986 Nobel Prize in Physics), due to the use of the much shorter electron wavelength λ_e. The latter is defined by the electron momentum P and **Planck**'s constant h via the **de Broglie** relationship, $\lambda_e = \frac{h}{P}$. In turn, electron momentum can be expressed via kinetic energy $\varepsilon_k = \frac{P^2}{2m} = eU$, which an electron accumulates when traveling within an electron microscope under an applied electric field, creating the potential difference U. Here, e and m stand for electron charge and mass, respectively. Finally, all this gives us the resulting numerical expression for an electron wavelength (in Å):

$$\lambda_e = \frac{0.38}{\sqrt{U(kV)}} \text{Å} \tag{10.56}$$

where U is measured in kV.

In scanning electron microscope (SEM), which, according to its working principle, is closer to optical microscopes than other types of electron microscopes, typically $U = 10–20$ kV. Correspondingly, $\lambda_e \approx 0.1$ Å, that is, 5×10^4 times shorter than typical wavelength of visible light, $\lambda_0 \approx 5{,}000$ Å. The actual gain in resolving power, however, is much smaller because of difficulties associated with focusing and related aberrations of electron lenses. As a result, maximal numerical aperture is rather small, $(NA)_e \approx 5 \times 10^{-3}–10^{-2}$, and the achievable gain in resolving power is about 300. This provides the best spatial resolution being on the nm-scale. Practically, high-resolution SEM instruments allow us to resolve the object features on a 10 nm scale. High-resolution transmission electron microscopes use high voltage $U = 300$ kV and offer much shorter electron wavelengths of $\lambda_e \approx 0.02$ Å. Correspondingly, sub-nanometer and even atomic resolution becomes available.

Returning to X-rays, we stress that the latter have wavelengths of about 0.1 nm = 1 Å, that is, 5,000 times shorter than those of visible light. Therefore, with X-rays, as well, one can expect much better spatial resolution than that achievable by using visible light optics. This is the driving force behind intensive development of X-ray microscopy at synchrotron beamlines. We discuss this issue in Section 12.4.

Chapter 11
Beyond diffraction limit

The second half of the twentieth century and, especially, its last few decades have been marked by groundbreaking inventions and intensive developments of novel microscopies that broke the diffraction limit in spatial resolution of optical devices established at the end of the nineteenth century. Below, we describe in more detail, confocal microscopy and near-field microscopy. These two techniques contributed a lot to the progress of contemporary life sciences and nanotechnology. Our description also includes a brief review of new ideas, based on the usage of negative materials already introduced in Section 6.4.

11.1 Confocal microscopy

As we discussed in Section 3.6.1, coherent illumination of an object by parallel light beams brings important benefits and, first of all, one-to-one correspondence between local wavefields in the object and image planes. When using such conventional microscopes, however, we lose a very important piece of information, that is, what happens in the sample depth. To circumvent this difficulty and solve the problem, the confocal optical microscope was invented. Its development began in the 1940s, but it took another 15 years before the first instrument was patented and built by **Marvin Minsky**.

Marvin Minsky, who came from the field of brain research, was trying to map the assembly of nerve cells but encountered serious problems related to diffuse scattering that made the image very blurry, with no clearly resolved details. He decided that ". . . to avoid all that scattered light," he had "to never allow any unnecessary light to enter in the first place." For this purpose, **Minsky** used the entrance aperture (pinhole), which spatially restricted the incident light (see Figure 11.1). The next element in his microscope was an objective lens aimed at focusing the light from the pinhole onto the desired point of an object. An object's points are chosen by moving the object in the *xy*-plane situated perpendicular to the optical axis, as well as along the axis of the optical system, thus changing the focal spot location in the sample's depth. The light scattered by the sample is intercepted by the second lens and guided into a small spot registered by a detector system. To further diminish the undesired scattered light from the object, the second aperture (exit aperture) is placed in front of the detector (see Figure 11.1). As **Minsky** said: "We end up with an elegant, symmetrical geometry: a pinhole and an objective lens on each side of the specimen."

The confocal scheme can also be used in the reflection mode (see Figure 11.2) described by **Minsky** as follows: "We could also employ a reflected light scheme by placing a single lens and pinhole on only one side of the specimen – and using a half-silvered mirror to separate the entering and exiting rays."

https://doi.org/10.1515/9783111140100-012

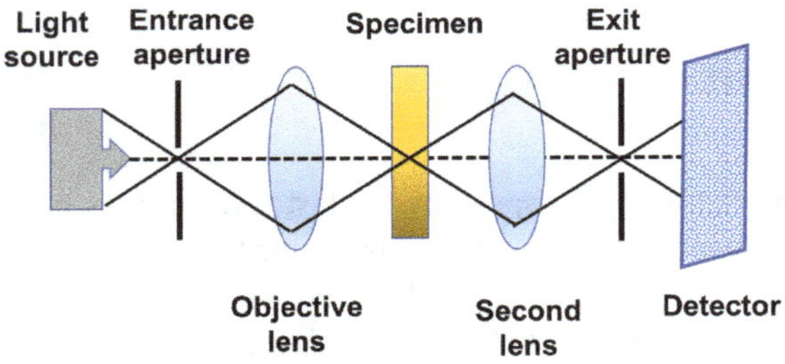

Figure 11.1: Principal scheme of **Minsky**'s transmission confocal microscope.

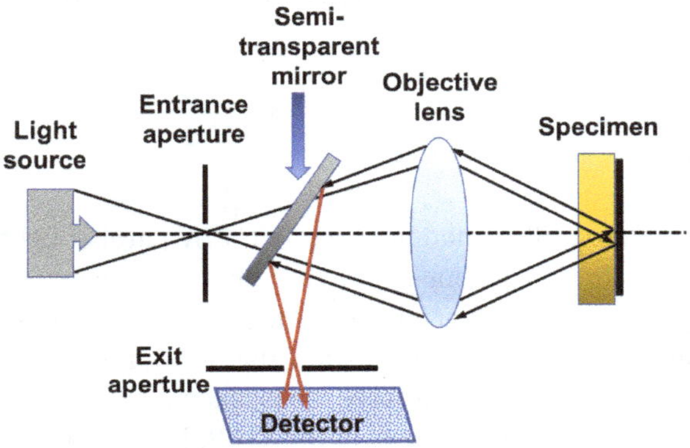

Figure 11.2: Principal scheme of **Minsky**'s reflection confocal microscope.

Confocal microscopy allows us to obtain three-dimensional information on the sample structure, which provides a sharper contrast and then, better resolution in the *xy*-plane because of the effective cut-off of the high-order diffraction rays by the exit aperture. The most important feature, however, is the ability to visualize and explore depth-dependent structural variations. An illustrative example is shown in Figure 11.3, which presents a Technion's logo produced by photolithography and subsequent etching.

Modern confocal microscopes use high-intensity laser sources in combination with high-speed scanning systems. Nowadays, the moving light spot on the sample is achieved by quickly redirecting the light beams rather than by the much slower physical movement of the sample. One way to do the former is by using a scanning **Nipkow** disc (or wheel) invented by **Paul Nipkow** at the end of the nineteenth century. It was used to realize the scanning mode in the first television systems. The scanning disk has a set (up to many thousands) of equidistant holes arranged in **Archimedean** spirals (see Figure 11.4). Upon rotation, the **Nipkow** disc transmits incident light through successive holes, thus creating moving pinholes employed to generate multiple images. For example, this method was applied in 1967 by **David Egger** and **Mojmir Petran** to image, for the first time, the brain cells and cell processes in living vertebrates.

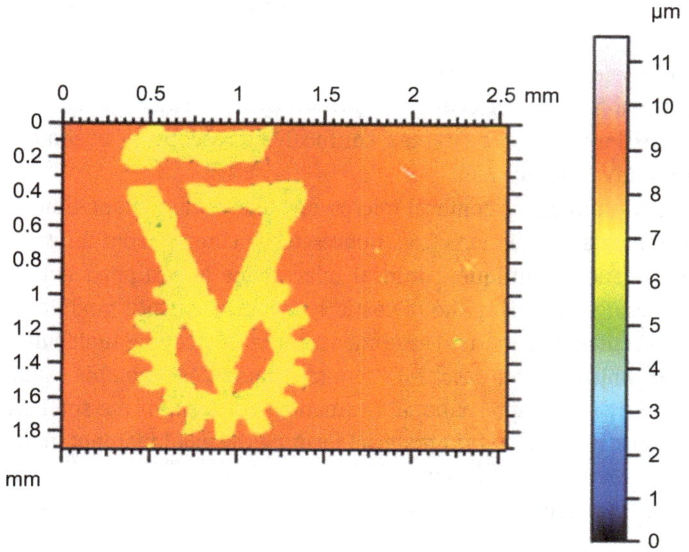

Figure 11.3: Illustrative example of the capabilities of 3D imaging by confocal microscope: A Technion's logo produced by photolithography procedures. The color scale from yellow to red corresponds to the increase in the depth of etching (see also the ruler at the right-hand side of the image) (courtesy of Boaz Pokroy, Technion-Israel Institute of Technology).

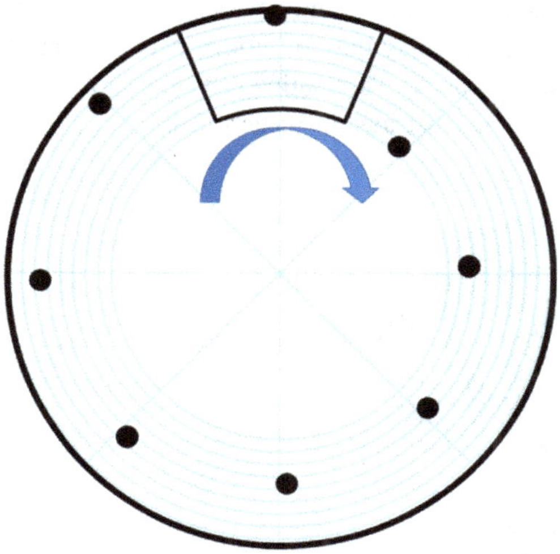

Figure 11.4: Design of the **Nipkow** disc with a set of holes arranged in an **Archimedean** spiral.

Another approach is realized in programmable array microscopes, which use electrically controlled spatial light modulators (SLM) to produce a set of moving pinholes. The SLM may contain microelectromechanical mirrors and liquid-crystal components. The image is usually acquired by a charge-coupled device (CCD). This combination drastically reduced the acquisition time and enabled confocal microscopy to become a highly powerful and useful investigative tool.

In addition to direct optical imaging, confocal microscopy allows easy adaptation towards detecting various secondary radiations (e.g., fluorescence) coming from different points of an object. In one such technique, confocal microscope is equipped with a **Raman** spectrometer, allowing registering the inelastic scattering intensity modulated by the frequencies of molecular vibrations. These measurements provide molecular-level information about the object's depth-resolved chemical content. As a notable example, we can mention the study by the international German-Israeli team of the spatially resolved composition of a protein/silica hybrid material forming the axial filament in the spicules of marine sponges. Despite the filament's tiny diameter (nearly 1.5 μm), it was possible to extract the depth-resolved protein/silica volume ratio within the filament as well as around it, and find that in the filament it is nearly 50:50. Combined with experimental results obtained by sophisticated X-ray diffraction and electron microscopy methods, this allowed scientists to conclude that the filament structure is composed of periodically arranged nm-sized amorphous protein clusters surrounded by mesoporous silica (see Figure 11.5). For the filament in the sponge *Monorhaphis chuni* shown in Figure 11.5, the structure symmetry is body-centered tetragonal (BCT) with lattice parameters $a = 9.8$ nm, $c = 10.8$ nm. This finding is of great importance because it shows that nature can produce highly periodic structures not only built of atomic networks, as in crystals, but also using nanometric amorphous blocks. Fascinating consequences of the highly periodic filament structure will be further illustrated in Section 12.4.

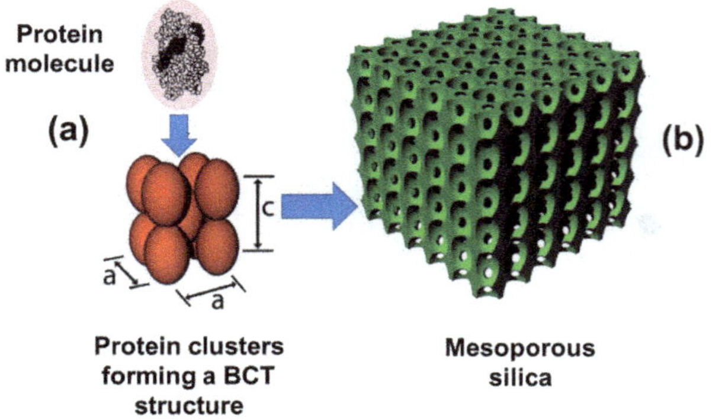

Figure 11.5: The 3D model of the highly ordered internal structure of the axial filament having body-centered tetragonal (BCT) symmetry and composed of amorphous protein nano-blocks (a) imbedded into a mesoporous silica structure (b).

Principally, a confocal fluorescent microscope is designed as shown in Figure 11.6.

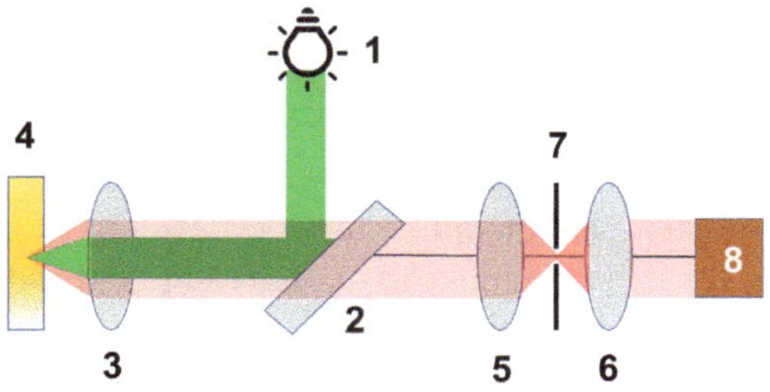

Figure 11.6: Principal scheme of a fluorescence confocal microscope: 1, light source; 2, semitransparent mirror (light deflector); 3, objective lens; 4, sample; 5, 6, auxiliary lenses; 7, pinhole aperture; and 8, detector.

An incident light with wavelength λ_1 (colored in green in Figure 11.6), issuing from the laser (1), is redirected towards the sample (4) using the light deflector (2) and, then, is focused by the objective lens (3) onto a certain point (x, y) of the sample (4). Point illumination of the sample activates fluorescent emission with wavelength $\lambda_2 > \lambda_1$ (colored in pink in Figure 11.6), which is converted into the parallel beam by the objective lens (3). After that, fluorescent radiation passes through the deflector (2) and is focused by the lens (5) onto the pinhole aperture (7) being confocal with the objective's focal point on the sample. Using the auxiliary lens (6), the fluorescent radiation is again converted into a parallel beam and is registered within the detector (8). Take special note of the crucial role of the pinhole aperture (7) in achieving enhanced spatial resolution of the confocal imaging. In simple words, the pinhole stops all rays issuing outside the focal point of the objective lens; therefore, only the light from the chosen illumination volume is detected.

In the X-ray wavelength range, a very popular technique is confocal X-ray fluorescent imaging of buried layers and inclusions in different materials systems. Recall that upon X-ray irradiation, all materials, besides scattering, partially absorb the incoming X-rays. The absorbed energy is released mostly in the form of electrons ejected from the electron shells located close to the nucleus. The resulting vacant energy states are filled by electrons from the upper shells. This process is accompanied by the emission of characteristic fluorescent radiation with fixed wavelengths determined by the energy differences between respective electron shells. Therefore, the X-ray fluorescence spectrum is the fingerprint of the specific set of elements composing the investigated sample.

Confocal X-ray fluorescence microscopes have been installed at synchrotron beamlines of several synchrotron sources, for example, at the European Synchrotron Radiation Facility (ESRF). X-rays are focused by the elliptic **Kirkpatrick-Baez** mirrors, polycapillary systems, or by pinholes, while samples are mechanically scanned using piezoelectric actuators. Typical studies include non-invasive three-dimensional imaging of inclusions, chemically different from the matrix, localization of quantum dots, and non-destructive layer-by-layer chemical analysis of old paintings.

To summarize, we can say that applications of confocal microscopy with visible light and X-rays to nano-objects and nanotechnology, in general, are very wide, covering three-dimensional nanostructures, two-dimensional structures such as thin films with nanopores, nanoholes and nanomembranes, one-dimensional carbon nanotubes, and zero-dimensional structures such as luminescent quantum dots.

Regarding life science applications, the most common is the usage of confocal fluorescence microscopy in the visible light range for studying biological samples treated with fluorescent dyes. This technique gained a great boost with the discovery of green fluorescent protein (GFP). The latter emits bright green light with a wavelength of 509 nm, when illuminated by blue or ultraviolet radiation. This protein was first extracted from the *Aequorea victoria* jellyfish (see Figure 11.7) and studied already in the 1960s by **Osamu Shimomura**. Later, such proteins were found in some other marine organisms.

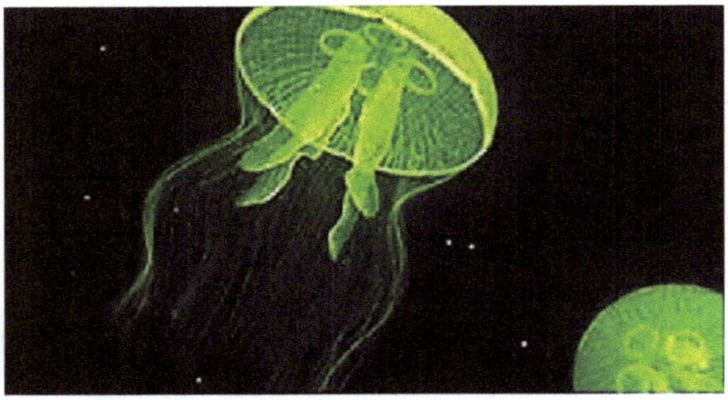

Figure 11.7: *Aequorea victoria* jellyfish colored in green due to the presence of the green fluorescent protein (GFP) (credit: Wikimedia Commons).

It took another thirty years until the gene expression of GFP was elucidated by research groups of **Martin Chaflie** and **Roger Tsien**. In 2008, **Osamu Shimomura**, **Martin Chalfie,** and **Roger Tsien** were awarded the Nobel Prize in Chemistry "for the discovery and development of the green fluorescent protein, GFP."

The accessibility of GFP and its derivatives revolutionized the field of confocal fluorescence microscopy and, especially, its large-scale use in cell biology. All this facilitated the development of automated living-cell fluorescence microscopy systems, which are widely employed to observe various processes in living cells "highlighted" by fluorescent proteins.

The most direct way of using GFP is to attach it to a biomolecule of interest. This allows us to selectively activate certain regions of a cell and trace the movements of proteins labeled with the GFP. In this way, it is possible to localize information on single molecules with an optical resolution of 10 nm. We stress that this does not mean that we are able to resolve the structural features on this scale. The most correct is to say that fluorescent proteins serve as beacons shining a light that help identify their positions within biological specimens with about 10 nm precision (see Figure 11.8). Note that these methods are extensively used in cancer research to label and track the propagation of cancer cells.

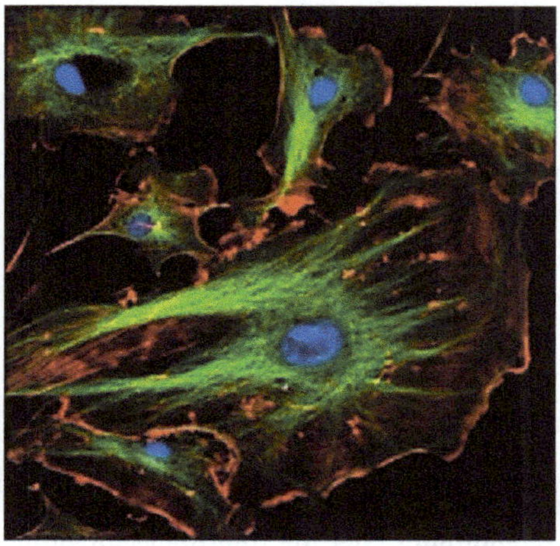

Figure 11.8: Endothelial cells imaged by fluorescent confocal microscopy. Blue color is used to indicate cell nuclei, microtubules are marked green, while actin filaments are labeled red (credit: Wikimedia Commons).

At the very end of the 1980s, **William Moerner** developed a new frequency-modulation laser technique for optical detection and spectroscopy of single molecules in solids. In the 1990s, **Eric Betzig** applied near-field microscopy to image individual molecules via their fluorescence activated by an external light source. High spatial resolution is achieved by combining individual images in which different molecules are triggered. This makes it possible to track processes occurring inside living cells. We will discuss the principles of near-field optical microscopy in Section 11.2.

Separately, in the 1990s, a fundamental analysis of nanoscale light imaging beyond the diffraction limit was performed by **Stefan Hell** and coworkers. They proposed a new type of scanning fluorescence far-field microscopy, which broke the **Abbe** diffraction limit. They did this by employing stimulated emission to inhibit the fluorescence process in the chosen regions of the illumination spot. Specifically, in this technique, which is called stimulated emission depletion (STED) fluorescence microscopy, two laser pulses are used. The first, with higher energy, is needed to create the excited molecular energy level of GFP (or another fluorescent molecule). If no special precautions are taken, the energy excess will be released as green fluorescence. The second laser pulse, however, having lower energy (red-shifted), yet enough to initiate stimulated emission, suppresses this process. The latter prevents normal fluorescence, that is, makes molecules dark in contrast to the bright molecules emitting green fluorescent light. Consequently, by keeping some molecules dark and others bright, the spatial resolution in the image plane is increased.

For this purpose, the red-shifted pulse (called a STED pulse) is shaped to be a ring with zero intensity in its center. If the intensity of the STED pulse exceeds the threshold required for depletion, the spot from which fluorescence may originate has a minimum area. The latter diminishes with the increasing STED pulse intensity. In other words, only molecules located near the ring center are allowed to emit, and in that way, one can separate optical signals originating in the neighboring areas. Scanning this narrowed spot across the specimen allows us to obtain images with superior resolution, breaking the diffraction limit. In contrast to near-field scanning optical microscopy, this method can produce three-dimensional images of transparent specimens.

In 2014, **Eric Betzig**, **Stefan Hell**, and **William Moerner** were awarded the Nobel Prize in Chemistry "for the development of super-resolved fluorescence microscopy,"

11.2 Near-field microscopy

There is another approach towards breaking the diffraction limit in spatial resolution of optical devices, which is called near-field microscopy. The story of near-field microscopy started almost 100 years ago, in 1928, when **Edward Hutchinson Synge** (encouraged by **Albert Einstein**) proposed a completely new idea for tiny object imaging. Using contemporary language, one can say that **Synge** proposed to use only evanescent waves that exist very close to the sample. Note that we analyzed in detail the evanescent light and X-ray waves, propagating under total internal and external reflection, in Sections 2.2 and 2.3, respectively. Recalling **Abbe**'s theory of image formation (see Section 3.6.1), we understand that resolving fine features in the object structure implies the existence of non-zero amplitudes of scattered waves with large wave vector projections q_x, q_y in the image plane (x, y).

In fact, let us consider a wave pocket $\boldsymbol{E}(\boldsymbol{r}, t)$, propagating from an object to the image plane close to the optical axis (the z-axis) of an optical system. Expanding the wave pocket into a **Fourier** series,

$$\boldsymbol{E}(\boldsymbol{r}, t) = \sum_{q_x, q_y} A(q_x, q_y) \exp\left[i(q_x x + q_y y + k_z z - \omega t)\right] \tag{11.1}$$

one can follow the optical transformation of each $\boldsymbol{q}$-component, all of which contribute to the image formation. High spatial resolution in the image plane (x, y) perpendicular to the z-axis means that **Fourier**-components with large q_x and q_y should transmit through the optical system without considerable attenuation. At this point, however, we confront a serious problem, which impedes the development of conventional optics towards better spatial resolution. In fact, based on eqns. (1.22) and (1.28) for zero conductivity, we find that:

$$k^2 = \frac{\omega^2}{c^2} n^2 \tag{11.2}$$

and, correspondingly,

$$k_z = \sqrt{\frac{\omega^2}{c^2} n^2 - \left[(q_x)^2 + (q_y)^2 \right]} \tag{11.3}$$

At high enough values of the transverse components, q_x, q_y, that is, when $[(q_x)^2 + (q_y)^2] > \frac{\omega^2}{c^2} n^2$, the magnitude of component k_z becomes an imaginary number, which means rapid attenuation of electromagnetic waves along the direction of its propagation (the z-direction), since $\exp(ik_z z) = \exp(-|k_z|z)$. These strongly attenuated (evanescent) waves have significant amplitudes only in the near-field regime, that is, very close to $z = 0$. In other words, by detecting evanescent waves very close to the object, we do not have fundamental limitations on the resolving power. Also note that in the spirit of the **Fourier**-transform, we can immediately say that large enough (in the sense of eq. (11.3)) characteristic transverse q-values, $q \approx \frac{2\pi}{D}$, become physically attainable when the aperture radius D diminishes well below the light wavelength λ. Furthermore, in the proximity to the aperture exit (at distances being on the order of its size), the radiation is well-collimated by the tiny aperture, in contrast to the divergent (diffracted) beam far away from it (see Figure 11.9). Thus, by placing a sample in this "near-field" region, we can illuminate it using a light probe whose dimensions are determined by the aperture size and not by the illuminating wavelength.

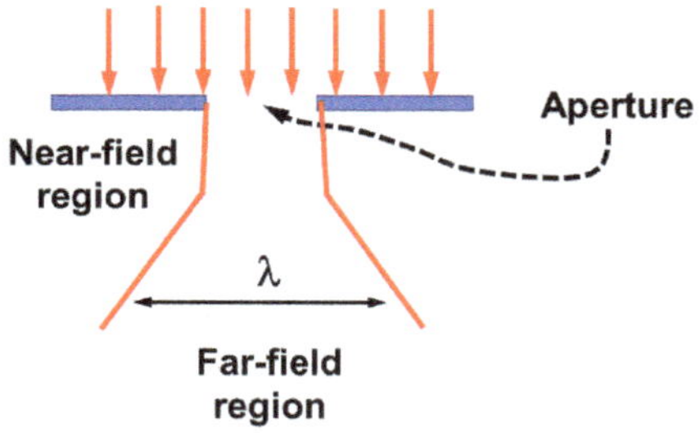

Figure 11.9: Illustration of the formation of the near-field region in the wavefront propagation.

All these considerations were presented, though qualitatively, in **Synge**'s seminal paper (1928), entitled "A suggested method for extending microscopy resolution into the ultra-microscopic regime," which he himself called an abstract. The paper mainly focuses on experimental problems that future researchers will encounter when trying to build such near-field microscopes. Most important among them, as indicated by the author, are fabricating tiny apertures, placing them very close to the investigated sample surface, and achieving mechanical stability of these elements and the entire assembly under sample scanning. It took more than fifty years until it became possible to overcome these difficulties. As a result, the first patent describing the design of the near-field visible light microscope was filed by **Dieter Pohl** in 1982.

In this instrument, a quartz rod with an open aperture of about 30 nm at one end was used to shine laser light onto the sample. The quartz rod was mounted on a piezoelectric actuator opposite a microscope slide carrying a test structure. The whole assembly was placed under a microscope focused on a transparent window of the test structure comprising fine transparent lines produced by e-beam lithography. In 1984, using a 488-nm-wavelength argon laser and moving the aperture along a test object (i.e., in the *xy*-plane), **Pohl**'s group at IBM (Zurich) claimed that they achieved spatial resolution of about $\lambda/20$, which is ten times better than the diffraction limit of $\lambda/2$. In the same year (1984), **Michael Isaacson** and his coworkers from Cornell University proposed a design for a reflecting near-field microscope and reported substantial light transmission through a tiny aperture with diameter of about $\lambda/16$. In the 1990s, the **Betzig** group, another Cornell University team, combined near-field microscopy with light fluorescence detection towards biological applications (see previous Section 11.1).

Nowadays, near-field scanning optical microscopy (NSOM) is a well-established technique, with several companies offering different types of instruments. In the standard NSOM setup (see Figure 11.10), the primary components are the laser light source, the scanning atomic force microscope (AFM)-like tip with a feedback mechanism, the detector, and the (x,y,z) piezoelectric sample stage.

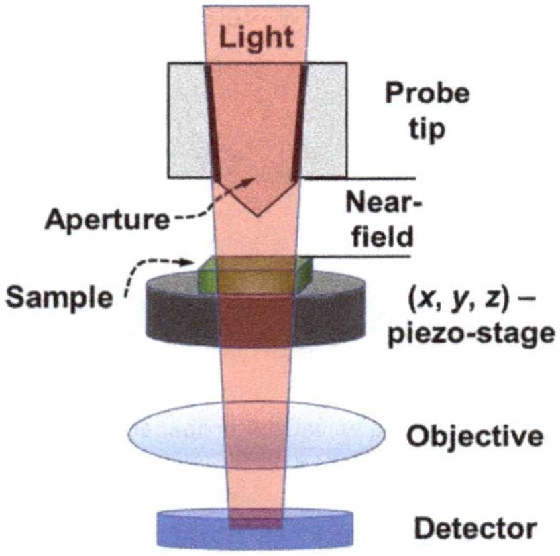

Figure 11.10: Principal scheme of an NSOM instrument.

The scanning tip is typically an optical fiber coated with metal, except at the tip, or just a standard AFM cantilever with a hole in the center of the pyramidal tip. The laser is focused into an optical fiber, while standard optical detectors, such as an avalanche photodiode, photomultiplier tube, or CCD, are employed to register the light in the near-field regime. NSOM instruments can be equipped with **Raman** spectrometers to supply additional information on local molecular structure. NSOM apparatuses are used in the transmission and reflection modes (see Figure 11.11). In the transmission mode, the sample is illuminated through the probe, and the light passing through the sample is collected and detected (Figure 11.11a). In the reflection mode, the sample is also illuminated through the probe, but the light reflected from the sample surface is collected and detected (Figure 11.11b).

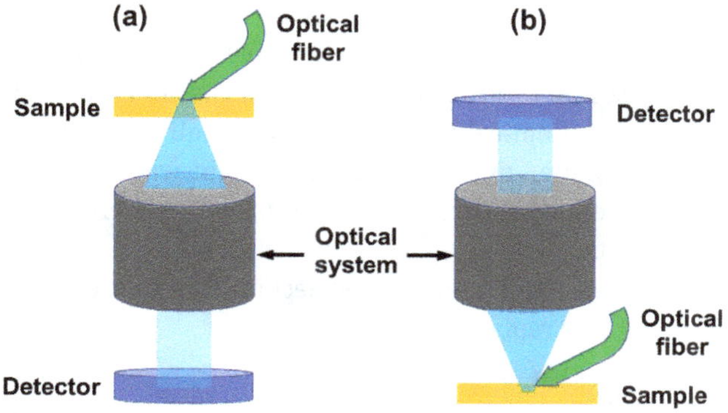

Figure 11.11: NSOM measurement modes: (a) transmission mode and (b) reflection mode.

In summary, to conduct an NSOM experiment, a point light source must be placed very close (on the nanometer scale) to the surface of interest. Then, the point light source is scanned over the surface without touching it, and the optical signal, coming from the regions located very close to the investigated surface, is collected. The resolution of an NSOM measurement is defined by the size of the point light source used (routinely about 50 nm and, when special measures are taken, even well below, towards 10 nm resolution).

Perhaps the strongest feature of NSOM is its unique versatility due to the sensitivity of evanescent waves to local electric and magnetic fields, dielectric permittivity, chemical content and, certainly, topographic landscape at the surface. All this helps investigate essential nanotechnology issues, which are crucial in the contemporary trend of miniaturization of microelectronic components towards the nm scale. In fact, modern microelectronics devices, based on nanometric semiconductor quantum wells and quantum dots, use the electron's quantum confinement to produce the desirable bandgap modifications. Probing the luminescence spectra, which originate in the confined charge carriers, by NSOM provides valuable information on the spatial distribution of quantum parameters – information that is obtained in a non-destructive way. Another example is NSOM application to study plasmonic and photonic nanostructures by mapping the internal electric fields and, on this basis, decide on device functionality.

Near-field microscopy, indeed, offers spatial resolution well below the light wavelength limit. It, however, suffers from a very small working distance and shallow depth of field, which impose some limitations on the optical design. Moreover, certain problems with reliable nanoprobe fabrication still exist. These drawbacks are eliminated when using negative materials (i.e., having negative refractive index), as is explained in Section 11.3.

11.3 New ideas with negative materials

In 2000, **John Pendry** published a seminal paper entitled "Negative refraction makes a perfect lens." He showed that lenses built of negative materials can break the diffraction limit for the resolving power of optical instruments. As we already said, in classical optics, the latter cannot be better than the half of the light wavelength λ, which is easy to understand using semiquantitative arguments, which were given around eq. (11.3) in the previous section.

Recalling what we know about negative materials (see Section 6.4), the situation is changed drastically because eq. (11.3) is replaced by:

$$k_z = -\sqrt{\frac{\omega^2}{c^2}n^2 - \left[(q_x)^2 + (q_y)^2\right]} \tag{11.4}$$

The [−] sign before the square root is a game changer since it provides the growth of the wave amplitude, $\exp(ik_z z) = \exp(|k_z|z)$, along the z-direction (instead of its attenuation) at $[(q_x)^2 + (q_y)^2 > \frac{\omega^2}{c^2}n^2$. A more detailed analysis shows that the wave amplitude is restored, exactly compensating for its decay when passing through a vacuum region of the same length. As a result, all **Fourier** components, including those having high magnitudes of transverse wave vectors, contribute to image formation. This fact, in principle, allows us to build "superlenses," which offer super-resolution far below the wavelength limit and beyond the near-field regime.

It is also worth mentioning another brilliant proposal of **Pendry** and coworkers in the field of negative materials, which is called "invisible cloak." This term means that it is possible, using negative materials, to make an object invisible under certain conditions. In other words, there is no light refraction or scattering in the object region, which could be recorded by any detecting system. More precisely, the object's invisibility implies that the wavefronts behind the object and ahead of it are indistinguishable (see Figure 11.12). The main idea to achieve this uses the fact that the phase of the propagating wave, $\exp(ink_z z)$, accumulated in the positive material can be canceled when passing through a negative material, which produces a wave phase of opposite sign, that is, $\exp(-ink_z z)$. We also already know that the wave amplitude decaying in positive material is amplified and restored when passing through negative material. Hence, we can say that positive and negative materials work optically as "matter" and "antimatter," cancelling out each other's action. Therefore, by combining

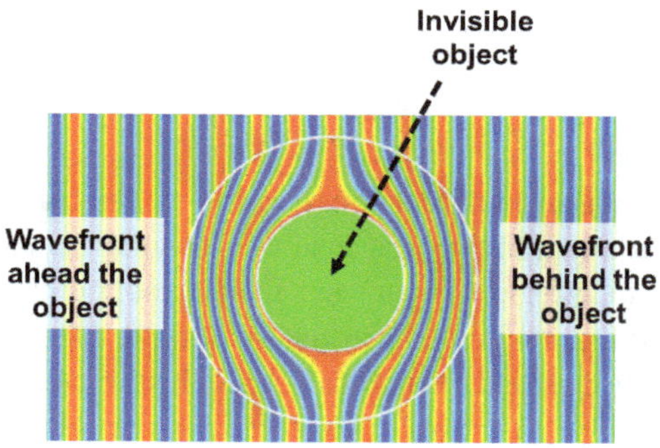

Figure 11.12: Illustration of the "invisible cloak" effect on propagating wavefronts.

equal portions of positive and negative materials, it is principally feasible to build an "invisible cloak." Certainly, in proximity to such objects, which may have rather complex shapes, the ray trajectories are curved (see Figure 11.12). At some distances near the object, however, the wavefronts will be indistinguishable behind and ahead of it, thus fully concealing the object.

Chapter 12
Fresnel diffraction

In Chapter 10, we discussed in detail **Fraunhofer** diffraction of plane waves being propagated through different objects. It is worth noting that **Fraunhofer** approximation of wave diffraction is valid in the far-field region. In contrast, the **Fresnel** diffraction analyzed here properly describes wave propagation in the near field, that is, close to the object. **Fresnel** diffraction played a fundamental role in the historical development of the wave theory of light and led to several important discoveries and practical inventions, some of which are discussed in this chapter.

12.1 X-ray scattering amplitude from an individual crystallographic plane: the Fresnel zone approach

As a first example, we calculate the X-ray scattering amplitude from individual crystallographic planes, the key parameter in the kinematic theory of X-ray diffraction that has remained analytically undefined in Section 10.2. A crystallographic plane is considered here as a monolayer of crystal unit cells, infinitely covering space in two directions (x, y).

X-rays come from a point source (point P in Figure 12.1), which delivers an incident beam with finite angular divergence formed by slits or other optical elements. The diffracted X-rays are registered by the detector, its receiving slit being represented (for the sake of simplicity) by point D in Figure 12.1. The detector is placed close to the plane, so attenuation of the scattering amplitude with distance, in the first approximation, is negligible. Accordingly, the main focus of the analysis is on the correct account of the phases of the X-ray waves scattered by different plane points $M(x, y)$.

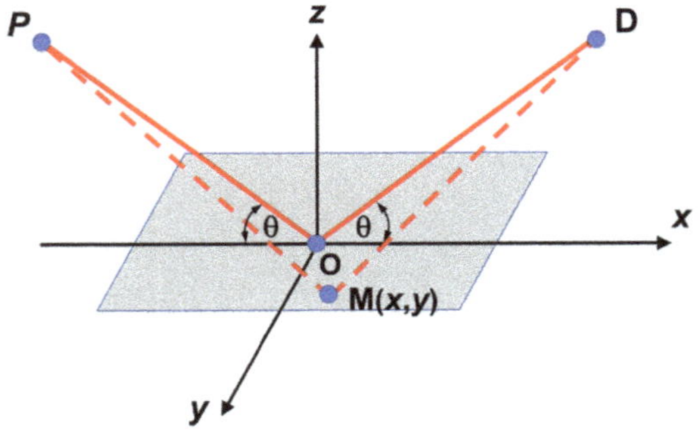

Figure 12.1: Calculating the scattering amplitude from an infinite crystallographic plane.

https://doi.org/10.1515/9783111140100-013

The central X-ray trajectory (marked as POD in Figure 12.1) is the shortest path of X-rays between the source and detector at specular reflection with equal incident and exit angles Θ, counted from the crystallographic plane. Due to the angular divergence, there are other X-ray trajectories between points P and D, for example, PMD in Figure 12.1, which are characterized by the phase difference, as compared to the central trajectory POD. As we will see later in this section, only a very small region around point O contributes significantly to the overall X-ray diffraction intensity in point D. This fact strengthens the previous statement that, in gross mode, we can neglect the attenuation of the wave amplitude with distance; the same is valid for the PM and MD paths.

For the sake of simplicity, we will deal with the phase problem mentioned in the framework of geometrical optics. To calculate the phase differences between X-ray trajectories, let us introduce, first, the coordinate system with the z-axis being normal to our plane, the x-axis being situated in the plane along the trace of the central trajectory POD, and the y-axis being perpendicular to both the z- and x-axes (see Figure 12.1). The distances PO and OD equal R_1 and R_2, respectively. Now, we can calculate the path PMD = PM + MD:

$$\begin{aligned} \mathrm{PM} + \mathrm{MD} &= \sqrt{R_1^2 \sin^2\Theta + y^2 + (R_1 \cos\Theta + x)^2} + \sqrt{R_2^2 \sin^2\Theta + y^2 + (R_2 \cos\Theta - x)^2} \\ &= R_1 \sqrt{1 + \frac{2x\cos\Theta}{R_1} + \frac{x^2 + y^2}{R_1^2}} + R_2 \sqrt{1 - \frac{2x\cos\Theta}{R_2} + \frac{x^2 + y^2}{R_2^2}} \end{aligned} \tag{12.1}$$

We assume that $x, y \ll R_1, R_2$ (justified below) and expand the square roots in eq. (12.1) in the **Taylor** series, keeping the linear and quadratic terms over x and y. This procedure yields:

$$\mathrm{PM} + \mathrm{MD} = R_1 + R_2 + \frac{x^2 \sin^2\Theta + y^2}{2R_1} + \frac{x^2 \sin^2\Theta + y^2}{2R_2} \tag{12.2}$$

Since the path length for the central trajectory is POD = PO + OD = $R_1 + R_2$, we can find the path difference, $\Delta P = \mathrm{PMD} - \mathrm{POD}$, and then the phase difference, $\Delta\varphi_d = \frac{2\pi}{\lambda}\Delta P$, between trajectories PMD and POD:

$$\Delta\varphi_d = \frac{\pi}{\lambda}\left(x^2 \sin^2\Theta + y^2\right)\left(\frac{1}{R_1} + \frac{1}{R_2}\right) \tag{12.3}$$

Considering the wavefield of scattered X-rays near the scattering plane, that is, at $R_2 = R \ll R_1$, we finally obtain:

$$\Delta\varphi_d = \frac{\pi}{R\lambda}\left(x^2 \sin^2\Theta + y^2\right) \tag{12.4}$$

Note that the phase difference is quadratic over coordinates (x,y), which is characteristic of **Fresnel** diffraction.

It follows from eq. (12.4) that the contours of the equal phase changes ($\Delta\varphi_d$ = const) in the *xy*-plane are ellipses centered at point O (see Figure 12.2).

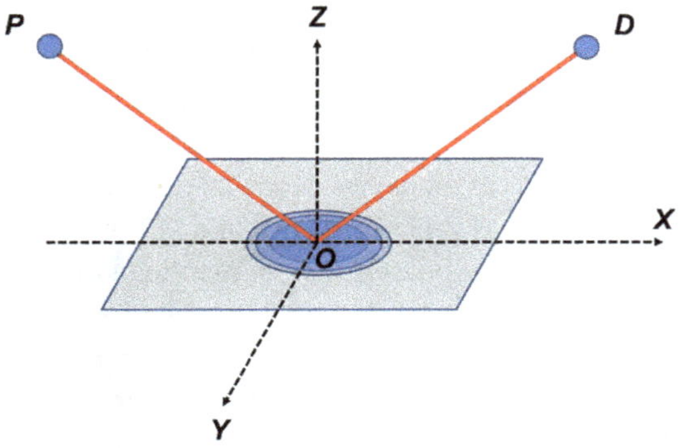

Figure 12.2: Sectioning of a crystallographic plane by elliptic **Fresnel** zones.

Let us demand that for neighboring concentric ellipses, the $\Delta\varphi_d$-values should differ by π. This requirement is satisfied if

$$\frac{\pi}{R\lambda}\left(x^2\sin^2\Theta + y^2\right) = m\pi \tag{12.5}$$

where integer numbers, $m = 1, 2, 3 \ldots$, enumerate the subsequent ellipses. Correspondingly, the equations of these ellipses have the following canonic shape:

$$\left(\frac{x\sin\Theta}{\sqrt{mR\lambda}}\right)^2 + \left(\frac{y}{\sqrt{mR\lambda}}\right)^2 = 1 \tag{12.6}$$

The halves of the ellipses' axes equal:

$$a_m = \frac{\sqrt{mR\lambda}}{sin\theta}; \quad b_m = \sqrt{mR\lambda} \tag{12.7}$$

Regions outlined by adjacent ellipses are called **Fresnel** zones (Figure 12.2). Since the waves scattered by neighboring **Fresnel** zones are in antiphase (i.e., phase difference equals π), their contributions almost cancel each other out. As we show in Section 12.2, the resultant scattering amplitude A_p from individual crystallographic planes is determined by half a contribution from the first **Fresnel** zone ($m = 1$). Using eq. (12.7), we find the area S_{F1} of the latter:

$$S_{F1} = \pi a_1 b_1 = \frac{\pi R\lambda}{\sin\Theta} \tag{12.8}$$

At large scattering angles ($\Theta \to 90^0$), **Fresnel** zones, indexed by integer numbers m, become circular (see eq. (12.7)) with radii,

$$r_m = \sqrt{mR\lambda} \tag{12.9}$$

Substituting $R \approx 1$ cm, $\lambda \approx 1$ Å, and $\sin\Theta \approx 0.3$ into eq. (12.8) yields $S_{F1} \approx 10^{-7}$ cm^2. In other words, the region around point O in Figure 12.2, which strongly contributes to the X-ray diffraction intensity, is about 10 μm^2, that is, its linear (x,y) sizes (a few microns) are indeed very small compared with distance R. This result justifies approximations made when deriving eq. (12.2).

Nevertheless, to accurately calculate the scattering amplitude A_p from an individual plane, we need to summate (with the respective phase differences (12.4)) the contributions from all crystal unit cells within the monolayer of the thickness equal to the interplanar spacing d:

$$A_p = \frac{1}{R}\iint\limits_{-\infty}^{\infty} \frac{dFr_0}{V_c}\exp(-i\Delta\varphi_d)dxdy = \frac{dFr_0}{RV_c}\iint\limits_{-\infty}^{\infty} dxdy\left\{\exp\left[-i\frac{\pi}{R\lambda}\left(x^2\sin^2\Theta + y^2\right)\right]\right\} \tag{12.10}$$

Here, the term $\frac{d}{V_C}dxdy$ is the number of unit cells within the in-plane element of the surface area $dxdy$, with V_c standing for unit cell volume. The term Fr_0 defines the scattering amplitude from individual unit cell and comprises the classical radius of an electron, $r_0 = 2.817 \cdot 10^{-13}$ cm (the scattering amplitude by one electron), and the structure factor describing X-ray scattering by a unit cell:

$$F = \sum_j f_j \exp(i\boldsymbol{Q}\boldsymbol{r}_j) \tag{12.11}$$

Here, $\boldsymbol{Q}$ is the diffraction vector, introduced in Section 10.2 (see eq. (10.16)), while f_j and $\boldsymbol{r}_j$ are the atomic scattering factors and coordinates of individual atoms comprising the unit cell. For completeness, we also introduced into eq. (12.10) the factor $\frac{1}{R}$, which recounts the reduction of the amplitude of the scattered spherical wave with the distance from its origin.

Using known expressions for **Fresnel** integrals:

$$C(x) = \int_0^x \cos\left(\frac{\pi}{2}u^2\right)du$$

$$S(x) = \int_0^x \sin\left(\frac{\pi}{2}u^2\right)du \tag{12.12}$$

$$C(\infty) = S(\infty) = \frac{1}{2}$$

we find:

$$A_p = -i\frac{Fr_0 d\lambda}{V_c \sin\Theta} \tag{12.13}$$

and

$$|A_p| = \frac{r_0 \lambda d|F|}{V_c \sin\Theta} \tag{12.14}$$

which indicates the strength of X-ray scattering by an individual plane. Applying eq. (10.13) for the characteristic attenuation depth Λ_e in the **Bragg** scattering geometry, one obtains:

$$\Lambda_e = \frac{d}{|A_p|} = \frac{V_c sin\Theta}{r_0 |F| \lambda} \tag{12.15}$$

We see that the diffraction-mediated penetration depth Λ_e is inversely proportional to the strength of the X-ray scattering amplitude $|A_p|$. The stronger the amplitude is, the shorter the penetration depth, and vice versa. Substituting typical numbers into eq. (12.15), we find that Λ_e is on a μm scale. Therefore, perfect crystals, a few micrometers thick (or thicker) already provide total X-ray reflection in the **Bragg** scattering geometry due to complete repumping of incident radiation into the diffracted beam. This process occurs within the very narrow angular interval around the **Bragg** angle Θ_B (eq. (10.14)), the interval defined by another fundamental parameter of the dynamical diffraction theory – the so-called extinction length τ (proportional to Λ_e). The latter is introduced and analyzed in detail in Chapters 14 and 15. The extinction length plays the role of effective thickness for thick crystals, replacing the physical thickness in the calculations of angular widths $\Gamma \sim \frac{1}{\tau}$ of diffraction profiles (in the spirit of eq. (10.9)). Note that typical values of Γ for perfect silicon crystals used in most such measurements are only few seconds of arc, which allows extra-small angular scatterings to be separated, which is necessary for different applications (e.g., for phase-contrast X-ray imaging introduced in Section 9.4). We will elaborate on this issue in more detail in Chapter 15.

12.2 Fresnel zone construction in transmission geometry

In Section 12.1, we presented the concept of **Fresnel** zones, which was applied to calculate the X-ray diffraction intensity from infinite crystallographic planes. The beauty of **Fresnel**'s idea was in dividing the entire plane into elliptic segments, which scatter in antiphase with respect to their neighbors, and therefore, almost compensate for each other's effects. As a result, the total contribution to the coherent scattering is rather small; it is only half of the contribution from the first **Fresnel** zone (see below), its area being defined by eq. (12.8).

For soft X-rays and visible light, when there is no diffraction on atomic networks (since $\lambda \gg d$), the **Fresnel** zone construction allows us to analyze wave propagation through different objects in transmission mode. Note that the radii of the **Fresnel** zones are scaled as $\sqrt{\lambda}$ (see eq. (12.9)) and, therefore, for visible light, the size of the first Fresnel zone is $\sqrt{5,000} \simeq 70$ times larger than that for hard X-rays, that is, it is hundreds of microns in size.

We assume that a spherical wave propagates in vacuum from point source S through some object (opaque screen, hole in the screen, etc.) located in point O (see Figure 12.3). The distance SO equals a.

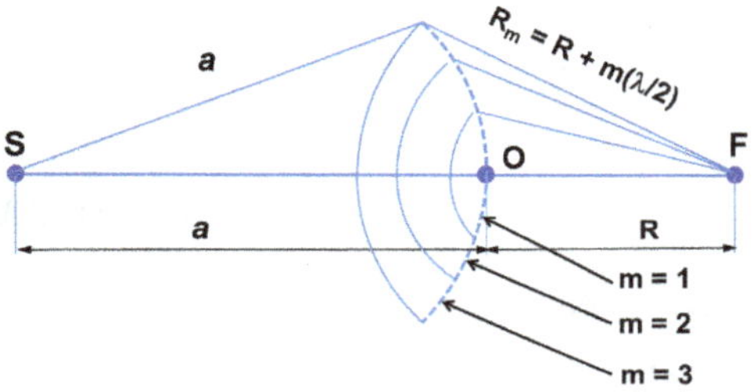

Figure 12.3: Calculating the sizes of subsequent **Fresnel** zones in transmission geometry.

The key question is how to calculate the X-ray intensity in some point F located on the optical axis SO at distance R behind the object. Again, we divide the equal-phase spherical surface near point O into the **Fresnel** zones, whose distances R_m to point F differ by integer number m of $\frac{\lambda}{2}$ (see Figure 12.3):

$$R_m = R + m\frac{\lambda}{2} \tag{12.16}$$

The zone border lines cut the wavefront surface into spherical segments having heights h_m (see Figure 12.4).

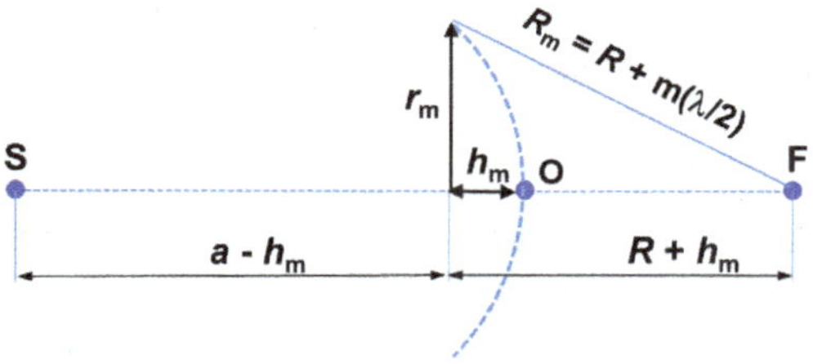

Figure 12.4: Illustration of deriving eq. (12.23) for the radius of the *m*th **Fresnel** zone.

The area S_m of the spherical segment of radius a with height h_m equals:

$$S_m = 2\pi a h_m \tag{12.17}$$

With the aid of Figure 12.4, we find:

$$(R_m)^2 = \left(R + m\frac{\lambda}{2}\right)^2 = (r_m)^2 + (R + h_m)^2 \tag{12.18}$$

and

$$a^2 = (r_m)^2 + (a - h_m)^2 \tag{12.19}$$

where r_m, as before, is the radius of the mth circular **Fresnel** zone boundary. Solving eqns. (12.18) and (12.19) yields:

$$h_m = \frac{Rm\lambda + (m\lambda/2)^2}{2(R + a)} \tag{12.20}$$

$$(r_m)^2 = 2ah_m - (h_m)^2 \tag{12.21}$$

In all practical cases, $m\lambda \ll 4R$ and $h_m \ll 2a$, and hence:

$$h_m \approx \frac{Rm\lambda}{2(R + a)} \tag{12.22}$$

$$r_m \approx \sqrt{\frac{Ram\lambda}{(R + a)}} \tag{12.23}$$

Using eqs. (12.17) and (12.22), we find the area of the mth spherical segment:

$$S_m = 2\pi a h_m = m\frac{\pi a R\lambda}{(R + a)} \tag{12.24}$$

as well as the area of the m-th **Fresnel** zone:

$$\Delta S_m = S_m - S_{m-1} = \frac{\pi a R\lambda}{(R + a)} \tag{12.25}$$

The contributions of neighboring zones A_m and $A_{\mathrm{m}-1}$ to the total scattering amplitude A_tot are in antiphase with each other. This implies that:

$$A_\mathrm{tot} = A_1 - A_2 + A_3 - A_4 + A_5 - \cdots \tag{12.26}$$

Note that according to eq. (12.25), the areas of the **Fresnel** zones, $\Delta S_\mathrm{m} = \mathrm{const}$, do not depend on the zone index m. Also, the distance R_m from the mth **Fresnel** zone to point F increases very little with m (see eq. (12.16)), since practically, $m\lambda \ll 2R$. These considerations allow us to conclude that the contributions A_m are decreased monotonically and very slowly when increasing zone index m, and we can (following the original **Fresnel**'s idea) assume that:

$$A_m = \frac{A_{m+1} + A_{m-1}}{2} \tag{12.27}$$

Using this representation for every even **Fresnel** zone in eq. (12.26) yields:

$$A_{\text{tot}} = A_1 - \left(\frac{A_1 + A_3}{2}\right) + A_3 - \left(\frac{A_3 + A_5}{2}\right) + A_5 - \left(\frac{A_5 + A_7}{2}\right) + A_7 - \cdots \tag{12.28}$$

We see that most of the terms in eq. (12.28) cancel each other out, and for a large enough number of open **Fresnel** zones, we finally obtain:

$$A_{\text{tot}} \approx \frac{A_1}{2} \tag{12.29}$$

In other words, as we already mentioned at the beginning of this chapter, the total interference effect of all open **Fresnel** zones is half of the contribution from the first **Fresnel** zone. As we estimated above, for visible light, the size of the first **Fresnel** zone is a few hundred microns. Therefore, light from point S to point F propagates within a rather narrow channel, that is, practically straightforward. By placing a non-transparent screen with an open hole having an area equal to that of the first **Fresnel** zone at point O, we double the wave amplitude at point F and increase the transmitted intensity by a factor of four.

Let us make more comprehensive estimations in the more general case in which the hole in the screen has radius r_{h}, and the distances between the screen and points S and F equal a and b, respectively (see Figure 12.5).

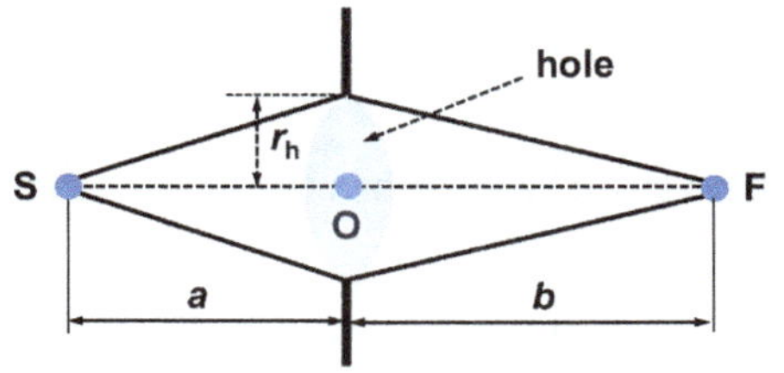

Figure 12.5: Calculating the **Fresnel** diffraction in transmission mode through a circular hole having radius r_{h}.

Suppose that the hole size r_{h} corresponds to the integer number m of the open **Fresnel** zones. Now we can use eq. (12.23) to relate parameters r_{h}, a, and b to the zone number m. Considering parameter h_{m} in Figure 12.4 to be small compared to a and $R \simeq b$, we find that:

$$r_{\text{h}} = r_{\text{m}} = \sqrt{\frac{abm\lambda}{(a+b)}} \tag{12.30}$$

and

$$m = \frac{r_h^2}{\lambda}\left(\frac{1}{a} + \frac{1}{b}\right) \tag{12.31}$$

In the spirit of eq. (12.26), the scattering amplitude at point F will be:

$$A_{tot} = A_1 - A_2 + A_3 - A_4 + A_5 - \cdots \pm A_m \tag{12.32}$$

where sign [+] or [−] before A_m stands for odd or even m-numbers, respectively. Again using presentation (12.28), we finally find,

$$A_{tot} = \frac{A_1}{2} \pm \frac{A_m}{2} \tag{12.33}$$

When moving up and down from point F perpendicular to the optical axis (see Figure 12.5), the analysis of the diffraction intensity distribution becomes more complicated, since the circular symmetry of the **Fresnel** zones is broken, and subsequent zones may be only partially open or closed. In any event, the obtained diffraction pattern appears as alternating bright and dark rings. According to eq. (12.33), the central spot will be brighter or darker for odd or even m-numbers, respectively.

In the same manner, **Fresnel** diffraction from a non-transparent disc of radius r_d can be analyzed (see Figure 12.6).

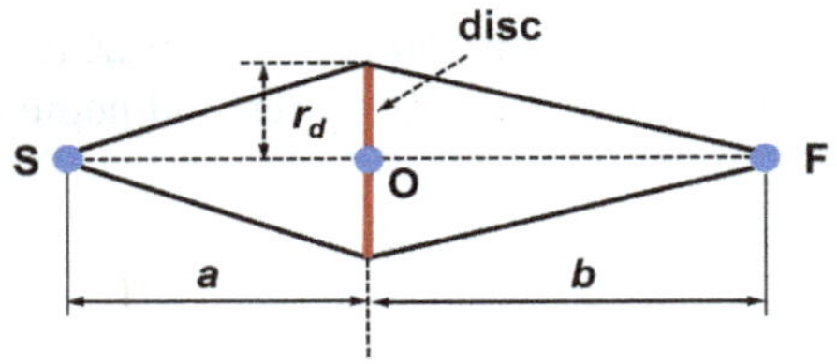

Figure 12.6: Calculating the **Fresnel** diffraction in transmission mode from a non-transparent disc having radius r_d.

If the disc is blocking number m of several first **Fresnel** zones, then the total scattering amplitude at point F will be:

$$A_{tot} = A_{m+1} - A_{m+2} + A_{m+3} - A_{m+4} + A_{m+5} \cdots \tag{12.34}$$

Regrouping the terms in the right-hand side of eq. (12.34), as was done in eq. (12.28), yields:

$$A_{tot} = \frac{A_{m+1}}{2} + \left(\frac{A_{m+1}}{2} - A_{m+2} + \frac{A_{m+3}}{2} + \cdots\right) \tag{12.35}$$

Since the sum in rounded brackets is close to zero, we finally obtain:

$$A_{\text{tot}} \simeq \frac{A_{m+1}}{2} \tag{12.36}$$

Therefore, one can conclude that a diffraction pattern again comprises alternating dark and bright rings, but in contrast to the hole case, the central spot at point F is always bright, independently of the m number.

This completely unexpected result, that is, the appearance of a bright spot in the classic shadow region, provoked a grave dispute between **Poisson** and **Fresnel** during a presentation by the latter of his diffraction theory to the French Academy of Sciences. **Siméon Poisson** claimed that this result is in opposition to common sense and, hence, disproves the entire **Fresnel** theory. However, **François Arago**, who also attended this meeting, immediately performed the critical experiment and demonstrated the presence of a bright spot in the center of the shadow region. It was the triumph of the wave theory of light.

12.3 Light and X-ray focusing by Fresnel zone plates and Bragg–Fresnel lenses

One can ask what will happen if one removes, by etching, every second **Fresnel** zone, which are in antiphase with its neighbors? Definitely, the remaining segments will all scatter in phase and the resultant scattering intensity will be enhanced many times, as in the case of conventional focusing device, for example, optical lens. Such devices, intended to work in the diffraction mode, are called **Bragg–Fresnel** lenses. Nowadays, they are routinely produced by photolithography techniques and used for focusing hard X-rays at synchrotron beam lines (see Figure 12.7).

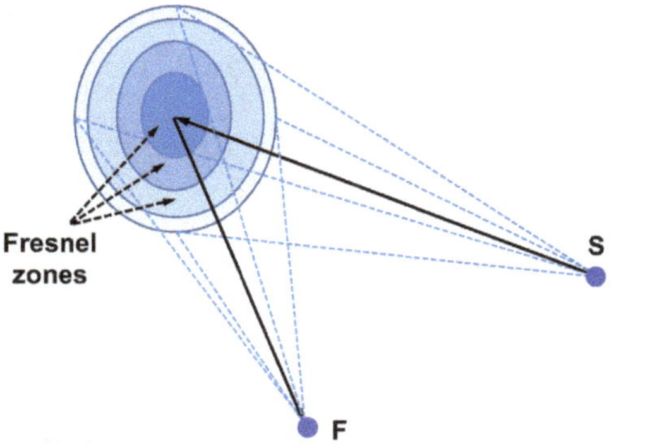

Figure 12.7: Illustration of the **Bragg–Fresnel** lens.

Considering eqs. (12.26) and (12.29), one can say that in the first approximation, the total scattering amplitude at point F will increase nearly by a factor of $\frac{A_1\left(\frac{N}{2}\right)}{\left(\frac{A_1}{2}\right)} = N$, where

N is the number of the last open **Fresnel** zone. This is true until $\frac{\lambda}{R} \ll \frac{2}{N}$. The size of a **Bragg–Fresnel** lens r_{N} (or r_{m} defined by eq. (12.9)) together with the wavelength λ used determines the length of focus R. Usually, $R \approx 1$ m, $\lambda \approx 1$ Å, $N \approx 100$, and then $r_{\mathrm{N}} \approx$ 0.1 mm = 100 μm (according to eq. (12.9)). Note that for a fixed lens size r_{N}, the focal length is $R \sim 1/\lambda$. Therefore, a particular **Bragg–Fresnel** lens can effectively be exploited when working with monochromatic X-rays, having a wavelength that fits eq. (12.9). In other words, **Bragg–Fresnel** lenses suffer from chromatic aberrations.

For soft X-rays and visible light, such kinds of devices can also be used in transmission mode. In that case, the focusing element consists of successive strongly- and weakly-absorbing circular zones and is called a **Fresnel** zone plate (FZP, see Figure 12.8). Note that circular **Fresnel** zones are also preferable from the fabrication point of view. Let us consider its operation in more detail.

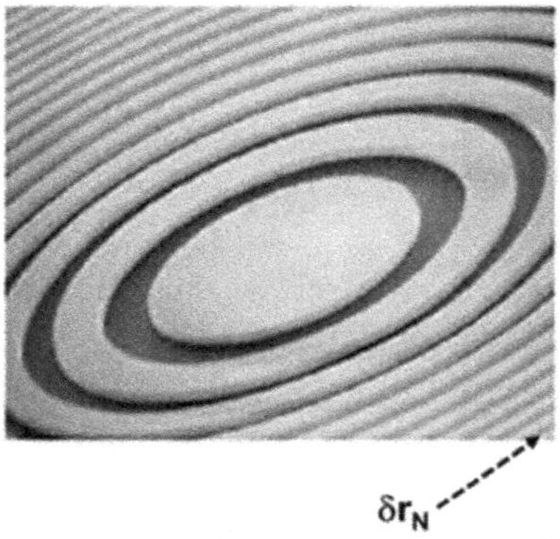

Figure 12.8: Fresnel zone plate (FZP) produced by nanolithography. The last open **Fresnel** zone δr_{N}, which determines the spatial resolution of FZP (see eq. (12.43)), is indicated by an arrow.

The key question, which we address here, is the spatial resolution achievable with FZPs. According to the results obtained in Chapter 10 (eq. (10.44)), the spatial resolution Δ of a circular focusing element is defined by its radius D and focal length F, as well as radiation wavelength λ:

$$\Delta = 0.61\frac{\lambda}{D}F \tag{12.37}$$

In our case, $D = r_N$ and $F = R$; therefore,

$$\Delta = 0.61\frac{\lambda}{r_{\mathrm{N}}}R \tag{12.38}$$

For FZP, the relationship between its radius and focal length is given by eq. (12.23). Taking into account that in real experiments at synchrotron beam lines $a \gg R$, one obtains:

$$r_N = \sqrt{RN\lambda} \tag{12.39}$$

that coincides with eq. (12.9). Substituting eq. (12.39) into (12.38) yields

$$\Delta = 1.22\frac{r_N}{2N} \tag{12.40}$$

Using eq. (12.39), we also find that

$$(r_N)^2 - (r_{N-1})^2 \approx (r_N - r_{N-1})2r_N = R\lambda \tag{12.41}$$

Therefore, the width of the last open **Fresnel** zone δr_N is:

$$\delta r_N = r_N - r_{N-1} = \frac{R\lambda}{2r_N} \tag{12.42}$$

Substituting eq. (12.42) into eq. (12.38), we finally obtain that the spatial resolution of FZP.

$$\Delta = 1.22\delta r_N \tag{12.43}$$

is determined by the width of the last open **Fresnel** zone (see Figure 12.8). This result imposes severe requirements on the photolithography procedures, which must provide the values of δr_N on a 10 nm scale. Nowadays, this is possible, thanks to the remarkable progress in nanotechnology; correspondingly, FZPs are installed and employed at many synchrotron sources to focus monochromatized radiation. Nevertheless, as we already said, they are not free of chromatic aberrations because the focal distance is inversely proportional to the radiation wavelength (see eq. (12.9)).

12.4 X-ray microscopy

X-ray imaging, aimed at uncovering structural features hidden to the human eye, began being used immediately after the discovery of X-rays by **Wilhelm Konrad Röntgen** in 1895. Perhaps the most famous in this sense is the early X-ray photograph of **Röntgen**'s wife's hand with a wedding ring, clearly showing all her finger bones due to the absorption contrast (see Figure 12.9).

Already during World War I, mobile X-ray laboratories were employed on the battlefields to help diagnose wounded soldiers. X-ray imaging for medical needs was continually developed and culminated in the invention of computer-assisted tomography (CT), which is now commonly used in hospitals across the world. This technique is based on absorption contrast and allows the reconstruction of the three-dimensional structure of different objects from large numbers of plane projections. For this great achievement, **Allan Cormack** and **Godfrey Hounsfield** were awarded the 1979 Nobel Prize in Physiology or Medicine: “for the development of computer assisted tomography.”

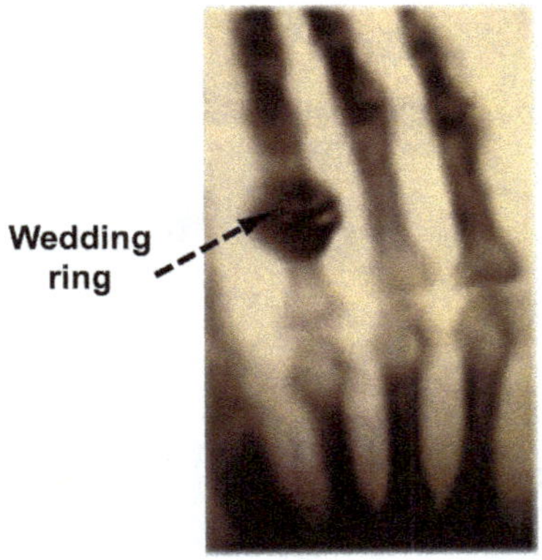

Figure 12.9: X-ray photograph of the hand of **Röntgen**'s wife, **Anna Bertha Röntgen**, taken on December 22, 1895.

During the entire period of X-ray history, the greatest expectation was to achieve X-ray imaging with spatial resolution much better than that of optical imaging, because X-ray wavelengths are about 5,000 times shorter than those for visible light. For decades, however, these expectations remained unmet due to the problems of manipulating X-ray beams, mainly because the refractive index in the X-ray field is very close to one, $n \approx 1$. Note that in conventional imaging methods, X-ray beams are formed by slits whose size is difficult to reduce below 200 micrometers. In fact, this is the best spatial resolution of advanced CT.

Another approach is to use X-ray diffraction in nearly perfect crystals for detecting intrinsic (defect-induced) or artificially generated deformation fields. In this method, the crystal is illuminated by a wide parallel X-ray beam to cover a larger crystal area. The spatial resolution is defined by the pixel size of the detector or the resolving power of the X-ray photographic film. In this way, many studies have been performed towards visualizing micron-sized dislocations, staking faults and other lattice defects. In the same fashion, using X-ray photographic films with a resolving power of 3,000 lines/mm (i.e., about 0.3 μm), it has been possible to visualize short-wavelength surface acoustic waves (SAW) and their interaction with dislocations. Applying the stroboscopic mode of imaging at the European Synchrotron Radiation Facility (ESRF), an Israeli research team successfully visualized the deformation field of the SAW, having a 6 μm wavelength (frequency 0.58 GHz) and propagating at the speed of about 4 km/s across the surface of a $LiNbO_3$ crystal (see Figure 12.10). The image contrast arises due to the alternating focusing and defocusing of X-rays by periodic corrugation of the crystal surface. Considering the local X-ray trajectories, one can say that the deformation "valleys" focus the diffracted X-rays, thus increasing the diffraction intensity in the appropriate regions of the image, while the deformation "hills" defocus X-rays, leading to the reduced intensity.

A real breakthrough in enhancing the resolving power of X-ray imaging methods, which happened in the 1980s–1990s, was primarily impacted by two factors: (1) construction of powerful synchrotron radiation sources supplying intense (and coherent) X-ray beams; and (2) radical progress in fabricating high-quality X-ray focusing elements

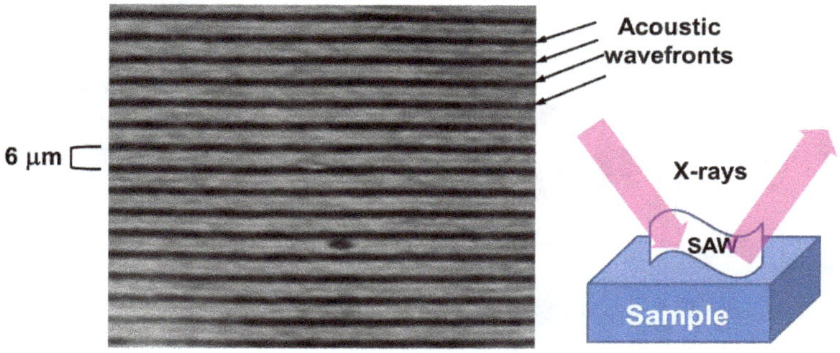

Figure 12.10: Visualization of surface acoustic waves (SAW) traveling across an $LiNbO_3$ crystal with the speed of a few km/s, using stroboscopic X-ray topography.

owing to the advances in nanotechnology. The commonly used major focusing elements, today, are the already mentioned compact lenses (Section 3.2.1), **Kirkpatrick–Baez** (K-B) reflecting mirrors and capillaries (Section 3.3.2), and FZP (Section 12.3). We remind the reader that only K-B mirrors are completely free of chromatic aberrations and, hence, are preferable when robust monochromatization is an issue.

Historically, X-ray microscopy began to be developed employing soft X-rays with energies between 100 and 1,000 eV, that is, having wavelengths in the range 1–10 nm. The driving force was to study biological objects comprising light atoms difficult to image using electron microscopy because of low scattering amplitudes. The wavelength range mostly used is between 2.3 and 4.4 nm (the so-called water window), that is, confined between the oxygen and carbon *K*-absorption edges. Note that water is considerably less absorbing than the carbon, which paves the way for imaging of hydrated living cells and cell aggregates.

Two types of X-ray instruments that can do this are very popular today: full-field or transmission X-ray microscope (TXM) and scanning transmission X-ray microscope (STXM). The TXM principal scheme is depicted in Figure 12.11. The sample is illuminated from the back through the condenser system, as in a conventional light microscope. The obtained plane projection is magnified by the objective lens and is registered by a two-dimensional detector. For the objective lens, the X-ray focusing elements of any kind can be used. In Figure 12.11, we show the application of FZP for this purpose.

In STXM, incident X-rays are emanating from point source (produced by source aperture, e.g., pinhole) and then are focused onto specific point of the sample by X-ray optical element, for example, FZP (see Figure 12.12). The sample is scanned in the focal plane of the FZP and the transmitted X-ray intensity is recorded as a function of the sample position. Spatial resolution achieved with such microscopes is in the range 20–50 nm, that is, in between the resolving powers of conventional light and electron microscopes.

To solve materials science problems, X-ray microscopes working with harder X-rays (having energies around 10 keV and higher) are used. In addition to conventional TXM

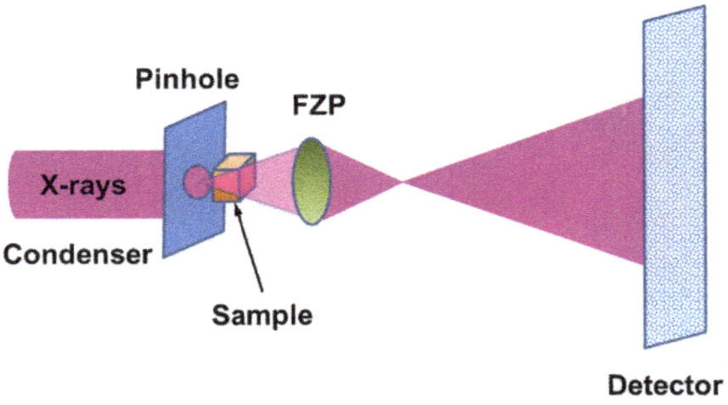

Figure 12.11: Principal scheme of TXM.

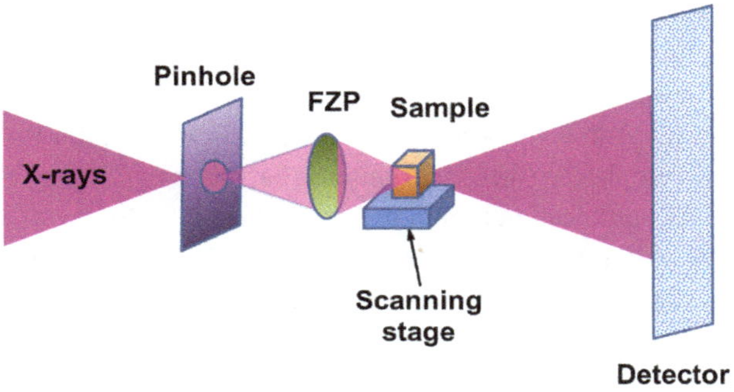

Figure 12.12: Principal scheme of STXM.

imaging, nanoscale X-ray microscopy techniques offer spectroscopic elemental and chemical mapping, microdiffraction-based structural analysis, 3D tomographic visualization of the sample interior, and coherent methods for nanomaterials imaging. These developments again became possible due to remarkable progress in high-brilliance, third- and fourth-generation synchrotron X-ray sources, as well as in high-resolution X-ray focusing optics, with characteristics approaching that used for soft X-rays.

Local elemental and chemical mappings are based on detecting X-ray fluorescence from nanoscale-sized sample regions irradiated by a focused X-ray beam. This technique, often called scanning X-ray fluorescence (SXRF) microscopy, besides elemental mapping, also provides information on the chemical state of atoms, as well as trace element detection with ppm sensitivity within irradiated regions being tens of nanometers in size.

Microdiffraction-based structural analysis exploits **Bragg** diffraction from nanoscale-sized volumes within the sample irradiated by the focused X-ray beam. The sample may be scanned, and because of this option, the respective technique is called scanning **Bragg** diffraction microscopy or simply scanning X-ray diffraction microscopy (SXDM). The strongest advantage of this technique is its sensitivity to different crystalline phases within the sample, which can be characterized separately by measuring multiple **Bragg** diffraction peaks. In that sense, SXDM is similar to scanning transmission electron microscopy. For SXDM, a wide-bandpass illumination with $\Delta\lambda/\lambda$ ~ 1% that requires aberration free K-B mirror systems for X-ray focusing is used. In this case, by using large-acceptance area detectors, multiple **Bragg** peaks can be simultaneously registered, allowing for a local determination of the unit cell parameters and then the full strain tensor for individual crystalline grains. The relative precision of the lattice strain determination in this mode of measurement is approximately 10^{-4}. Switching the measurement mode to the single **Bragg** peak analysis, the precision is improved by a factor of ten, reaching ~10^{-5}, which is 100 times better than that achieved with high-resolution transmission electron microscopy. The most influential contributions towards the advancement the three-dimensional X-ray diffraction microscopy were made by two groups – one from Oak Ridge National Laboratory led by **Bennett (Ben) Larson** and the second from Denmark Technical University led by **Henning Poulsen**. Specifically, they developed special triangulation methods, which allowed tracking lattice rotations and strains in individual grains composing polycrystalline materials, as well as following the grain growth kinetics.

Nowadays, micro- and nano-tomography are very popular tools installed at most synchrotron sources, allowing us to visualize the 3D internal structure of investigated samples. In gross mode, this kind of imaging is similar to CT, in which 3D images are composed from a set of planar projections with the aid of sophisticated computer programs. In nano-tomography measurements, absorption-contrast or phase-contrast planar projections can be acquired. In contrast to conventional CT, X-ray focusing is used down to few tens of nm in size. Correspondingly, the locality of the obtained information is greatly increased.

As illustrative example, we present the experimental results obtained by an international German-Israeli research team at ESRF when studying several marine sponges, whose skeletons are composed of amorphous silica (see Figure 12.13).

Note that marine sponges are one of the oldest multicellular organisms, with a fossil record that stretches back more than half a billion years. *Demospongiae* and *Hexactinellida*, two classes of sponges, synthesize mineralized silica-based skeletal elements, called glass spicules, which provide the animals with structural support and mechanical strength and help protect them from their environment. The spicules are microns to millimeters long and exhibit a diversity of highly regular three-dimensional branched morphologies that are an example of symmetry in biological systems (see Figure 12.13a). It has long been a mystery how these marine organisms form branched glass architectures in cold water, taking into account that the man-made technology of glass forming and shaping requires heat treatment at high temperatures of about 1,000 °C.

Using the advanced X-ray methods available at ESRF, that is, nano-tomography and focused X-ray diffraction, helped unravel the mystery and uncover the principles of spicule morphogenesis. It turned out that during spicule formation, the process of silica deposition is templated by an axial organic filament already mentioned in

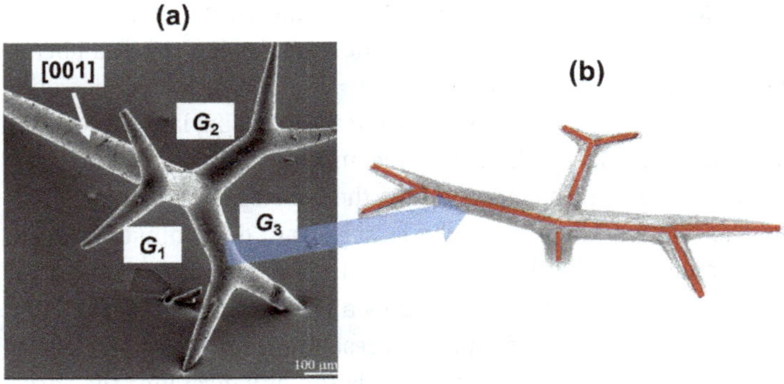

Figure 12.13: The morphology of demosponge spicules *Tethya aurantium* and their inner axial filaments (colored in red), as revealed by scanning electron microscopy (a) and X-ray nano-tomography (b). The growth direction of the main shaft is [001]; the tripod branches are marked by vectors $\boldsymbol{G}_1$, $\boldsymbol{G}_2$, and $\boldsymbol{G}_3$ in (a). The correspondence between one of the tripod branches shown in (a) and (b) is indicated by the blue arrow.

Section 11.1. The filament, up to 2 µm in diameter, is predominantly composed of enzymatically active proteins, silicatein and its derivatives, which catalyze bio-fabrication of silica, the process being genetically controlled by specialized cells called sclerocytes. The protein nano-blocks (about 5 nm in size) in the axial filament are arranged in a crystal-like three-dimensional structure having hexagonal or body-centered tetragonal symmetry (BCT), whereas the pores within this structure are filled by amorphous silica (see Figure 11.5 in Chapter 11). Using nano-tomography, it was possible to trace the filament across the spicules (red lines in Figure 12.13b) and prove that the glass branching is governed by the symmetry of the protein crystal structure. In the shown spicules of *Tethya aurantium*, the filament structure has hexagonal symmetry with lattice parameters $a = 5.95$ nm, $c = 11.89$ nm. The filament in the main shaft of the spicule (the longer element in Figure 12.13a) grows along the [001] direction, while the primary branching, which leads to the characteristic tetrapod shape clearly seen in Figure 12.13a, proceeds on crystallographic planes with the following Miller indices: $\boldsymbol{G}_1$ (101), $\boldsymbol{G}_2$ $(0\bar{1}1)$, $\boldsymbol{G}_3$ $(\bar{1}11)$. Therefore, by using the crystalline-like axial filament of certain symmetry, nature mastered the fabrication of extremely complex glass structures at low temperatures that is far beyond the abilities of current human technology.

In the last two decades or so, a completely new approach to the imaging of nano-objects has been under intensive development, that is, coherent diffractive imaging (CDI), which, in the case of **Bragg** diffraction, is also known as ptychography. This approach changed the imaging paradigm in two aspects: (*i*) employing coherent incident beams, and (*ii*) relying on computational algorithms, rather than on X-ray optics, in the processing of coherently scattered X-ray diffraction patterns and converting them into an image of the sample. Many research groups participated in this development, but the influential contribution for CDI philosophy and development of phase retrieval algorithms by **John**

Rodenburg should be specially mentioned. Note that CDI got a strong boost after construction of third- and fourth-generation synchrotron sources, which are characterized by greatly reduced emittance of the electron beam circulating along the storage ring and subsequent gain in coherent X-ray flux. Further enhancement of X-ray beam coherency and brilliance is achieved with free electron lasers.

There are several setups for CDI. The simplest one is schematically depicted in Figure 12.14. A coherent X-ray beam comes from the synchrotron source and, after passing the restricted pinhole aperture, illuminates some region of the sample. The produced far-field (**Fraunhofer** regime) diffraction pattern, "modulated" by the sample's shape, is registered by the 2D detector.

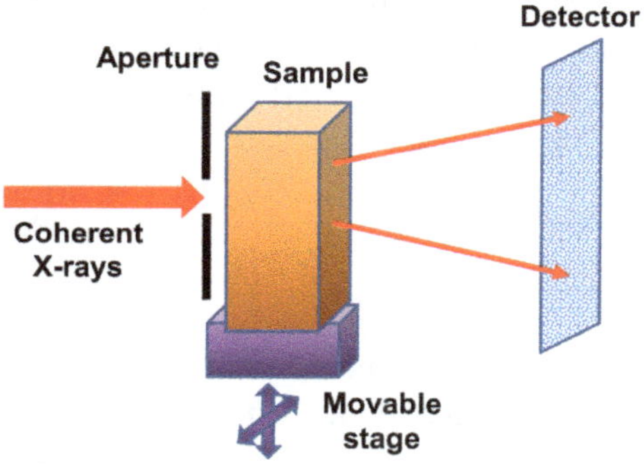

Figure 12.14: Scheme of the ptychography setup.

Then, the sample is shifted under the beam by a certain distance, and the measurement repeats. We stress that the shift should provide some overlap between the first and second illumination positions (see Figure 12.15).

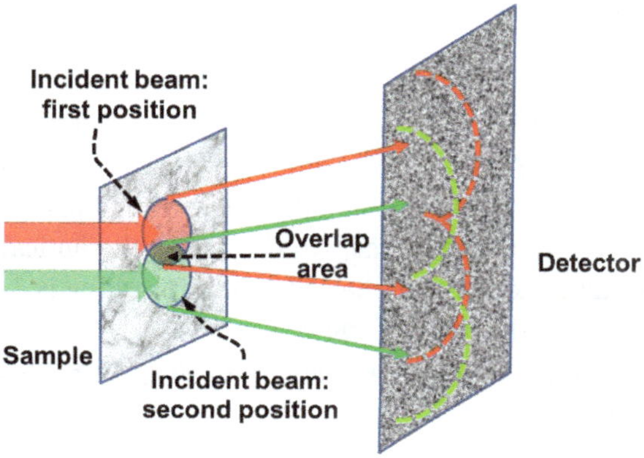

Figure 12.15: Illustration of the measurement steps involved in the ptychographic procedure.

Collecting several diffraction patterns, it is possible to retrieve the phases and moduli of respective scattering amplitudes (i.e., complex refractive index) by applying specially developed iterative computer algorithms, which start from the presumed object shape in real space. The spatial resolution of CDI is on a 10 nm-scale. The ultimate spatial resolution, however, is not limited by the X-ray optics used (if any), but by the wavelength of radiation only. For these reasons, CDI is being continuously used (and further developed) in the efforts towards resolving nano-sized features in materials density, strain, or other properties, which generate the X-ray contrast mechanisms.

The ideas of CDI are implemented in electron microscopy as well. Recently, the highest resolution of an electron microscopy (0.039 nm) was demonstrated by researchers in Cornell University and was even registered in the Guinness World Records. The Cornell team showed that ptychography can really be used to overcome the resolution limit of conventional electron microscopy, which stands on lens-based imaging.

It is also worth mentioning the successful attempts of the **Pierre Thibault** group in combining ptychography principles with STXM.

12.5 The Talbot effect

One more very interesting and practically important phenomenon related to **Fresnel** diffraction, is the **Talbot** effect. It is named after **Henry Fox Talbot** who first observed it in 1836. The **Talbot** effect appears as periodic self-imaging of a diffraction grating at regular distances z_T from the grating location, the distances being equal to:

$$z_T = \frac{\lambda}{1 - \sqrt{1 - \left(\frac{\lambda}{d}\right)^2}} \tag{12.44}$$

where d is the grating period and λ is the radiation wavelength. This formula was first derived in 1881 by **Lord Rayleigh**. If $\lambda \ll d$ (as in most practical cases), eq. (12.44) transforms into

$$z_T = 2\frac{d^2}{\lambda} \tag{12.45}$$

Let us consider wave propagation from source S through a diffraction grating placed at point O and then up to the observation (image) plane, which crosses the optical axis at point P (see Figure 12.16). The distance SO from the source to the grating is T, whereas the distance OP from the grating to the image plane is z.

A one-dimensional periodic grating with spatial periodicity in the y-direction equal to d can be represented via a **Fourier** series:

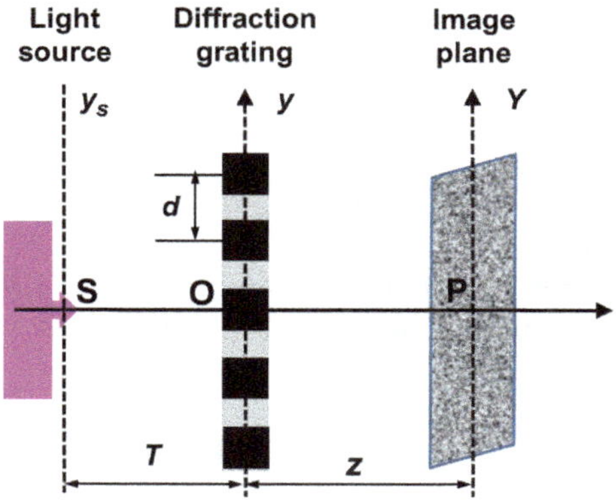

Figure 12.16: Scheme of the **Talbot** experiment.

$$G(y) = \sum_{-\infty}^{\infty} a_n \exp\left(2\pi i \frac{ny}{d}\right) \tag{12.46}$$

where a_n is the strength of the nth **Fourier** harmonic. In general, the diffracted amplitude $E(Y)$ in the image plane (along the Y-coordinate) is the convolution of the amplitude $G(y)$ of the object (diffraction grating, in this case) with the wavefield produced by the source $S(y_s)$, that is, along the y_s-coordinate:

$$E(Y) = \frac{\exp\left[\frac{2\pi i(T+z)}{\lambda}\right]}{i\lambda Tz} \int dy_s S(y_s) \int dy G(y) \exp\left[i\pi \frac{(y-y_s)^2}{\lambda T}\right] \exp\left[i\pi \frac{(Y-y)^2}{\lambda z}\right] \tag{12.47}$$

As usual in **Fresnel** diffraction, the phase factors, $\exp\left[i\pi \frac{(y-y_s)^2}{\lambda T}\right]$ and $\exp\left[i\pi \frac{(Y-y)^2}{\lambda z}\right]$, quadratically depend on the coordinates in the object or image planes and are inversely proportional to the product of the respective distance and wavelength λ (see also eq. (12.4)). Concentrating on the phase terms containing both coordinates z and y, we can say that diffraction amplitude is proportional to

$$E(Y) \propto \int dy G(y) \exp\left[i\pi \frac{(Y-y)^2}{\lambda z}\right] \tag{12.48}$$

Substituting eq. (12.46) into eq. (12.48) yields:

$$E(Y) \propto \sum_{-\infty}^{\infty} a_n \int dy \exp\left[\frac{2\pi i}{\lambda}\left(\frac{ny\lambda}{d} + \frac{Y^2}{2z} - \frac{Yy}{z} + \frac{y^2}{2z}\right)\right] \tag{12.49}$$

To perform integration in eq. (12.49), it is convenient to complete the square in circular brackets, which leads to the following expression for the phase factor:

$$\frac{2\pi}{\lambda}\left[\frac{Y^2}{2z} + \left(\frac{y}{\sqrt{2z}}\right)^2 + 2\left(\frac{y}{\sqrt{2z}}\right)\left(\frac{n\lambda}{d} - \frac{Y}{z}\right)\frac{\sqrt{2z}}{2} + \frac{2z}{4}\left(\frac{n\lambda}{d} - \frac{Y}{z}\right)^2 - \frac{2z}{4}\left(\frac{n\lambda}{d} - \frac{Y}{z}\right)^2\right] \tag{12.50}$$

Separating the completed square, which now appears in the squared brackets, one obtains:

$$\frac{2\pi}{\lambda}\left\{\left[\frac{y}{\sqrt{2z}}+\left(\frac{n\lambda}{d}-\frac{Y}{z}\right)\frac{\sqrt{2z}}{2}\right]^2-\frac{zn^2\lambda^2}{2d^2}+\frac{n\lambda}{d}Y\right\} \tag{12.51}$$

After integration, which converts the completed square into some constant, the diffraction field in the observation plane is defined by two phase factors:

$$E(Y) \propto \sum_{-\infty}^{\infty} a_n \exp\left[-2\pi i\left(\frac{zn^2\lambda}{2d^2}\right)\right] \exp\left[2\pi i\left(\frac{n}{d}Y\right)\right] \tag{12.52}$$

The second exponent shows the same periodicity in the Y-direction as the diffraction grating itself. We stress that this periodicity does not depend on distance z. The most interesting to us is the first exponential term, which does reveal the z-dependence in the phase changes of the diffraction orders. At a certain distance z_T, which is called the **Talbot** distance, all diffraction orders are in phase and, hence, are enhanced when satisfying the constructive interference condition,

$$\frac{z_T\lambda}{2d^2} = m \tag{12.53}$$

or

$$z_T = 2m\frac{d^2}{\lambda} \tag{12.54}$$

which coincides with eq. (12.45) for $m = 1$. At these distances, the self-imaging of the diffraction grating, often called a **Talbot** carpet, occurs (see Figure 12.17).

On the left side of Figure 12.17, we see the light diffracting through a grating, and this pattern is exactly reproduced on the right side of the picture (one **Talbot** length, $z_T = 2\frac{d^2}{\lambda}$, away from the grating). Halfway between these two edges, we observe the

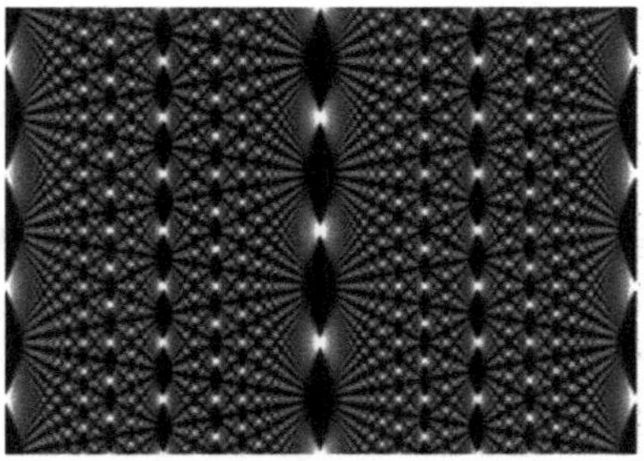

z = 0 **z = ½ ·z$_T$** **z = z$_T$**

Figure 12.17: Illustration of **Talbot** self-imaging (**Talbot** carpet).

image shifted along the Y-axis, and at regular fractions of the **Talbot** length, the sub-images also are clearly seen. Halfway images appear at distances,

$$z = m\frac{d^2}{\lambda} \tag{12.55}$$

with m being an odd integer. These images result from the π-phase shift of odd-number diffraction orders with respect to the zero- and even-number diffraction orders.

Due to the unique features of the **Talbot** self-images, they are widely used in contemporary optics. For example, placing a phase object after the diffraction grating causes a shift in the diffraction pattern, which is used for phase retrieval and then for phase-contrast imaging, as was explained in Section 9.4. The use of the **Talbot** effect for X-ray phase-contrast imaging will be discussed in more detail, in Section 12.5.1.

As a promising direction, the plasmonic **Talbot** effect should be mentioned. Surface plasmons (SPs) are coherent electron oscillations that are formed at the interface between any two materials where the real part of the dielectric function changes its sign across the interface (e.g., a metal part in air). SPs have lower energy than bulk plasmons, which are related to the longitudinal electron gas oscillations about positive ion cores. When SPs are coupled with a photon, the resulting hybridized excitation is called surface plasmon polariton (SPP). This SPP can propagate along the surface of a metal, until its energy is dissipated via absorption in the metal or through radiation losses into free space. Recent research has shown that SPPs may also be promising objects to observe the **Talbot** effect. Accelerated by advanced nanofabrication technologies, this direction has already inspired many interesting proposals towards observing plasmon self-imaging with metallic/dielectric materials having holes, metallic dots, dielectric dots, and metal waveguide arrays.

Another encouraging application field of the **Talbot** effect is related to metamaterials (see Section 6.4). Some metamaterials are composed of structural elements smaller than the wavelength of the waves they should affect. The strength of metamaterials is in unconventional processing of evanescent waves, for example, towards imaging beyond the diffraction limit (see Section 11.3). We remind the reader that in a conventional optical material, evanescent waves decay exponentially with the propagation distance, and hence, no **Talbot** self-imaging occurs if the periodicity of the object is much smaller than the light wavelength. However, in metamaterials comprising negative elements, the situation changes drastically, since evanescent waves can be converted back to propagating ones and collected in the far-field region (see Section 11.3). Taking advantage of this phenomenon, the super **Talbot** effect in an anisotropic metamaterial was proposed towards bulk-plasmon-based imaging of structural features much smaller than the light wavelength.

12.5.1 Employing diffraction gratings for phase-contrast X-ray imaging

As already mentioned in Section 9.4, over the last two decades, there has been substantial progress in the implementation of grating-based **Talbot** interferometers in the field of X-ray imaging. The simplest X-ray **Talbot** interferometer comprises two gratings: a phase grating (G1) with periodicity d_1 and an absorption/transmission grating (G2) with periodicity d_2 (see Figure 12.18).

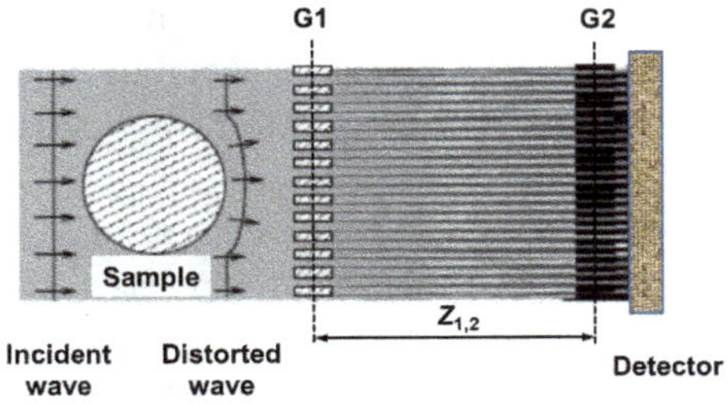

Figure 12.18: Scheme of the phase-contrast imaging method using the **Talbot** effect and double-grating X-ray interferometer.

After the first grating (G1), the incident beam forms periodic interference fringes in the plane of the grating (G2). A phase object (sample), placed into the incident beam, causes phase perturbations of the wavefront, leading to the shifts of these interference fringes. The absorption grating (G2), situated in front of the detector, transforms local fringe positions into intensity variations. By detecting fringe shifts, one can reconstruct the shape of the wavefront and generate the image.

It is possible to show that if the phase grating (G1) is characterized by a phase shift of π, the period of the grating (G2) should be $d_2 = \frac{d_1}{2}$. If the phase shift equals π, the first order of diffraction is enhanced at the expense of the 0-order. To achieve maximal contrast, the distance $Z_{1,2}$ between the two gratings should satisfy the condition:

$$Z_{1,2} = (m - 1/2)\frac{d_2^2}{\lambda} = (m - 1/2)\frac{d_1^2}{4\lambda} \tag{12.56}$$

with $m = 1, 2, 3 \ldots$ Note that eq. (12.56) has the same structure as eqns. (12.54) and (12.55), being different in the m-dependent numerical factor only, viz. $(m - 1/2)/4$.

Further contrast enhancement is achieved using three-grating interferometers (introduced in Section 9.4, see Figure 9.8), in which an additional grating is placed between the X-ray source and the object, towards improving the X-ray beam coherence. Note that three-grating design is similar to that used in the **Bonse-Hart** interferometer (see Section 8.4). Double- and triple-grating interferometers provide high quality phase-contrast X-ray images of weakly absorbing biological samples, polymers, and fiber composites. As an illustrative example, the inner structure of a small spider is shown in Figure 12.19. The X-ray images of the spider "architecture," taken by the **Pfeiffer** group at Swiss Light Source and ESRF, revealed fine features with spatial resolution on a 10-micron scale, the features that are hard to access with alternative techniques.

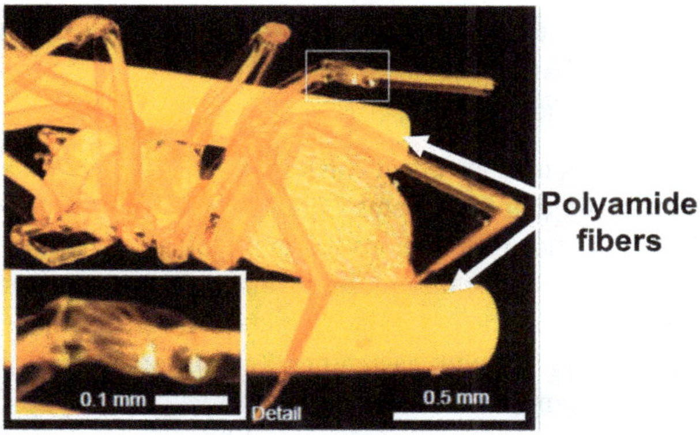

Figure 12.19: Example of phase-contrast imaging of the spider's body (supported by two polyamide fibers) revealing fine features of its internal structure. The latter is visualized due to the spatial distribution of the refractive index. Reprinted with permission from Optics Express 13, 6296 (2005) © The Optical Society.

A **Momose** group made another impressive demonstration of the capabilities of **Talbot** self-imaging by capturing the image of a mouse tail (not shown here) at the synchrotron source Super Photon ring – 8 GeV (SPring–8, Japan). It is important to note that within the same image, the bones together with soft tissues, such as skin, muscle, ligament, and cartilage, were well resolved. These and other pioneering experiments have facilitated further developments of the **Talbot** self-imaging towards extended medical and biological applications.

Chapter 13
Optics of dynamical diffraction

Next three chapters are devoted to fundamental optical phenomena that evolve in the framework of dynamical diffraction of quantum beams (X-rays, electrons, and neutrons). In contrast to kinematical diffraction (see Sections 10.2 and 12.1), which is valid for small crystals only, dynamical diffraction takes place in rather thick perfect or nearly perfect crystals. In this case, the diffracted and transmitted waves may have comparable amplitudes and strongly interact with each other. To analyze dynamical diffraction, we first must uncover deep interrelation between diffraction in crystals and translational symmetry, introduce the concept of reciprocal lattice, and understand how many diffracted waves can simultaneously propagate within a medium.

13.1 Wave propagation in periodic media from symmetry point of view

For this purpose, let us consider following ideas of **Léon Brillouin**, the propagation of plane waves across a periodic medium. A plane wave is defined as:

$$Y = Y_0 \exp[i(\boldsymbol{kr} - \omega t)] \tag{13.1}$$

where Y stands for a physical parameter that oscillates in space ($\boldsymbol{r}$) and time (t), while Y_0, $\boldsymbol{k}$, and ω are the wave amplitude, wave vector, and angular frequency, respectively. The term in circular brackets in eq. (13.1),

$$\varphi = \boldsymbol{kr} - \omega t \tag{13.2}$$

is the phase of the plane wave. At any instant t, the surface of the steady phase, φ = constant, is defined by the condition, $\boldsymbol{kr}$ = const. The latter is, in fact, the equation of a geometrical plane perpendicular to the direction of wave propagation $\boldsymbol{k}$ and, therefore, this type of wave has accordingly been so named (plane wave).

Considering first a homogeneous medium, we can say that a plane wave having wave vector $\boldsymbol{k}_i$ at a certain point in its trajectory will continue to propagate with the same wave vector because of the momentum conservation law. Note that wave vector $\boldsymbol{k}$ is linearly linked to momentum $\boldsymbol{P}$ via the reduced **Planck** constant $\hbar$, that is, $\boldsymbol{P} = \hbar\boldsymbol{k}$. We also remind that the momentum conservation law is a direct consequence of the specific symmetry of a homogeneous medium known as the homogeneity of space (the **Emmy Nöther**'s theorem).

The situation is drastically changed for a non-homogeneous medium, in which the momentum conservation law, generally, is not valid because of the breaking of the aforementioned symmetry. As a result, in such a medium, one can find wave vectors $\boldsymbol{k}_d$ differing from the initial wave vector $\boldsymbol{k}_i$. It implies that, in principle, additional

https://doi.org/10.1515/9783111140100-014

waves (scattered or diffracted waves) can propagate within such a medium, besides the incident one.

We will focus here on the particular non-homogeneous medium with translational symmetry, the medium which comprises scattering centers in certain points $\boldsymbol{r}_s$ only:

$$\boldsymbol{r}_s = n_1\boldsymbol{a}_1 + n_2\boldsymbol{a}_2 + n_3\boldsymbol{a}_3 \tag{13.3}$$

with the rest of the space being empty (see Figure 13.1). Here, in eq. (13.3), the vectors $\boldsymbol{a}_1$, $\boldsymbol{a}_2$, $\boldsymbol{a}_3$ are three non-coplanar translation vectors, while n_1, n_2, and n_3 are integer numbers (positive, negative, or zero). Currently, this is our model of a crystal.

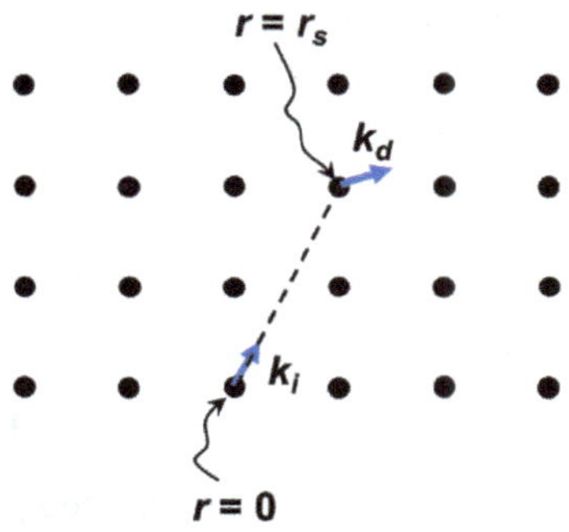

Figure 13.1: Illustration of an X-ray scattering event in a periodic medium.

Based on only the translational symmetry, we can say that in an infinite medium with no absorption, the magnitude of plane wave Y should be the same in close proximity to any lattice node described by eq. (13.3). It means that the amplitude Y_0 is the same at all points $\boldsymbol{r}_s$, whereas the phase φ can differ by an integer number m of 2π. Let us suppose that the plane wave has wave vector $\boldsymbol{k}_i$ at starting point $\boldsymbol{r}_0 = 0$ and $t_0 = 0$. Then, according to eq. (13.2), $\varphi(0) = 0$. Correspondingly, at point $\boldsymbol{r}_s$, the phase $\varphi(\boldsymbol{r}_s)$ of plane wave should be as given by the equation, $\varphi(\boldsymbol{r}_s) = \boldsymbol{k}_d\boldsymbol{r}_s - \omega t = 2\pi m$. Note that the change of the wave vector from $\boldsymbol{k}_i$ to $\boldsymbol{k}_d$ physically means that the wave obeys scattering in point $\boldsymbol{r}_s$ (see Figure 13.1). Here as well as below, only elastic scattering (with no energy change) is considered. Therefore,

$$|\boldsymbol{k}_d| = |\boldsymbol{k}_i| = |\boldsymbol{k}| = \frac{2\pi}{\lambda} \tag{13.4}$$

where λ stands for the radiation wavelength. Note that we already used this equation in Chapter 10 (see eq. (10.17)).

To further proceed, we recall the linear dispersion law for electromagnetic waves in vacuum (see eq. (1.20)), that is, the linear relationship between an absolute value of wave vector $|\boldsymbol{k}|$ and angular frequency ω:

$$\omega = c|\boldsymbol{k}| \tag{13.5}$$

where c is the speed of wave propagation. With the aid of eqs. (13.4) and (13.5), one can express the time interval t for a wave traveling between points $\boldsymbol{r}_0 = 0$ and $\boldsymbol{r}_s$:

$$t = \frac{\boldsymbol{k}_i \boldsymbol{r}_s}{|\boldsymbol{k}_i| c} \tag{13.6}$$

By using eqs. (13.2), (13.4)–(13.6), we calculate the phase of plane wave $\varphi(\boldsymbol{r}_s)$ after scattering in point $\boldsymbol{r}_s$ to be

$$\varphi(\boldsymbol{r}_s) = \boldsymbol{k}_d \boldsymbol{r}_s - \omega t = (\boldsymbol{k}_d - \boldsymbol{k}_i)\boldsymbol{r}_s \tag{13.7}$$

Since the initial phase $\varphi(0) = 0$, eq. (13.7) determines the phase difference φ as the result of wave scattering. The difference vector between wave vectors in the final ($\boldsymbol{k}_d$) and initial ($\boldsymbol{k}_i$) wave states is called as a wave vector transfer (to a crystal) or diffraction (scattering) vector $\boldsymbol{Q}$:

$$\boldsymbol{Q} = \boldsymbol{k}_d - \boldsymbol{k}_i \tag{13.8}$$

Substituting eq. (13.8) in eq. (13.7) finally yields the phase difference φ as:

$$\varphi = \varphi(\boldsymbol{r}_s) = \boldsymbol{Q}\boldsymbol{r}_s \tag{13.9}$$

According to eq. (13.8), different values of $\boldsymbol{k}_d$ are indeed permitted from the symmetry point of view, but only those that provide a scalar product in eq. (13.9), that is, a scalar product of a certain diffraction vector $\boldsymbol{Q}_{\mathrm{B}}$ with the set of lattice vectors $\boldsymbol{r}_s$ (13.3) to be equal to integer number m of 2π:

$$\varphi = \boldsymbol{Q}_{\mathrm{B}} \boldsymbol{r}_s = 2\pi m \tag{13.10}$$

To avoid the usage of factor 2π in eq. (13.10), another vector $\boldsymbol{G}$ is introduced as:

$$\boldsymbol{G} = \frac{\boldsymbol{Q}_B}{2\pi} \tag{13.11}$$

for which the condition (13.10) is rewritten as:

$$\boldsymbol{G} \cdot \boldsymbol{r}_s = m \tag{13.12}$$

By substituting eq. (13.3) in eq. (13.12), we finally obtain the necessary diffraction condition:

$$\boldsymbol{G} \cdot (n_1 \boldsymbol{a}_1 + n_2 \boldsymbol{a}_2 + n_3 \boldsymbol{a}_3) = m \tag{13.13}$$

13.2 The concept of reciprocal lattice and quasi-momentum conservation law

To find the set of allowed vectors $\boldsymbol{G}$ satisfying eq. (13.13), the reciprocal space is introduced, which is based on three non-coplanar vectors $\boldsymbol{b}_1$, $\boldsymbol{b}_2$, and $\boldsymbol{b}_3$. Real space and reciprocal space are related to each other by the orthogonality conditions:

$$\boldsymbol{a}_i\boldsymbol{b}_j = \delta_{ij} \tag{13.14}$$

where δ_{ij} is the **Kronecker** symbol equal to 1 for $i=j$ or 0 for $i \neq j$ $(i,j = 1, 2, 3)$. To build reciprocal space from real space, we use the following mathematical procedure:

$$\begin{aligned} \boldsymbol{b}_1 &= [\boldsymbol{a}_2 \times \boldsymbol{a}_3]/V_c \\ \boldsymbol{b}_2 &= [\boldsymbol{a}_3 \times \boldsymbol{a}_1]/V_c \\ \boldsymbol{b}_3 &= [\boldsymbol{a}_1 \times \boldsymbol{a}_2]/V_c \end{aligned} \tag{13.15}$$

where V_c stands for the volume of the parallelepiped (unit cell) built in real space on vectors $\boldsymbol{a}_1$, $\boldsymbol{a}_2$, $\boldsymbol{a}_3$:

$$V_C = \boldsymbol{a}_1 \,.\, [\boldsymbol{a}_2 \times \boldsymbol{a}_3] \tag{13.16}$$

By using eq. (13.16), it is easy to directly check that the procedure (13.15) provides the orthogonality conditions (13.14). For example, $(\boldsymbol{a}_1 \cdot \boldsymbol{b}_1) = \boldsymbol{a}_1 \cdot [\boldsymbol{a}_2 \times \boldsymbol{a}_3]/V_c = V_c/V_c = 1$, whereas $(\boldsymbol{a}_2 \cdot \boldsymbol{b}_1) = \boldsymbol{a}_2 \cdot [\boldsymbol{a}_2 \times \boldsymbol{a}_3]/\, V_c = 0$.

In the reciprocal space, the allowed vectors $\boldsymbol{G}$ are linear combinations of the basic vectors $\boldsymbol{b}_1$, $\boldsymbol{b}_2$, $\boldsymbol{b}_3$

$$\boldsymbol{G} = h\boldsymbol{b}_1 + k\boldsymbol{b}_2 + l\boldsymbol{b}_3 \tag{13.17}$$

with integer projections (hkl) known as **Miller** indices. The ends of vectors $\boldsymbol{G}$ being constructed from the common origin (000) form the nodes of a reciprocal lattice (see Figure 13.2), within which the wave vectors $\boldsymbol{k}_i$ and $\boldsymbol{k}_d$ (in fact, with lengths divided by 2π) also "live."

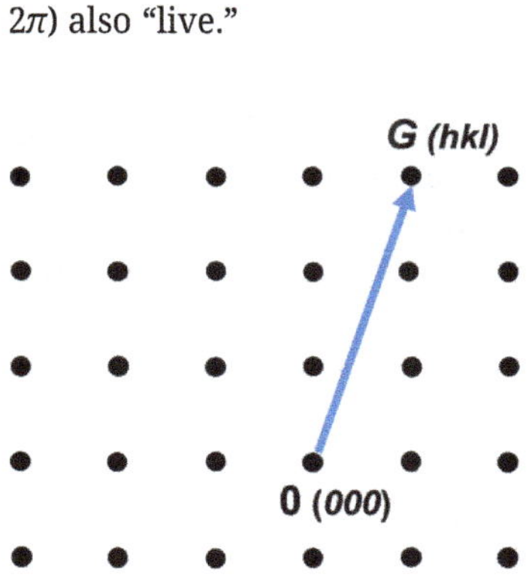

Figure 13.2: Reciprocal lattice as the set of reciprocal space vectors $\boldsymbol{G}(hkl) = h\boldsymbol{b}_1 + k\boldsymbol{b}_2 + l\boldsymbol{b}_3$. Their ends, plotted from the common origin (zero point (000)), form the nodes (black dots) of the reciprocal lattice.

The symmetry of reciprocal lattice is tightly related to the symmetry of the respective crystal lattice. For all vectors $\boldsymbol{G}$, which are called vectors of reciprocal lattice, eq. (13.13) is automatically fulfilled due to the orthogonality conditions (13.14). So, in a medium with translational symmetry, only those wave vectors $\boldsymbol{k}_d$ may exist that are in the following interrelation with the initial wave vector $\boldsymbol{k}_i$:

$$\boldsymbol{k}_d - \boldsymbol{k}_i = \boldsymbol{Q}_B = 2\pi\boldsymbol{G} \tag{13.18}$$

where $\boldsymbol{G}$-vectors are given by eq. (13.17). Sometimes, eq. (13.18) is called as quasi-momentum (or quasi-wave vector) conservation law in the medium with translational symmetry, which should be used, instead of the momentum (wave vector) conservation law ($\boldsymbol{k}_d = \boldsymbol{k}_i$) in a homogeneous medium. We stress that due to the quasi-momentum conservation law, all waves with wave vectors satisfying eq. (13.18) are symmetry equivalent. This important point and the related consequences will be elaborated in more detail in Chapters 14 and 15.

Graphical representation of eq. (13.18) leading to the **Bragg** law is given in Figure 13.3. According to eq. (13.18), the necessary condition for the diffraction process is indeed the quasi-momentum (or quasi-wave vector) conservation law, which determines specific angles $2\Theta_B$ between wave vectors $\boldsymbol{k}_d$ and $\boldsymbol{k}_i$, when the diffraction intensity could, in principle, be observed (see Figure 13.3). Solving the wave vector triangle in Figure 13.3, together with eq. (13.4), yields:

$$2|\boldsymbol{k}|\sin\Theta_B = \frac{4\pi\sin\Theta_B}{\lambda} = 2\pi|\boldsymbol{G}| = |\boldsymbol{Q}_B| \tag{13.19}$$

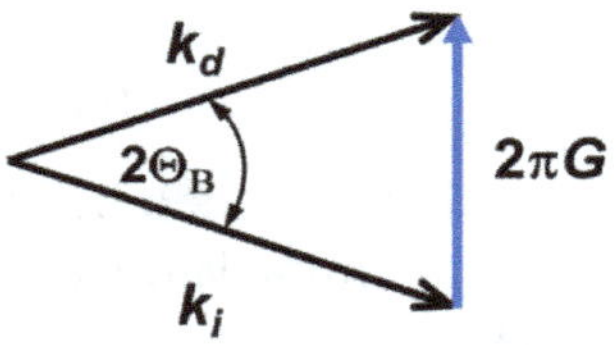

Figure 13.3: Graphical representation of eq. (13.18).

Note that each vector of reciprocal lattice $\boldsymbol{G} = h\boldsymbol{b}_1 + k\boldsymbol{b}_2 + l\boldsymbol{b}_3$ is perpendicular to a certain crystallographic plane in real space. This connection is directly given by eq. (13.12), which defines the geometric plane for the ends of certain vectors $\boldsymbol{r}_s$, the plane being perpendicular to the specific vector $\boldsymbol{G}$ (see Figure 13.4).

Using eq. (13.19) and introducing a set of parallel planes of this type that are separated by the d-spacing

$$d = \frac{1}{|\boldsymbol{G}|} \tag{13.20}$$

we finally obtain the **Bragg** law already introduced in Sections 7.2 and 10.2,

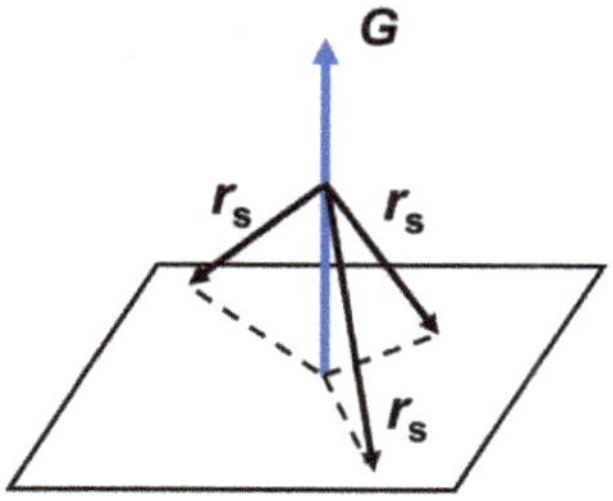

Figure 13.4: Schematic illustration of eq. (13.12).

$$2d \sin \Theta_B = \lambda \tag{13.21}$$

It provides interrelation between the possible directions for the diffracted wave propagation (via **Bragg** angles Θ_B) and the inter-planar spacings (d-spacings) in crystals. By using eqs. (13.15)–(13.17), and (13.20), one can calculate the d-spacings in crystals as functions of their lattice parameters and **Miller** indices, for all possible symmetry systems in real space. Therefore, measuring the diffraction peak positions $2\Theta_B$ and calculating the lattice d-spacings via the **Bragg** law (13.21) provides a vital tool for solving crystal structures by diffraction methods.

13.3 Diffraction condition and Ewald sphere

Next important question that we must answer is how many diffracted waves can simultaneously propagate through a crystal lattice. For this purpose, the so-called **Ewald** construction within the reciprocal lattice is used (see Figure 13.5).

In this construction, the wave vector $\boldsymbol{k}_i$ of the incident wave (in fact, reduced in length by a factor 2π) is placed within the reciprocal lattice of the investigated crystal. Since the wave vector can be shifted in space without changing its direction (i.e., remaining parallel to itself), we choose vector $\boldsymbol{k}_i/2\pi$ to be ended at some node of the reciprocal lattice, which is now taken as the **0**-node.

After that, the starting point A of vector $\boldsymbol{k}_i/2\pi$ is also well defined (see Figure 13.5). Since we are interested in elastic scattering (see eq. (13.4)), the ends of all possible wave vectors $\boldsymbol{k}_d$ (divided by 2π) should be located on the surface of the sphere of radius $|\boldsymbol{k}_i|/2\pi = |\boldsymbol{k}_d|/2\pi = |\boldsymbol{k}|/2\pi = 1/\lambda$, drawn from the common center located at point A (see Figure 13.5). This sphere is called the **Ewald** sphere. In Figure 13.5, we see the circular cross section of the **Ewald** sphere cut by the scattering plane formed by vectors $\boldsymbol{k}_i$ and $\boldsymbol{k}_d$. With the aid of Figure 13.5, we can say that the diffraction condition (13.18) simply means that the **Ewald** sphere intersects at least one additional node $\boldsymbol{G}$, besides the trivial **0**-node; the latter being always located on the **Ewald** sphere, according to the chosen construction procedure. Correspondingly, the alignment procedure of the crystal to fit the diffraction conditions means proper rotation of a single

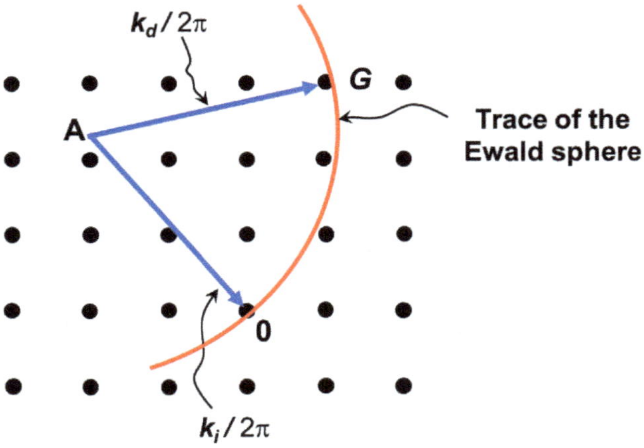

Figure 13.5: Reciprocal lattice (black dots) and the **Ewald**'s sphere construction. In the present case, **Ewald** sphere is crossing only two nodes, that is, the **0**-node and ***G***-node. Correspondingly, we have two propagating X-ray waves, with wave vectors indicated by $\boldsymbol{k}_i$ and $\boldsymbol{k}_d$.

crystal (and correspondingly, the rotation of its reciprocal lattice) until at least one additional node $\boldsymbol{G}$ touches the surface of the **Ewald** sphere.

The latter claim requires further clarification. In fact, if the length of wave vector $\boldsymbol{k}_i/2\pi$ is much smaller than the length of the smallest vector of reciprocal lattice $\boldsymbol{G}_{\min}$, the **Ewald** sphere does not intersect any node of the reciprocal lattice (except a trivial intersection at the **0**-node, see Figure 13.6). In other words, in this case, diffraction within the atomic network cannot be realized.

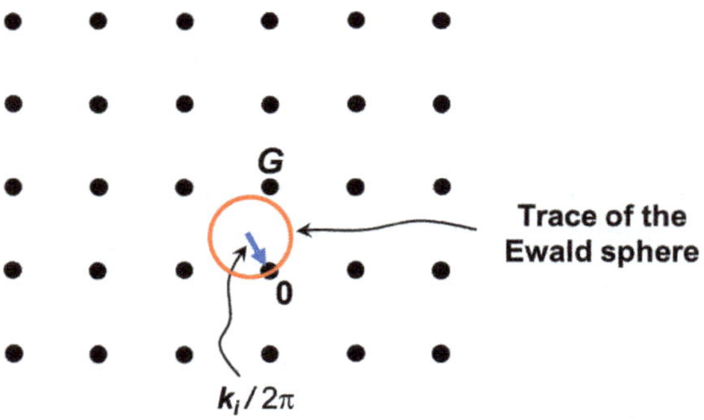

Figure 13.6: **Ewald**'s sphere construction in the case described by eq. (13.22).

Using eqs. (13.4) and (13.20), we can formulate a quantitative criterion for such a situation:

$$\lambda \gg d_{\max} \tag{13.22}$$

where $d_{\max}$ is the maximal d-spacing within a unit cell of a crystal. By using eq. (13.21), one can find the stronger constraint:

$$\lambda > 2d_{\max} \tag{13.23}$$

For example, the wavelength range for visible light fits the criterion (13.22). As was explained in section 10.4, visible light does obey diffraction when it meets obstacles (holes, slits, screens, etc.) with geometrical sizes comparable with λ. However, according to eq. (13.22), visible light does not “feel” the periodicity of the atomic network.

Another extreme is realized when $|\boldsymbol{k}_i|/2\pi \gg |\boldsymbol{G}_{\min}|$ or

$$\lambda \ll d_{\max} \tag{13.24}$$

In this case, many nodes of the reciprocal lattice can simultaneously be intersected by the **Ewald** sphere (see Figure 13.7). It means that numerous diffracted waves, differing in wave vectors $\boldsymbol{k}_d$ can concurrently propagate within the crystal and, therefore, the diffraction process must be treated in the framework of the so-called multi-wave approximation. This is typical for electron diffraction in crystals.

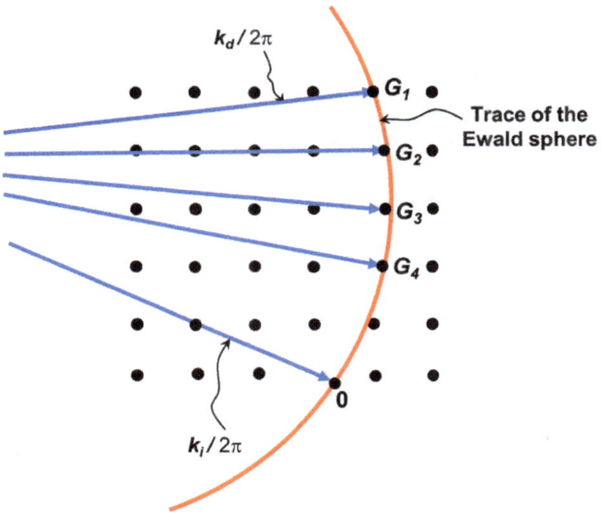

Figure 13.7: Ewald’s sphere construction in the case described by eq. (13.24).

For X-rays, as a rule, the following condition is fulfilled:

$$\lambda \leq d_{\max} \tag{13.25}$$

which allows the propagation of a single diffracted wave in a crystal, in addition to the incident wave, that is, the location of two nodes, **0** and ***G***, on the **Ewald** sphere, as depicted in Figure 13.5. In this case, the diffraction process, generally, should be handled in the two-beam approximation, which is quantitatively treated in the next Chapters 14 and 15. Note that in carefully designed experiments, multiple-wave diffraction can also be achieved with X-rays and then considered appropriately within a multi-beam approximation.

It is important to stress that under **Bragg** condition (13.21), the incident and the diffracted waves in an infinite crystal are identical (degeneracy) quantum-mechanical states, since they are linked to each other by the quasi-wave vector conservation law (13.18), a direct consequence of translational symmetry. The degeneracy point (marked by letter L in Figure 13.8) is located at the intersection of the isoenergetic surfaces (red lines in Figure 13.8) for the incident ($\boldsymbol{k}_i$) and the diffracted ($\boldsymbol{k}_d$) waves. Being projected onto vector $\boldsymbol{G}$ of the reciprocal lattice, the degeneracy point is in the middle between the **0**- and $\boldsymbol{G}$-nodes of the reciprocal lattice (see Figure 13.8).

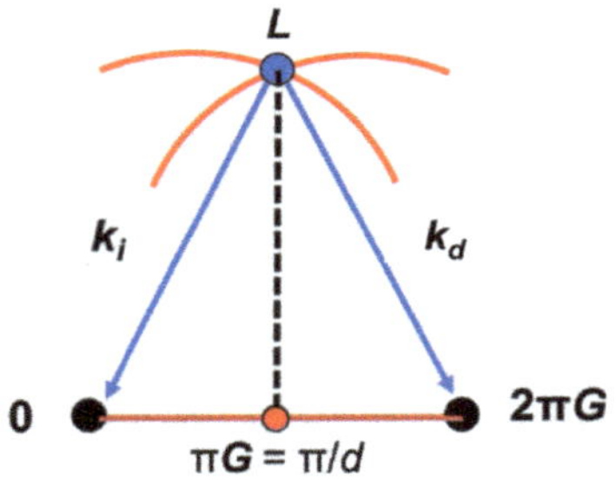

Figure 13.8: The traces of isoenergetic surfaces (red curves) in the reciprocal space for the incident ($\boldsymbol{k}_i$) and diffracted ($\boldsymbol{k}_d$) waves. The point of degeneracy of quantum states (i.e., the intersection point of red curves) is marked by the letter L (the **Lorentz** point).

This point is called the **Lorentz** point in the X-ray diffraction theory or the **Brillouin** zone boundary in solid state physics. The degeneracy of states is removed by strong interaction between these two waves via periodic lattice potential (in case of electrons) or periodic dielectric permittivity (in case of X-rays) that leads to novel interference phenomena, forbidden quantum states within a crystal, and total reflection of quanta in narrow angular range around the **Bragg** angles. All these will be quantitatively analyzed in Chapters 14 and 15.

Chapter 14
Dynamical diffraction of quantum beams: basic principles

The physics of dynamical diffraction is basically very similar for particles (electrons and neutrons) and electromagnetic waves (X-rays) since the analyses are performed in the framework of the perturbation theory. Nevertheless, one important exception still exists: for classical particles, we use scalar fields (wave functions) and **Schrödinger** equation, whereas for X-rays, we use vector fields and **Maxwell** equations. Correspondingly, in the case of X-ray diffraction, theoretical description should consider different X-ray polarizations (see next Chapter 15). Before doing this, let us start with a mathematically easier case of electron (or neutron) dynamical diffraction, which illustrates well the most essential aspects of the diffraction physics involved.

Motion of non-relativistic electrons in crystals obeys the non-stationary **Schrödinger** equation:

$$i\hbar\frac{\partial\psi}{\partial t} = -\frac{\hbar^2}{2m}\nabla^2\psi + V(\boldsymbol{r})\psi \tag{14.1}$$

where $\psi(\boldsymbol{r}, t)$ is the electron wave function, $V(\boldsymbol{r})$ is periodic lattice potential, m is the mass of electron, $\hbar$ is the reduced **Planck** constant, and

$$\nabla^2 = \frac{\partial^2}{\partial x^2} + \frac{\partial^2}{\partial y^2} + \frac{\partial^2}{\partial z^2} \tag{14.2}$$

is the **Laplace** operator already introduced in Chapter 1. Let us first find the solution of eq. (14.1) in free space, that is, when $V(\boldsymbol{r})$=0. It is natural to assume that in free space, an electron exists in a form of plane wave with amplitude ψ_0

$$\psi = \psi_0\exp(i\boldsymbol{kr} - i\omega t) \tag{14.3}$$

which propagates with constant wave vector $\boldsymbol{k}$ and angular frequency ω. Substituting eq. (14.3) in eq. (14.1) yields the relationship between the magnitude of the wave vector $k = |\boldsymbol{k}|$ and frequency ω:

$$\hbar\omega = \frac{\hbar^2}{2m}k^2 \tag{14.4}$$

that is, the dispersion law for electron wave in vacuum.

In the presence of periodic lattice potential $V(\boldsymbol{r})$, we are trying to find the solution of **Schrödinger** equation (14.1) as linear combination of plane waves with different wave vectors $\boldsymbol{k}$ and amplitudes $\psi(\boldsymbol{k}, \omega)$:

https://doi.org/10.1515/9783111140100-015

$$\psi = \sum_{k} \psi\,(\boldsymbol{k}, \omega)\exp(i\boldsymbol{kr} - i\omega t) \tag{14.5}$$

In turn, the periodic lattice potential $V(\boldsymbol{r})$ can be expanded in the **Fourier** series using the set of vectors of reciprocal lattice $\boldsymbol{G}$ or, more precisely, the set of diffraction vectors $\boldsymbol{Q} = 2\pi\boldsymbol{G}$ (see eq. (13.11)):

$$V(\boldsymbol{r}) = \sum_{Q} V\,(\boldsymbol{Q})\exp(i\boldsymbol{Qr}) \tag{14.6}$$

Fourier-coefficients $V(\boldsymbol{Q})$ are given by the expression:

$$V(\boldsymbol{Q}) = \frac{1}{V_c}\int V(\boldsymbol{r})\exp(-i\boldsymbol{Qr})d^3\boldsymbol{r} \tag{14.7}$$

in which integration is performed over the volume of unit cell V_c. Substituting plane wave, $\psi(\boldsymbol{k}, \omega)\exp(i\boldsymbol{kr} - i\omega t)$, from the set (14.5) into the **Schrödinger** equation (14.1) yields:

$$\psi(\boldsymbol{k}, \omega)\exp(i\boldsymbol{kr} - i\omega t)\left(\hbar\omega - \frac{\hbar^2}{2m}\boldsymbol{k}^2\right) - \sum_{Q} V(\boldsymbol{Q})\psi(\boldsymbol{k}, \omega)\exp[i(\boldsymbol{k} + \boldsymbol{Q})\boldsymbol{r} - i\omega t] = 0 \tag{14.8}$$

Introducing the amplitude of the scattered wave as:

$$\psi(\boldsymbol{k} + \boldsymbol{Q}, \omega) = \psi(\boldsymbol{k}, \omega)\exp(i\boldsymbol{Qr}) \tag{14.9}$$

one finally obtains the system of algebraic equations

$$\left(\hbar\omega - \frac{\hbar^2}{2m}\boldsymbol{k}^2\right)\psi(\boldsymbol{k}, \omega) - \sum_{Q} V(\boldsymbol{Q})\,\psi(\boldsymbol{k} + \boldsymbol{Q}, \omega) = 0 \tag{14.10}$$

which connect the amplitude of the incident wave $\psi(\boldsymbol{k}, \omega)$ with the amplitudes of an infinite number of scattered waves $\psi(\boldsymbol{k} + \boldsymbol{Q}, \omega)$, each linked to the specific diffraction vector $\boldsymbol{Q}$. As usual, we will consider elastic scattering only and, hence, frequency ω does not change during the wave scattering processes.

Further analysis depends on how many nodes of the reciprocal lattice, that is, vectors $\boldsymbol{G} = \boldsymbol{Q}/2\pi$, are simultaneously crossed by the **Ewald**'s sphere, introduced in Chapter 13. Let us start with the simple case that there are no additional nodes (except **0**-node) at the surface of the **Ewald**'s sphere. It means that only one strong wave is propagating through the crystal, which is the incident wave with wave vector $\boldsymbol{k}$. Correspondingly, there is no diffraction, that is, $\boldsymbol{Q} = 0$. Applying this consideration to eq. (14.10) yields

$$\hbar\omega = \frac{\hbar^2}{2m}k^2 + V(0) = \frac{\hbar^2}{2m}k_m^2 \tag{14.11}$$

which represents the dispersion law in a homogeneous medium, which differs from that in vacuum. The magnitude k_m of wave vector in a medium is related to that in vacuum (eq. (14.4)) by the following expression:

$$k_m = k\sqrt{1 + \frac{V(0)}{\varepsilon_k}} \tag{14.12}$$

where $\varepsilon_k = \frac{\hbar^2}{2m}k^2$ is the electron kinetic energy in vacuum. By means of eq. (14.12), we immediately obtain the refractive index n for electron waves in a homogeneous medium with respect to vacuum:

$$n = \frac{k_m}{k} = \sqrt{1 + \frac{V(0)}{\varepsilon_k}} \tag{14.13}$$

For electron waves, $V(0) > 0$ and $n > 1$, that is, materials are optically denser than vacuum (as for visible light).

Note that, for neutrons, the refractive index is described by the same expression (14.13). However, due to the specific features of nuclear scattering, the **Fourier** coefficient $V(0)$ may be positive or negative, depending on the internal structure of a particular nucleus. It means that some materials (with negative values of $V(0)$) will be optically less dense than vacuum, which leads to the possibility of total external reflection for neutrons entering from the vacuum side toward such a medium. Moreover, by cooling the neutrons, that is, by reducing their kinetic energy ε_k, we can arrange the refractive index, $n = \sqrt{1 - \frac{|V(0)|}{\varepsilon_k}}$, to be an imaginary number. It means that such ultracold neutrons cannot penetrate material from the vacuum side at any angle of incidence. Therefore, they will stay for a while (limited only by the mean neutron lifetime of about 15 min) within the evacuated volume surrounded by the walls built of a material for which $\frac{|V(0)|}{\varepsilon_k} > 1$. This interesting effect was theoretically predicted by **Yaakov Zeldovich** in 1959, and is used to study the quantum characteristics of neutrons.

For X-rays, as we know, the real part of the refractive index is a little smaller than 1 for all materials. Therefore, for X-ray waves, materials are less optically dense compared to vacuum. It means that at small angles of incidence (a few tenths of a degree), with respect to the sample surface, X-rays experience total external reflection when entering materials from the vacuum side (see Section 2.3). This phenomenon allows us to concentrate the X-ray energy in ultrathin layers beneath the sample surface that is used in grazing incidence diffraction (GID) and X-ray reflectivity measurements (see **Appendix 5.B**).

14.1 The two-beam approximation

If additional nodes of the reciprocal lattice are crossed by the surface of the **Ewald** sphere, diffraction phenomena occur. If the total number of such nodes is N, then dynamical diffraction is described by the system of N algebraic eqs. (14.10), relating the amplitudes of one incident and $(N-1)$ diffracted waves. In the simplest case, to organize diffraction, we need two nodes (**0**-node and some ***G***-node) being intersected by the **Ewald**'s sphere (see Figure 13.5). It implies that the system of N-equations (14.10) is transformed to only two equations, which are the basis of the so called two-beam approximation. Note that this easy case allows us to uncover all essential features of dynamic diffraction.

The first equation is obtained by attributing the wave vector $\boldsymbol{k}$ to the incident wave, while the wave vector $\boldsymbol{k}+\boldsymbol{Q}$ to the diffracted wave:

$$\left[\hbar\omega - \frac{\hbar^2}{2m}\boldsymbol{k}^2 - V(0)\right]\psi(\boldsymbol{k},\omega) - V(-\boldsymbol{Q})\psi(\boldsymbol{k}+\boldsymbol{Q},\omega) = 0 \tag{14.14}$$

Since in two-beam approximation, the quantum states of the incident and the diffracted waves are related to each other by the vector of the reciprocal lattice and, in that sense, are symmetry equivalent, we can equally attribute wave vector $(\boldsymbol{k}+\boldsymbol{Q})$ to the incident wave, while wave vector $\boldsymbol{k}$ to the diffracted wave. Now, these two vectors are connected by vector $(-\boldsymbol{Q})$ (see Figure 14.1) that gives us the second equation:

$$\left[\hbar\omega - \frac{\hbar^2}{2m}(\boldsymbol{k}+\boldsymbol{Q})^2 - V(0)\right]\psi(\boldsymbol{k}+\boldsymbol{Q},\omega) - V(\boldsymbol{Q})\psi(\boldsymbol{k},\omega) = 0 \tag{14.15}$$

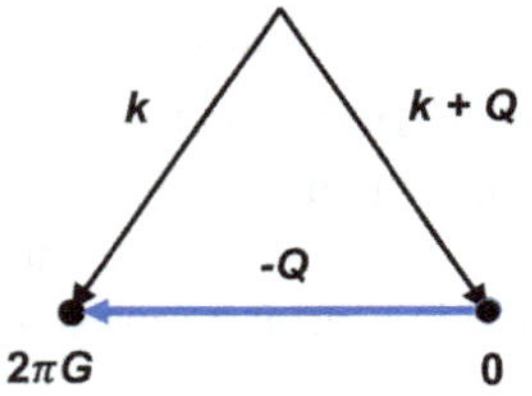

Figure 14.1: Wave vector construction used for deriving eq. (14.15).

Setting

$$\begin{aligned}\psi(\boldsymbol{k},\omega) &= \psi_0\\ \psi(\boldsymbol{k}+\boldsymbol{Q},\omega) &= \psi_Q\end{aligned} \tag{14.16}$$

we finally find the system of two equations for two unknown amplitudes, ψ_0 and ψ_Q, the equations being interrelated via the lattice potential V:

$$\left[\hbar\omega - \frac{\hbar^2}{2m}\boldsymbol{k}^2 - V(0)\right]\psi_0 - V(-\boldsymbol{Q})\psi_Q = 0$$

$$-V(\boldsymbol{Q})\psi_0 + \left[\hbar\omega - \frac{\hbar^2}{2m}(\boldsymbol{k}+\boldsymbol{Q})^2 - V(0)\right]\psi_Q = 0 \tag{14.17}$$

Non-trivial (non-zero) solution is realized when the determinant, built of the equations' coefficients, is equal to zero:

$$\det = \begin{vmatrix} \hbar\omega - [(\hbar^2\boldsymbol{k}^2)/(2m)] - V(0); & -V(-\boldsymbol{Q}) \\ -V(\boldsymbol{Q}); & \hbar\omega - [\hbar^2(\boldsymbol{k}+\boldsymbol{Q})^2/(2m)] - V(0) \end{vmatrix} = 0 \tag{14.18}$$

This is the secular equation for a new dispersion law, which almost everywhere practically coincides with the regular (quadratic) one (14.11) for a homogeneous medium. The situation is drastically changed in the vicinity of the degeneration point in the reciprocal space:

$$\boldsymbol{k}^2 = (\boldsymbol{k}+\boldsymbol{Q})^2 \tag{14.19}$$

in which, both waves have identical quantum-mechanical characteristics. For valence electrons within a crystal, the energy $\mathcal{E} = \hbar\omega$ is not fixed, and the degeneration of quantum states mentioned is removed by opening the energy gap in a specific point of the reciprocal space (i.e., for specific wave vector), defined by eq. (14.19). In the one-dimensional case, that is, when vector $\boldsymbol{k}$ is changed along the direction of the scattering vector $\boldsymbol{Q}$ (or vector of reciprocal lattice, $\boldsymbol{G} = \boldsymbol{Q}/2\pi$), eq. (14.19) yields (also considering eq. (13.20)):

$$|k| = \frac{Q}{2} = \frac{2\pi G}{2} = \frac{\pi}{d} \tag{14.20}$$

In other words, the degeneration point is in the middle of the diffraction vector or, as is said in solid state physics, at the **Brillouin** zone boundary $\frac{\pi}{d}$. At this point, eq. (14.18) for electron energy transforms into:

$$(\mathcal{E} - \mathcal{E}_0)^2 = V(\boldsymbol{Q})V(-\boldsymbol{Q}) \tag{14.21}$$

with $\mathcal{E}_0 = \dfrac{\hbar^2 Q^2}{8m} + V(0) = \dfrac{\pi^2\hbar^2 G^2}{2m} + V(0)$

If the lattice potential $V(\boldsymbol{r})$ is described by real function, then according to eq. (14.7), $V(-\boldsymbol{Q}) = V^*(\boldsymbol{Q})$ and $V(\boldsymbol{Q})V(-\boldsymbol{Q}) = |V(\boldsymbol{Q})|^2$. The same is valid for center-symmetric crystals. If so, the solution of eq. (14.21) in point (14.20) has the following form:

$$\mathcal{E} = \mathcal{E}_0 \pm |V(\boldsymbol{Q})| \tag{14.22}$$

with the characteristic gap $\Delta\mathcal{E} = 2\,|V(\boldsymbol{Q})|$, which separates the allowed energy states for electron waves in a crystal (see Figure 14.2).

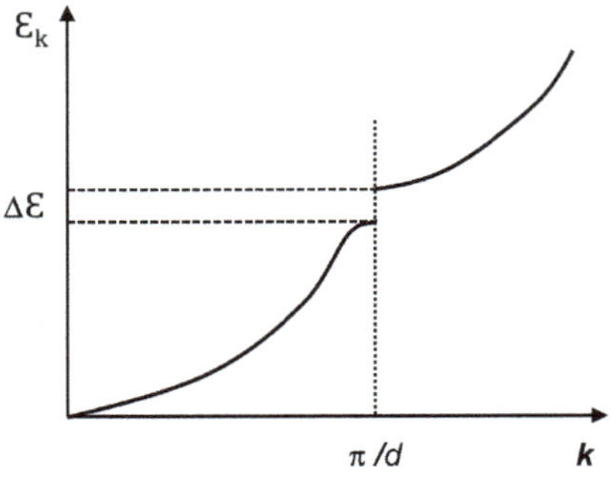

Figure 14.2: Dispersion curve, i.e., the dependence of the electron kinetic energy $\mathcal{E}_k$, as a function of the wave vector magnitude $\boldsymbol{k}$ for electron waves in a crystal. The energy gap $\Delta\mathcal{E}$ at the **Brillouin** zone boundary $k = \frac{\pi}{d}$ arises due to the electron wave diffraction on a periodic lattice potential.

We see that the energy gap is directly determined by the **Fourier** component of the lattice potential $|V(\boldsymbol{Q})|$. No electron states exist within the gap and, correspondingly, this energy interval is called a forbidden zone. Depending on the zone width, that is, the magnitude of the respective **Fourier** component of the lattice potential $|V(\boldsymbol{Q})|$ at the **Brillouin** zone boundary, materials are classified as semiconductors or insulators. Therefore, we can say that the forbidden zone (the bandgap) arises due to the diffraction of valence electrons on the periodic lattice potential created by ions in crystals.

For an electron wave entering a crystal from the vacuum side, the situation is essentially different. Assuming only elastic scattering processes, the energy of electrons is fixed, and we must solve eq. (14.18) with respect to the set of permitted wave vectors $\boldsymbol{k}$. In other words, we must find the shape of the so-called isoenergetic dispersion surface (i.e., the surface of equal energy) in three-dimensional reciprocal space near the degeneration point (14.19). Practically, we always reduce this problem to a two-dimensional one by considering the cross section of the isoenergetic surface created by the scattering plane, the latter being defined by wave vectors $\boldsymbol{k}$ and $\boldsymbol{k} + \boldsymbol{Q}$. In this plane, the degeneration point L (the **Lorentz** point) is located on the normal to the diffraction vector $\boldsymbol{Q}$, drawn through its middle (see eq. (14.20) and Figure 14.3).

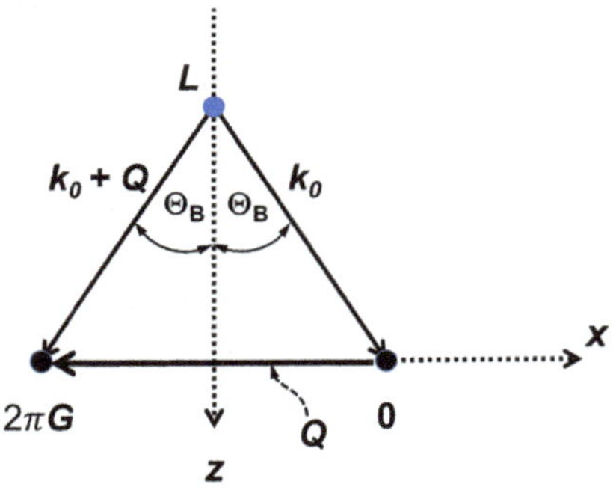

Figure 14.3: Coordinate system (x, z) used for deriving eq. (14.27). Wave vectors $\boldsymbol{k}_0$ and $\boldsymbol{k}_0 + \boldsymbol{Q}$ connect the **Lorentz** point in the reciprocal space (marked by the letter L) to the **0**- and $\boldsymbol{G}$-nodes, respectively. The angle between the wave vectors $\boldsymbol{k}_0$ and $\boldsymbol{k}_0 + \boldsymbol{Q}$ is double the **Bragg** angle $2\Theta_B$.

Let us introduce a new parameter in eq. (14.18), which in the degeneration point equals:

$$\hbar\omega - V(0) = \frac{\hbar^2 \boldsymbol{k}_0^2}{2m} = \frac{\hbar^2(\boldsymbol{k}_0 + \boldsymbol{Q})^2}{2m} \tag{14.23}$$

Note that wave vectors $\boldsymbol{k}_0$ and $\boldsymbol{k}_0 + \boldsymbol{Q}$ connect the **Lorentz** point to the **0**- and **G**-nodes of the reciprocal lattice, respectively (see Figure 14.3). Considering eq. (14.23), we can rewrite eq. (14.17) as follows:

$$\begin{gathered}(\boldsymbol{k}_0^2 - \boldsymbol{k}^2)\psi_0 - \frac{2mV(-\boldsymbol{Q})}{\hbar^2}\psi_Q = 0\\ -\frac{2mV(\boldsymbol{Q})}{\hbar^2}\psi_0 + \left[(\boldsymbol{k}_0 + \boldsymbol{Q})^2 - (\boldsymbol{k} + \boldsymbol{Q})^2\right]\psi_Q = 0\end{gathered} \tag{14.24}$$

Near the **Lorentz** point L, we can apply the following approximation for wave vectors:

$$\begin{gathered}\boldsymbol{k} - \boldsymbol{k}_0 = \delta\boldsymbol{k}_z + \delta\boldsymbol{k}_x; \quad (\boldsymbol{k} + \boldsymbol{Q}) - (\boldsymbol{k}_0 + \boldsymbol{Q}) = \delta\boldsymbol{k}_z + \delta\boldsymbol{k}_x\\ \boldsymbol{k}_0 + \boldsymbol{k} \approx 2\boldsymbol{k}_0; \quad (\boldsymbol{k}_0 + \boldsymbol{Q}) + (\boldsymbol{k} + \boldsymbol{Q}) \approx 2\boldsymbol{k}_0 + 2\boldsymbol{Q}\end{gathered} \tag{14.25}$$

where $\delta\boldsymbol{k}_x$ and $\delta\boldsymbol{k}_z$ are the wave vector differences (deviations from vector $\boldsymbol{k}_0$ connecting the **Lorentz** point to the **0**-node), respectively, along and perpendicular to the appropriate vector of the reciprocal lattice $\boldsymbol{G} = \boldsymbol{Q}/2\pi$. Substituting eq. (14.25) into eq. (14.24) yields:

$$\begin{gathered}(\delta\boldsymbol{k}_z + \delta\boldsymbol{k}_x)\boldsymbol{k}_0\psi_0 + \frac{mV(-\boldsymbol{Q})}{\hbar^2}\psi_Q = 0\\ \frac{mV(\boldsymbol{Q})}{\hbar^2}\psi_0 + (\delta\boldsymbol{k}_z + \delta\boldsymbol{k}_x)(\boldsymbol{k}_0 + \boldsymbol{Q})\psi_Q = 0\end{gathered} \tag{14.26}$$

With the aid of Figure 14.3, we find that the angles between vector $\boldsymbol{k}_0$ (which is very close to vector $\boldsymbol{k}$) and the z- and x-axes, are Θ_B and $(90° - \Theta_B)$, respectively. At the same time, these angles for vector $(\boldsymbol{k}_0 + \boldsymbol{Q})$ are Θ_B and $(90° + \Theta_B)$. Correspondingly, eq. (14.26) transforms into:

$$\begin{gathered}(\delta k_z + \delta k_x \tan\Theta_B)\psi_0 + \frac{mV(-\boldsymbol{Q})}{|\boldsymbol{k}_0|\hbar^2\cos\Theta_B}\psi_Q = 0\\ (\delta k_z - \delta k_x \tan\Theta_B)\psi_Q + \frac{mV(\boldsymbol{Q})}{|\boldsymbol{k}_0|\hbar^2\cos\Theta_B}\psi_0 = 0\end{gathered} \tag{14.27}$$

The secular equation of the system (14.27) provides the shape of the isoenergetic surface in the proximity to the **Lorentz** point L, as

$$\left(\delta k_z^2 - \delta k_x^2 \tan^2 \Theta_B\right) = \left(\frac{\Delta k_0}{2}\right)^2 \tag{14.28}$$

with

$$\Delta k_0 = \frac{|\boldsymbol{k}_0|\sqrt{V(\boldsymbol{Q})V(-\boldsymbol{Q})}}{\left(\frac{\hbar^2 \boldsymbol{k}_0{}^2}{2m}\right)\cos\Theta_B} = \frac{|\boldsymbol{k}_0|\sqrt{V(\boldsymbol{Q})V(-\boldsymbol{Q})}}{\mathcal{E}_k \cos\Theta_B} \tag{14.29}$$

For center-symmetric crystals, $V^*(\boldsymbol{Q}) = V(-\boldsymbol{Q})$, and

$$\Delta k_0 = \frac{|\boldsymbol{k}_0| \cdot |V(\boldsymbol{Q})|}{\mathcal{E}_k \cos\Theta_B} \tag{14.30}$$

where $\mathcal{E}_k$ again is the particle kinetic energy.

Equation (14.28) describes a two-branch hyperbolic isoenergetic surface in the reciprocal space (see Figure 14.4), with the branches being separated by the characteristic gap Δk_0 (see eq. (14.29) or eq. (14.30)) in the wave vector (reciprocal) space. As in the case of the energy gap $\Delta\mathcal{E}$, the wave vector gap Δk_0 is also proportional to the **Fourier** component of the lattice potential $|V(\boldsymbol{Q})|$ (see eq. (14.30)). Larger potential means stronger interaction between the incident and the diffracted waves and, hence, a larger separation gap Δk_0.

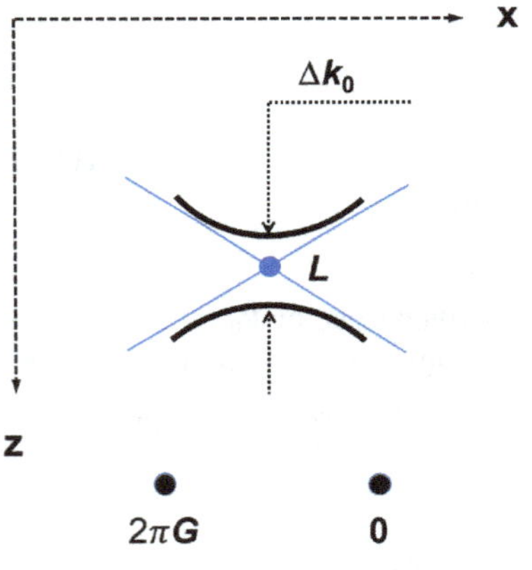

Figure 14.4: Two-branch hyperbolic isoenergetic surface near the **Lorentz** point (L) in the reciprocal space. These two branches are separated by the characteristic gap Δk_0.

To find the amplitudes of the incident and the diffracted waves, let us first simplify eq. (14.28) by introducing the dimensionless parameters:

$$\frac{\delta k_z}{\left(\frac{\Delta k_0}{2}\right)} = q \tag{14.31}$$

$$\frac{\delta k_x \tan \Theta_B}{\left(\frac{\Delta k_0}{2}\right)} = p$$

With the aid of definitions (14.31), the secular equation (14.28) transforms into a simple hyperbolic equation:

$$q^2 - p^2 = 1 \tag{14.32}$$

whereas the system of eqs. (14.27) is reduced to:

$$\begin{aligned} (q+p)\psi_0 + \psi_Q &= 0 \\ \psi_0 + (q-p)\psi_Q &= 0 \end{aligned} \tag{14.33}$$

To obtain eq. (14.33), we use eq. (14.29), and suppose again (for the sake of simplicity) that $V^*(\boldsymbol{Q}) = V(-\boldsymbol{Q})$, which is true, for example, for center-symmetric crystals.

14.2 Novel interference phenomena: Pendellösung and Borrmann effects

Each excitation point (called also as tie point) on the isoenergetic surface gives rise to a pair of waves, the incident and the diffracted ones, that propagate, respectively, toward the **0**-and ***G***-nodes of the reciprocal lattice with wave vectors $\boldsymbol{k}_i$ and $\boldsymbol{k}_d$ (see Figure 14.5).

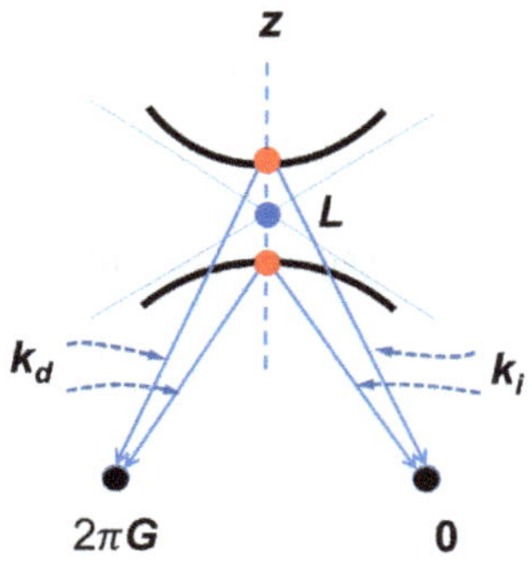

Figure 14.5: Illustration of the excitation (tie) points (colored in red) at the apexes of the hyperbolic branches in the exact **Bragg** position. The **Lorentz** point (blue) is marked by the letter L, while the wave vectors of the incident and diffracted waves by $\boldsymbol{k}_i$ and $\boldsymbol{k}_d$, respectively.

Exactly at the **Bragg** angle Θ_B, the tie points are situated on the straight line parallel to the z-axis and crossing the **Lorentz** point L in which $p = 0$. Respectively, $q = \pm 1$ (see eq. (14.32)), that is, in this case, the excitation points are the apexes of the hyperbola (see Figure 14.5). The upper apex corresponds to $q = 1$, whereas the bottom apex to $q = -1$. Substituting $p = 0$ and $q = 1$ into eq. (14.33), we find the following relationship between the amplitudes ψ_{01} and ψ_{Q1} for the incident and the diffracted waves originated in this excitation point (upper apex of the hyperbola):

$$\psi_{01} = -\psi_{Q1} \tag{14.34}$$

For the bottom apex ($p = 0$, $q = -1$), one respectively obtains:

$$\psi_{02} = \psi_{Q2} \tag{14.35}$$

Further analysis depends on the crystal shape and the specific scattering geometry, which both dictate boundary conditions for the incident and/or the diffracted waves. For example, for flat crystalline plate in the symmetric transmission geometry, which is called the **Laue** geometry (see Figure 14.6), there is no diffracted wave at the entrance crystal surface, situated at $z = 0$, that is,

$$\psi_{Q1} + \psi_{Q2} = 0 \tag{14.36}$$

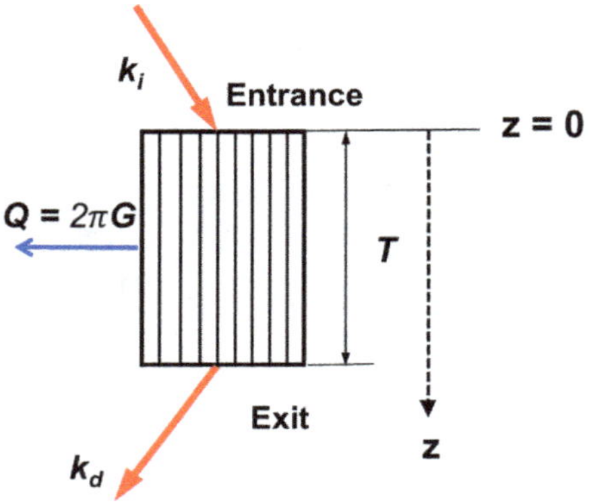

Figure 14.6: Diffraction scheme in the symmetric transmission (**Laue**) scattering geometry. The crystal thickness is indicated by letter T. Wave vectors of the incident and the diffracted waves are marked as $\boldsymbol{k}_i$ and $\boldsymbol{k}_d$, respectively.

Setting the amplitude of the incident wave at $z = 0$ as A_0 yields the second equation:

$$\psi_{01} + \psi_{02} = A_0 \tag{14.37}$$

Solving eqs. (14.34)–(14.37), one obtains:

$$\psi_{01} = \psi_{02} = A_0/2 \tag{14.38}$$

$$\psi_{Q1} = -A_0/2 \tag{14.39}$$

$$\psi_{Q2} = A_0/2 \tag{14.40}$$

Recalling the definitions of wave vectors via parameters p and q, that is, eqs. (14.25) and (14.31), we can express the diffracted ψ_D and the transmitted ψ_T waves at the exact **Bragg** diffraction position as:

$$\begin{aligned}\psi_{\mathrm{D}} &= \psi_{Q1}\exp\left[-i\omega t + i(\boldsymbol{k}_0+\boldsymbol{Q})\boldsymbol{r} + i\frac{\Delta k_0 z}{2}\right] + \psi_{Q2}\exp\left[-i\omega t + i(\boldsymbol{k}_0+\boldsymbol{Q})\boldsymbol{r} - i\frac{\Delta k_0 z}{2}\right]\\ \psi_{\mathrm{T}} &= \psi_{01}\exp\left(-i\omega t + i\boldsymbol{k}_0\boldsymbol{r} + i\frac{\Delta k_0 z}{2}\right) + \psi_{02}\exp\left(-i\omega t + i\boldsymbol{k}_0\boldsymbol{r} - i\frac{\Delta k_0 z}{2}\right)\end{aligned} \tag{14.41}$$

Substituting the amplitudes (14.38)–(14.40) into eq. (14.41) yields:

$$\psi_{\mathrm{D}} = -iA_0\sin\left(\frac{\Delta k_0 z}{2}\right)\exp[-i\omega t + i(\boldsymbol{k}_0+\boldsymbol{Q})\boldsymbol{r}] \tag{14.42}$$

$$\psi_{\mathrm{T}} = A_0\cos\left(\frac{\Delta k_0 z}{2}\right)\exp(-i\omega t + i\boldsymbol{k}_0\boldsymbol{r}) \tag{14.43}$$

Correspondingly, the intensities of the diffracted and the transmitted waves can be expressed as follows:

$$|\psi_{\mathrm{D}}|^2 = A_0^2\sin^2\left(\frac{\Delta k_0 z}{2}\right) \tag{14.44}$$

$$|\psi_{\mathrm{T}}|^2 = A_0^2\cos^2\left(\frac{\Delta k_0 z}{2}\right) \tag{14.45}$$

We see that both intensities oscillate as the waves penetrate the crystal depth z. The respective functions (14.44) and (14.45) are in antiphase with respect to each other. Neglecting absorption, we find that the total intensity of the two waves, described by eqs. (14.44) and (14.45), is constant, that is, $|\psi_{\mathrm{D}}|^2 + |\psi_{\mathrm{T}}|^2 = A_0^2$. In other words, the diffraction intensity reaches its maximum, A_0^2, when the intensity of the transmitted wave equals zero, and vice versa. Such a behavior is called the **Pendellösung** effect. Neglecting absorption, we can state that the complete repumping of the incident beam intensity into the diffraction intensity and back, takes place at the characteristic depth $z = \tau$:

$$\tau = \frac{2\pi}{\Delta k_0} \tag{14.46}$$

This fundamental quantity is called the extinction length, which is of primary importance to the whole field of dynamical diffraction. We will systematically use the concept of extinction length in this Chapter and in the next Chapter 15.

Another fascinating phenomenon arises when one considers the energy flow within a crystal under diffraction conditions. Since the isoenergetic surface is representative for quantum states with equal energies, the energy flow for waves originated in each excitation point is directed along the normal to the isoenergetic surface at that point (see Figure 14.7). In the exact **Bragg** position Θ_{B}, which corresponds to the excited hyperbolic apexes, the energy flow is evidently directed along the diffractive planes (see Figure 14.8).

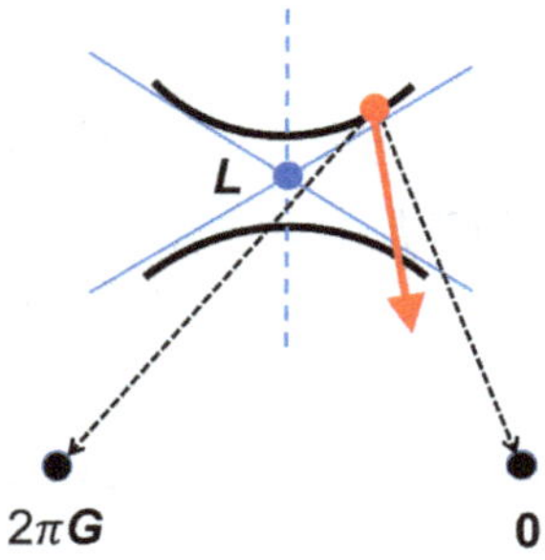

Figure 14.7: At any angular position of the crystal, the energy flow (red solid arrow) is directed along the normal to the isoenergetic surface at the respective excitation point (colored in red). The **Lorentz** point is marked by the letter *L*.

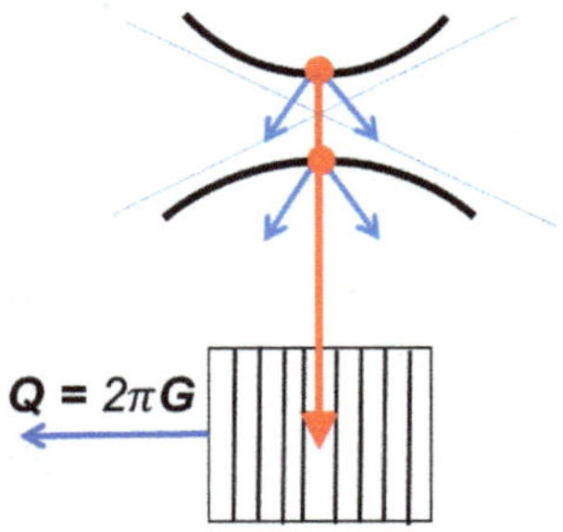

Figure 14.8: Scheme of the symmetric **Laue** scattering geometry. In the exact **Bragg** position, which corresponds to the excitation of the hyperbolic apexes (red dots in the upper panel), the energy flow (red solid arrow) is along the diffractive planes (shown in the bottom panel).

Let us consider the structure of the related wavefield in more detail. In fact, in the exact **Bragg** position, the quantum states of the incident and the diffracted waves are identical, since the corresponding wave vectors are connected by a certain vector of the reciprocal lattice (see eq. (13.18)). It means that we can take the sum of the incident and the diffracted waves for constructing the resulting wavefield responsible for factual energy flow. For the upper apex of the hyperbola ($p = 0$, $q = 1$), this wavefield is:

$$\psi_1 = \psi_{01}\exp\left(-i\omega t + i\boldsymbol{k}_0\boldsymbol{r} + i\frac{\Delta k_0 z}{2}\right) + \psi_{Q1}\exp\left[-i\omega t + i(\boldsymbol{k}_0 + \boldsymbol{Q})\boldsymbol{r} + i\frac{\Delta k_0 z}{2}\right] \tag{14.47}$$

By the aid of eqs. (14.38)–(14.40), it transforms into:

$$\begin{aligned}\psi_1 &= \frac{A_0}{2}\exp\left(-i\omega t + i\frac{\Delta k_0 z}{2}\right)\{\exp(i\boldsymbol{k}_0\boldsymbol{r}) - \exp[i(\boldsymbol{k}_0 + \boldsymbol{Q})\boldsymbol{r}]\} = \\ &= \frac{A_0}{2}\exp\left(-i\omega t + i\frac{\Delta k_0 z}{2} + i\boldsymbol{k}_0\boldsymbol{r} + i\frac{\boldsymbol{Q}}{2}\boldsymbol{r}\right)\left[\exp\left(-i\frac{\boldsymbol{Q}}{2}\boldsymbol{r}\right) - \exp\left(i\frac{\boldsymbol{Q}}{2}\boldsymbol{r}\right)\right] = \\ &= -A_0 i\exp\left(-i\omega t + i\frac{\Delta k_0 z}{2} + i\boldsymbol{k}_0\boldsymbol{r} + i\frac{\boldsymbol{Q}}{2}\boldsymbol{r}\right)\sin\left(\frac{\boldsymbol{Q}\boldsymbol{r}}{2}\right)\end{aligned} \tag{14.48}$$

The intensity of this wavefield is:

$$|\psi_1|^2 = A_0^2\sin^2\left(\frac{\boldsymbol{Q}\boldsymbol{r}}{2}\right) = A_0^2\sin^2(\pi\boldsymbol{G}\boldsymbol{r}) \tag{14.49}$$

Applying the same procedure to the bottom apex of the hyperbola ($p = 0$, $q = -1$) gives the second wavefield:

$$\psi_2 = \frac{A_0}{2}\exp\left(-i\omega t - i\frac{\Delta k_0 z}{2}\right)\{\exp(i\boldsymbol{k}_0\boldsymbol{r}) + \exp[i(\boldsymbol{k}_0 + \boldsymbol{Q})\boldsymbol{r}]\} = \tag{14.50}$$

$$= A_0\ \exp\left(-i\omega t - i\frac{\Delta k_0 z}{2} + i\,\boldsymbol{k}_0\,\boldsymbol{r} + i\frac{\boldsymbol{Q}}{2}\boldsymbol{r}\right)\cos\left(\frac{\boldsymbol{Q}\boldsymbol{r}}{2}\right)$$

with the intensity

$$|\psi_2|^2 = A_0{}^2\cos^2\left(\frac{\boldsymbol{Q}\boldsymbol{r}}{2}\right) = A_0{}^2\cos^2(\pi\boldsymbol{G}\boldsymbol{r}) \tag{14.51}$$

Analysis of eqs. (14.49) and (14.51) allows us to conclude that the intensities of wavefields ψ_1 and ψ_2 reveal spatial periodicity along the vector of the reciprocal lattice $\boldsymbol{G}$, that is, in the x-direction (see Figure 14.4). The period x_p is:

$$x_p = \frac{1}{|\boldsymbol{G}|} = d \tag{14.52}$$

that is, it equals the d-spacing between parallel crystal planes, oriented normally to the respective vector of the reciprocal lattice $\boldsymbol{G}$. Such wavefield periodicity is the physical basis of the lattice image contrast, which allows us to visualize the atomic columns in the transmission electron microscopy (TEM).

Actually, eqs. (14.49) and (14.51) describe standing waves, which are spatially arranged in antiphase with respect to each other (see Figure 14.9). In fact, the wavefield ψ_2 has the maximum intensity exactly at the atomic plane positions (i.e., at $x = md$, where $m = 0, 1, 2 \ldots$, see eq. (14.51) and Figure 14.9a), whereas the wavefield ψ_1 – exactly between the planes (i.e., at $x = (^1/_2 + md)$, see eq. (14.49) and Figure 14.9b). Correspondingly, the wavefield ψ_2 will strongly interact with atoms located in these planes, while the wavefield ψ_1 – very feebly. Weak interaction also implies weak absorption of the wavefield ψ_1, which is the origin of its anomalous transmission through relatively thick crystals, called the **Borrmann** effect in X-ray (and neutron) diffraction. Anomalous transmission, for example, is employed in X-ray interferometers (see, Section 8.4), utilizing rather thick perfect crystals.

On the other hand, strong interaction of the wavefield ψ_2 with a crystal, including its strong absorption, causes enhanced emission of secondary radiations, that is, X-ray fluorescence or **Auger** electrons. Registration of secondary radiations in dynamical diffraction experiments is the basis of the so-called X-ray standing wave technique, which allows us to study lattice imperfections in ultrathin layers beneath the crystal surface.

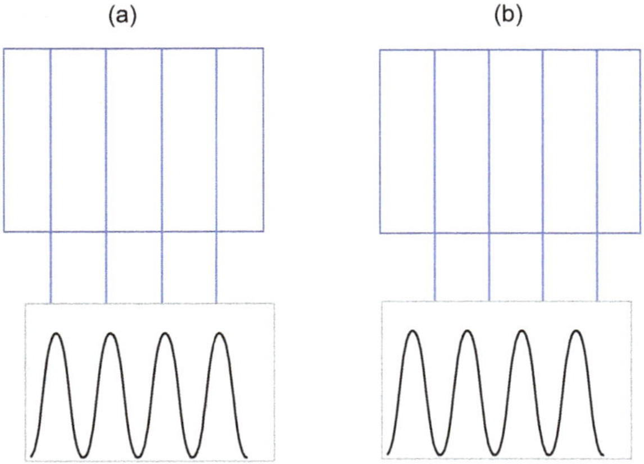

Figure 14.9: Illustration of the X-ray standing waves and the related **Borrmann** effect: (a) – wavefield ψ_2, eq. (14.51); (b) – wavefield ψ_1, eq. (14.49).

14.3 Diffraction profile in the Laue scattering geometry: extinction length and thickness fringes

For calculating the diffraction profile (i.e., angular dependence of diffraction intensity near exact **Bragg** position Θ_B) in dynamical diffraction theory, for every angular position of a crystal, we have to coherently summate the two diffraction waves originated in the excitation (tie) points located on the two branches of the isoenergetic dispersion surface (see Figure 14.10). To find these tie points for an arbitrary angular deviation $\Delta\Theta$ from **Bragg** position Θ_B, we take help from the conservation law for the projection of the wave vector along the entrance crystal surface. In other words, when the quanta are entering the crystal, the wave vector change occurs only normally to the surface. If so, we find the active tie points at the intersection between the normal mentioned and the branches of dispersion surface (see Figure 14.10). For a particular angular deviation, $\Delta\Theta = (\Theta - \Theta_B)$, the normal used is shifted from the **Lorentz** point L along the x-axis by an amount, $\delta k_x = p\Delta k_0/(2\tan\Theta_B)$ (see eq. (14.31)). As we will show below, the parameter p is indeed proportional to the angular deviation $\Delta\Theta = (\Theta - \Theta_B)$.

Starting again from the system of eqs. (14.33) and neglecting absorption, we find that

$$\psi_Q = -(q+p)\psi_0 \tag{14.53}$$

At the same time, the secular equation (14.32) yields:

$$q = \pm\sqrt{1+p^2} \tag{14.54}$$

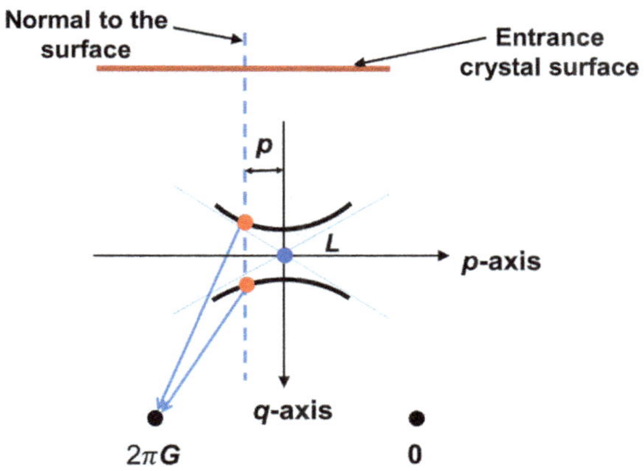

Figure 14.10: Finding the excitation points (red dots) for arbitrary angular deviation from the exact **Bragg** position in the **Laue** scattering geometry. The **Lorentz** point is marked by the letter *L*.

We remind that the upper and lower branches of the dispersion surface in Figure 14.10 correspond, respectively, to the signs [+] and [−] in eq. (14.54). Substituting eq. (14.54) into eq. (14.53) provides the following relationships between the amplitudes of the diffracted and transmitted waves for, respectively, the upper (1) and lower (2) excitation points:

$$\psi_{Q1} = -\left(p + \sqrt{1+p^2}\right)\psi_{01} \tag{14.55}$$

and

$$\psi_{Q2} = \left(\sqrt{1+p^2} - 1\right)\psi_{02} \tag{14.56}$$

Using the boundary conditions for the **Laue** scattering geometry, that is, eqs. (14.36) and (14.37), yields:

$$\begin{aligned} \psi_{Q1} &= -\frac{A_0}{2\sqrt{1+p^2}} \\ \psi_{Q2} &= \frac{A_0}{2\sqrt{1+p^2}} \end{aligned} \tag{14.57}$$

The next step is to summate the two diffracted waves with the amplitudes ψ_{Q1} and ψ_{Q2}, taking account of the phase differences between them. Therefore, the diffraction field ψ_D is expressed as

$$\psi_{\mathrm{D}} = \frac{A_0}{2\sqrt{1+p^2}} \exp[i(\boldsymbol{k}_0 + \boldsymbol{Q})\boldsymbol{r} - i\omega t] \cdot \exp(i\delta k_x x) \cdot \left[\exp\left(\frac{iz\Delta k_0}{2}\sqrt{1+p^2}\right) - \exp\left(-\frac{iz\Delta k_0}{2}\sqrt{1+p^2}\right)\right] \tag{14.58}$$

At the exit surface of the crystal $z = T$ (where T is the thickness of crystalline plate), the diffraction intensity (neglecting absorption effects) is

$$|\psi_{\mathrm{D}}|^2 = \frac{A_0{}^2}{1+p^2} \sin^2\left[\left(\frac{\Delta k_0 T}{2}\right)\sqrt{1+p^2}\right] \tag{14.59}$$

Recalling the relationship (14.46) between the gap Δk_0 and the extinction length τ allows us to rewrite eq. (14.59) in the following form:

$$|\psi_{\mathrm{D}}|^2 = \frac{A_0{}^2}{1+p^2} \sin^2\left[\left(\frac{\pi T}{\tau}\right)\sqrt{1+p^2}\right] \tag{14.60}$$

We see that the diffraction intensity (eq. (14.60)) oscillates as a function of parameter p, the latter being linearly proportional to the angular deviation $\Delta\Theta$ from the exact **Bragg** position Θ_{B}. Such oscillating behavior is shown in Figure 14.11. At large thicknesses $\left(T \gg \frac{\tau}{\pi}\right)$ or/and far away from diffraction maximum ($p \gg 1$), the oscillations are very fast. The averaged (over the oscillations mentioned) diffraction intensity

$$< |\psi_{\mathrm{D}}|^2 > = \frac{(A_0{}^2/2)}{1+p^2} \tag{14.61}$$

is described by **Lorentzian (Cauchy)** function (see Figure 14.11), having full width at half maximum (FWHM) $\Gamma_{\mathrm{L}} = 2$ on the p-scale.

To complete this analysis, we need to derive the exact interrelation between the parameter p and angular deviation $\Delta\Theta$. By the aid of Figure 14.12, we find that

$$\sin\Theta_{\mathrm{B}} = \frac{Q}{2k_0}$$
$$\sin(\Theta_{\mathrm{B}} - \Delta\Theta) \approx \frac{\left(\frac{Q}{2}\right) - \delta k_x}{k_0} \tag{14.62}$$

and then

$$\Delta\Theta = \frac{\delta k_x}{k_0 \cos\Theta_{\mathrm{B}}} = p\frac{\Delta k_0}{Q} = p\frac{d}{\tau} \tag{14.63}$$

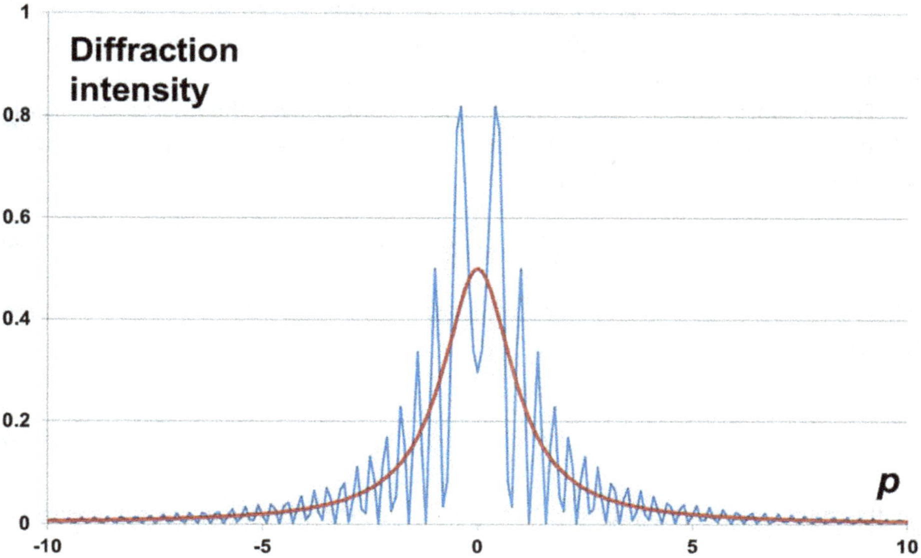

Figure 14.11: Diffraction intensity profile with thickness fringes in the **Laue** scattering geometry. The averaged intensity (red curve) is described by the **Lorentzian** function.

Using eq. (14.63), one obtains the FWHM Γ_L of the **Lorentzian** function (14.61) on the angular Θ-scale (i.e., substituting $p = 2$ into eq. (14.63)):

$$\Gamma_L = 2\frac{\Delta k_0}{Q} = 2\frac{d}{\tau} \tag{14.64}$$

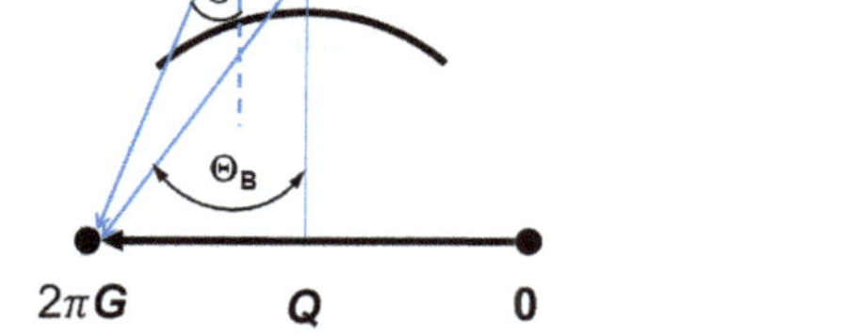

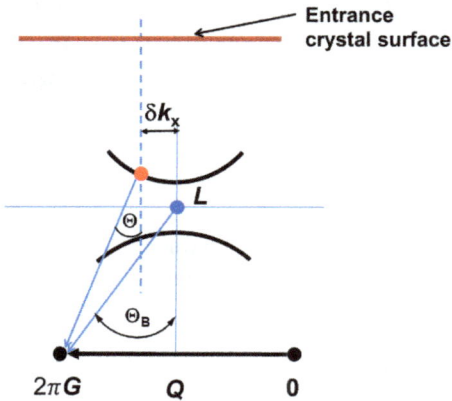

Figure 14.12: Illustration of deriving eq. (14.63). The **Lorentz** point is marked by the letter *L*, while the excitation (tie) point – by red dot.

It is worth to stress this fundamental result, that is, the angular width of the diffraction profile in dynamical diffraction is defined by the ratio of the d-spacing used in the specific measurement, and the extinction length τ.

Furthermore, one can find the oscillation periods over the diffraction profile (14.60). Let us consider, for example, the periodicity of zero-intensity points, the latter satisfying the following condition:

$$\left(\frac{\pi T}{\tau}\right)\sqrt{1+p^2} = m\pi \tag{14.65}$$

where m is integer number. It follows from eq. (14.65), that zero intensity is expected for real p -values:

$$p = \sqrt{\left(\frac{m}{T/\tau}\right)^2 - 1} \tag{14.66}$$

On the tails of the intensity distribution, that is, for $m \gg T/\tau$, the oscillation period is

$$\Delta p = p_{m+1} - p_m = \frac{\tau}{T} \tag{14.67}$$

By means of eq. (14.63), we can convert the parameter Δp in eq. (14.67) to angular interval $\Delta\Theta_{osc}$

$$\Delta\Theta_{osc} = \frac{d}{T} \tag{14.68}$$

which is inversely proportional to the plate thickness T. For this reason, such oscillations are often called as thickness fringes.

The last question in this section that we intend to address is the integrated diffraction intensity I_D, that is, the area under intensity distribution (14.60). Analytic expression for I_D can rather easily be obtained for a thick crystal, in which he following condition is fulfilled:

$$\frac{\pi T}{\tau} \gg 1 \tag{14.69}$$

In this case, oscillations are very frequent, and we can use the averaged **Lorentzian** function (14.61) for our analysis. Considering eq. (14.63), the integrated (over angular deviations $\Delta\Theta$) diffraction intensity equals:

$$I_D = \frac{A_0^2}{2}\int_{-\infty}^{\infty}\frac{dp}{1+p^2} = \pi\frac{A_0^2}{2}\frac{\Delta k_0}{Q} = \pi\frac{A_0^2}{2}\frac{d}{\tau} \tag{14.70}$$

Correspondingly, the integrated reflectivity in the **Laue** case $(\mathcal{R}_\mathrm{L})_{\mathrm{in}} = \dfrac{I_D}{A_0{}^2}$ is:

$$(\mathcal{R}_L)_{in} = \frac{\pi d}{2\tau} = \frac{d}{4}\Delta k_0 \tag{14.71}$$

We stress that the integrated diffraction intensity (reflectivity) in thick crystal does not depend on the crystal thickness T. At the same time, it is inversely proportional to the extinction length τ, that is, linearly increasing with the strength of interaction between the radiation and the crystal lattice. We remind that the strength of interaction expresses itself via the evolving gap value $\Delta k_0 = \frac{2\pi}{\tau}$ (see eq. (14.46)). The results obtained here are perhaps most fundamental in dynamical diffraction theory.

14.4 Diffraction profile in the Bragg scattering geometry: total reflection near the Bragg angle

Let us consider dynamical diffraction in a thick, plate-like single crystal, in which reflecting atomic planes are parallel to the entrance surface of the crystal. The same surface serves as the exit one for the diffracted beam. These two conditions define the so-called symmetric **Bragg** scattering geometry (see Figure 14.13). The definition of a thick crystal in the **Bragg** scattering geometry will be given at the end of this section.

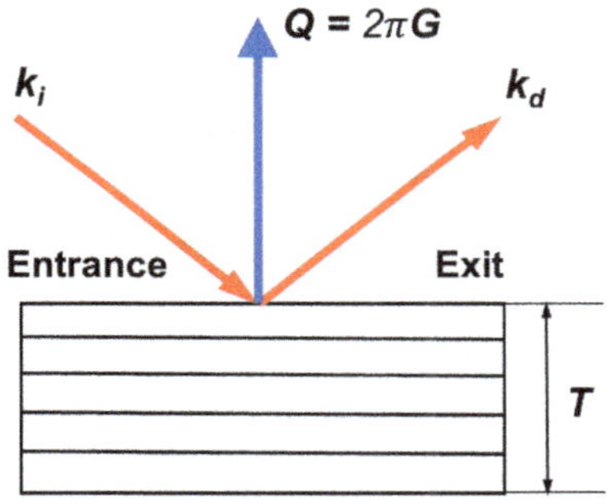

Figure 14.13: Scheme of the symmetric **Bragg** scattering geometry. The crystal thickness is indicated by the letter T. Wave vectors of the incident and the diffracted waves are marked as $\boldsymbol{k}_i$ and $\boldsymbol{k}_d$, respectively.

It is evident that within a crystal, we can use the same equations for dynamical diffraction as before, that is, eqs. (14.32) and (14.33), since the crystal lattice is the same. Drastic change occurs in the orientation of the coordinate system (x, z) with respect to crystal surface (see Figure 14.14) and, hence, in the boundary conditions.

In fact, in both the scattering geometries (**Laue** and **Bragg**), vectors $\delta \boldsymbol{k}_x$ and $\delta \boldsymbol{k}_z$ are, respectively, parallel and perpendicular to the vector of the reciprocal lattice $\boldsymbol{G}$. At the same time, their orientation relations with respect to the entrance crystal surface are very different in the **Laue** or **Bragg** case (see Figure 14.14). Specifically, in the **Laue** case vector $\delta \boldsymbol{k}_z$ is perpendicular to the entrance crystal surface (horizontal red solid line in Figure 14.14a), whereas in the **Bragg** case, this is true for vector $\delta \boldsymbol{k}_x$, since vector $\delta \boldsymbol{k}_z$ is parallel to the surface (vertical red solid line in Figure 14.14b). This is a

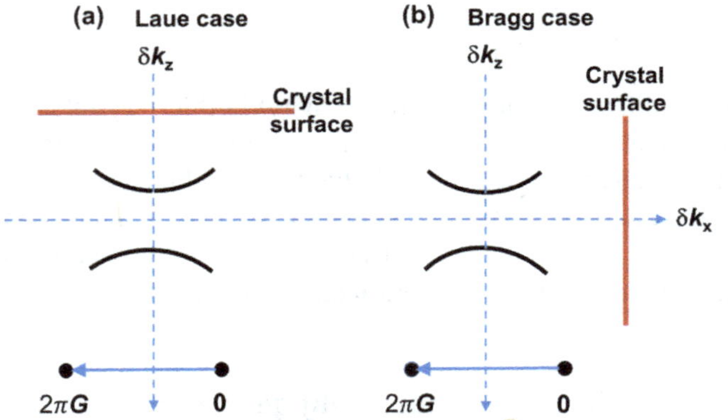

Figure 14.14: Orientation of the two branch isoenergetic dispersion surface (black solid curves) within a crystal and the coordinate axes (x, z) with respect to the entrance crystal surface (red solid line): (a) symmetric **Laue** case; (b) symmetric **Bragg** case.

very important dissimilarity since we are generally interested in the changes in wave vectors proceeding normally to the crystal surface. As was already mentioned, projection of a wave vector along the surface does not change at all when specular scattering is considered. Therefore, to mathematically analyze the diffraction problem in the **Bragg** case, we must simply mutually switch between vectors $\delta\mathbf{k}_x$ and $\delta\mathbf{k}_z$ in the respective equations. This sentence is graphically illustrated by means of Figure 14.15, which should be compared with Figure 14.10 for the **Laue** scattering geometry. In practical terms, we must now solve eq. (14.32) with respect to parameter p, rather than parameter q:

$$p = \pm\sqrt{q^2 - 1} \tag{14.72}$$

The latter result radically changes the entire diffraction picture.

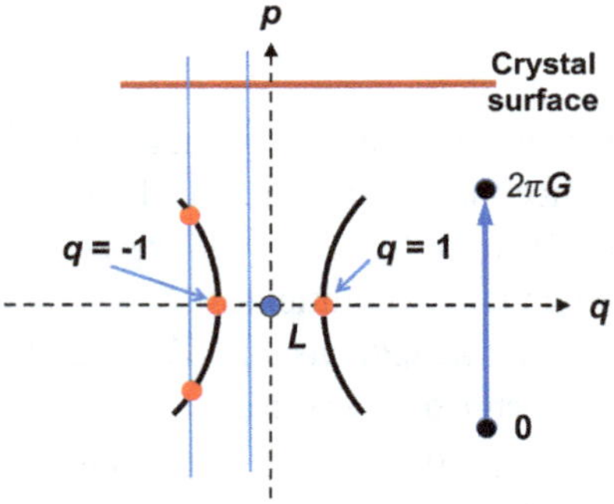

Figure 14.15: Illustration of deriving eq. (14.72) for the **Bragg** scattering geometry. The trace of the entrance crystal surface is indicated by red horizontal line. The **Lorentz** point is marked by the letter L, while the excitation points – by red dots.

First, it follows from eq. (14.72) that at $|q|<1$, parameter p becomes an imaginary number. Physically (see Figure 14.15), it means that at $|q|<1$, the normal to the crystal surface does not intersect the branches of the isoenergetic dispersion surface and, correspondingly, there are no active tie points that give rise to the diffracted waves. We can say that the respective quantum states (diffracted waves) do not exist within a thick crystal. In some sense, the situation is analogous to the formation of forbidden zones for electrons in crystals (see text related to eq. (14.22) and Figure 14.2). More exactly, it means that for a certain range of incident angles near the **Bragg** angle Θ_B, the quantum beam is completely expelled from the crystal and, hence, the wave reflectivity $\mathcal{R}_B$, that is, the ratio between the intensities of the diffracted and the incident waves equals one (neglecting absorption).

Let us check this important conclusion analytically. The solution of eqs. (14.33) and (14.72) is

$$\mathcal{R}_B = \frac{|\psi_Q|^2}{|\psi_0|^2} = |q \pm \sqrt{q^2-1}|^2 \tag{14.73}$$

For $|q| \le 1$, function (14.73) transforms into

$$\mathcal{R}_B = \frac{|\psi_Q|^2}{|\psi_0|^2} = |q \pm i\sqrt{1-q^2}|^2 = 1 \tag{14.74}$$

that indeed means the total reflection of incident wave, as was mentioned above.

For $|q| > 1$, that is, beyond the interval of total reflection, the normal to the crystal surface already crosses one of the branches of the dispersion surface (left or right side from the **Lorentz** point L), exciting a pair of tie points located, in this case, at the same branch (see Figure 14.15). Only one point from a pair should be considered at any angle of incidence, when constructing the diffraction field, since the second point will provide a physically meaningless solution. The latter means an exponential growth of diffraction intensity with depth when the solution is analytically continued into the total reflection region. On the contrary, true solution being analytically continued into the total reflection region provides exponential decay of the diffraction intensity whenever the diffracted wave penetrates the crystal depth.

In practical terms, when calculating the diffraction intensity, in eq. (14.73), we need to take the sign [–] for $q > 1$ and the sign [+] for $q < -1$. Such a choice provides the appropriate **Lorentzian**-like asymptotes for the reflectivity profile $\frac{|\psi_Q|^2}{|\psi_0|^2}$ at $|q| \gg 1$:

$$\mathcal{R}_B = \frac{|\psi_Q|^2}{|\psi_0|^2} = |q \pm \sqrt{q^2-1}|^2 \approx \frac{1}{4q^2} \tag{14.75}$$

Considering eqs. (14.74) and (14.75), we can plot the diffraction profile (the reflectivity profile) in the **Bragg** scattering geometry (see Figure 14.16).

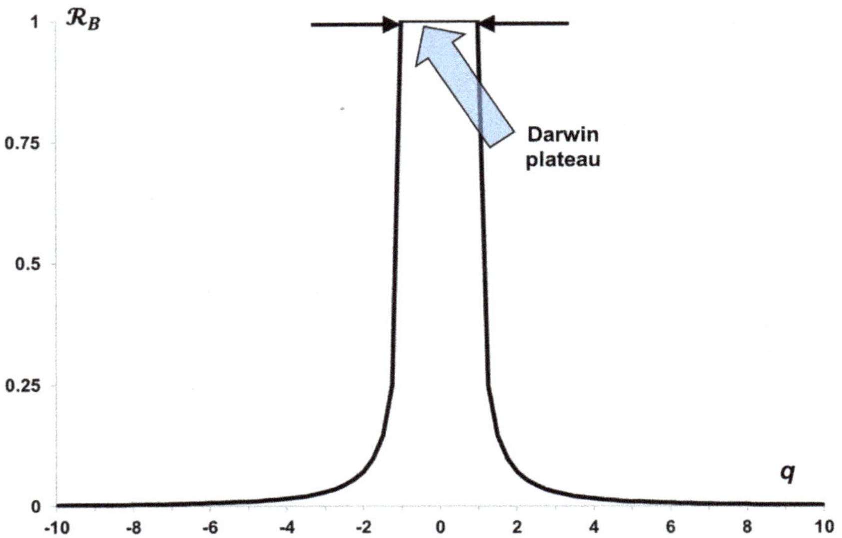

Figure 14.16: Diffraction (reflectivity) profile with **Darwin** plateau in the symmetric **Bragg** scattering geometry (no absorption).

Let us calculate an angular range of total reflection (so-called **Darwin** "plateau," i.e., the flat part of diffraction profile), which is confined between $-1 \leq q \leq 1$ in Figure 14.16. By the aid of Figure 14.17, we find that

$$k_0(\cos\Theta_B - \cos\Theta) \approx k_0[\cos\Theta_B - \cos(\Theta_B + \Delta\Theta] \approx k_0 \sin\Theta_B\Delta\Theta \approx \frac{\Delta k_0}{2} \tag{14.76}$$

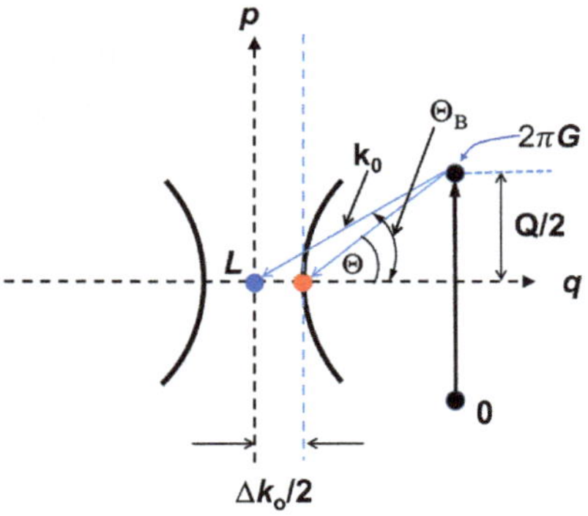

Figure 14.17: Illustration of deriving eq. (14.77) for the **Bragg** scattering geometry. The **Lorentz** point is marked by the letter L, while the specific excitation point – by red dot.

It follows from Figure 14.17 and eq. (14.76) that half width of the **Darwin** plateau $\Delta\Theta$ is:

$$\Delta\Theta = \frac{\Delta k_0}{2k_0 \sin\Theta_B} = \frac{\Delta k_0}{Q} = \frac{d}{\tau} \tag{14.77}$$

Correspondingly, the full width of the **Darwin** plateau, Γ_D:

$$\Gamma_D = 2\Delta\Theta = 2\frac{d}{\tau} \tag{14.78}$$

is again inversely proportional to the extinction length, and exactly equal to the FWHM of the diffraction profile in the **Laue** case (see eq. (14.64)). Numerical estimations of the extinction length and the FWHM of the diffraction profiles in the case of X-ray diffraction will be given in the next Chapter 15.

By using eqs. (14.74) and (14.75), we can calculate integrated reflectivity in the **Bragg** case $(\mathcal{R}_B)_{in}$, that is, the area under the diffraction profile shown in Figure 14.16:

$$(\mathcal{R}_B)_{in} = \frac{8}{3}\frac{d}{\tau} \tag{14.79}$$

This value is 1.7 times larger than $(\mathcal{R}_L)_{in}$ in the **Laue** case (see eq. (14.71)). In fact, the ratio between these two reflectivity coefficients, described by eqs. (14.71) and (14.79), is:

$$\frac{(\mathcal{R}_B)_{in}}{(\mathcal{R}_L)_{in}} = \frac{16}{3\pi} \tag{14.80}$$

An important additional question that we would like to address here is the penetration depth of wavefields in the **Bragg** scattering geometry in the exact **Bragg** position. In other words, at which depth is the diffracted wave really expelled from the crystal toward total reflection and, correspondingly, the transmitted wave is heavily attenuated? We recall that in this geometry, the projection δk_x of vector $\boldsymbol{k} - \boldsymbol{k}_0$ (see eq. (14.25)) is perpendicular to the crystal surface, that is, it is directed into the crystal depth (see Figure 14.14b). In the exact **Bragg** position $q = 0$ and $p = \pm i$ (see eq. (14.72)). By choosing the sign [+] to get the depth-decaying wavefield for depths $z > 0$ and using eq. (14.31), one finds the depth-depending part of the diffracted and transmitted waves in the form:

$$\psi_{D,T} \sim \exp(i\delta k_x z) = \exp\left(-\frac{\pi}{\tau \tan\Theta_B} z\right) \tag{14.81}$$

We see that the characteristic depth Λ_e for the exponential wave decay in the **Bragg** scattering geometry equals

$$\Lambda_e = \frac{\tau}{\pi}\tan\Theta_B \tag{14.82}$$

which can be substantially smaller than the extinction length τ (for moderate **Bragg** angles, $\Theta_B \leq 45°$).

Now, by the aid of parameter Λ_e, one can understand the exact meaning of thick crystal in the **Bragg** scattering geometry. Physically, the crystal should be thick enough to prevent the appearance of some X-ray intensity at the bottom surface of our crystalline plate. Note that only under this condition, we can expect total reflection through the entrance surface near the **Bragg** angle Θ_B in a non-absorbing crystal. To reach the target, the crystal thickness T should be much larger than the characteristic depth (14.82), that is, $T \gg \Lambda_e$. It means that:

$$\frac{\pi T}{\tau \tan \Theta_B} \gg 1 \tag{14.83}$$

We stress that for **Bragg** angles $\Theta_B < 45°$, eq. (14.83) provides smaller crystal thicknesses than those that are characteristic for thick crystal definition in the **Laue** case (compare eqs. (14.69) and (14.83)).

Chapter 15
Specific features of dynamical X-ray diffraction

As we comprehensively discussed in Chapter 5, electromagnetic waves, including X-rays, are characterized by different polarization states. This key issue must be incorporated into the dynamical theory of X-ray diffraction. To do this, we use **Maxwell** equations that describe the propagation of electromagnetic waves through a medium. These time-dependent and coordinate-dependent differential equations, already introduced in Chapter 1, interconnect the electric field ($\boldsymbol{E}$, $\boldsymbol{D}$) components and the magnetic field ($\boldsymbol{H}$, $\boldsymbol{B}$) components of the propagating electromagnetic waves:

$$\nabla \mathrm{x} \boldsymbol{E} = -\frac{\partial \boldsymbol{B}}{\partial t} \tag{15.1}$$

$$\nabla \cdot \boldsymbol{B} = 0 \tag{15.2}$$

$$\nabla \mathrm{x} \boldsymbol{H} = \boldsymbol{J} + \frac{\partial \boldsymbol{D}}{\partial t} \tag{15.3}$$

$$\nabla \cdot \boldsymbol{D} = \rho_f \tag{15.4}$$

Here, $\boldsymbol{E}$ is again the electric field, $\boldsymbol{D}$ is the electric displacement field, $\boldsymbol{H}$ is the magnetic field, $\boldsymbol{B}$ is the magnetic induction, $\boldsymbol{J}$ is the displacement current density, and ρ_f is the volume density of free electric charges. We remind the reader that the symbol (∇x) represents the differential vector operator called a rotor (rot) or curl. For example,

$$\nabla \mathrm{x} \boldsymbol{E} = \left(\frac{\partial E_z}{\partial y} - \frac{\partial E_y}{\partial z}\right)\boldsymbol{e}_x + \left(\frac{\partial E_x}{\partial z} - \frac{\partial E_z}{\partial x}\right)\boldsymbol{e}_y + \left(\frac{\partial E_y}{\partial x} - \frac{\partial E_x}{\partial y}\right)\boldsymbol{e}_z \tag{15.5}$$

where $\boldsymbol{e}_x$, $\boldsymbol{e}_y$, $\boldsymbol{e}_z$ are the unit vectors along the respective axes (x, y, x) of the **Descartes** coordinate system. In turn, the symbol ($\nabla\cdot$) represents the differential scalar operator, called divergence (div). For example,

$$\nabla \cdot \boldsymbol{D} = \frac{\partial D_x}{\partial x} + \frac{\partial D_y}{\partial y} + \frac{\partial D_z}{\partial z} \tag{15.6}$$

We also need to define an additional differential vector operator ($\nabla\Phi$), which is called a gradient operator (grad); applied to scalar field Φ, it produces a vector field:

$$\nabla \Phi = \frac{\partial \Phi}{\partial x}\boldsymbol{e}_x + \frac{\partial \Phi}{\partial y}\boldsymbol{e}_y + \frac{\partial \Phi}{\partial z}\boldsymbol{e}_z \tag{15.7}$$

Within an isotropic medium $\boldsymbol{B} = \mu_0 \mu_{\mathrm{m}} \boldsymbol{H}$, where μ_0 is the magnetic permeability of a vacuum and μ_{m} is the magnetic constant of a medium. In this chapter, we apply our

https://doi.org/10.1515/9783111140100-016

analyses only to non-magnetic materials, in which the parameter μ_m is very close to 1 and does not exhibit any change during X-ray propagation. Therefore, one can set:

$$\boldsymbol{B} = \mu_0 \boldsymbol{H} \tag{15.8}$$

As we learned in Chapter 6, the electric displacement field $\boldsymbol{D}$ is related to the electric field $\boldsymbol{E}$ via dielectric polarizability χ of a medium:

$$\boldsymbol{D} = \varepsilon_0(1+\chi)\boldsymbol{E} = \varepsilon_0\varepsilon_m\boldsymbol{E} \tag{15.9}$$

where ε_0 is the fundamental constant defining dielectric permittivity of vacuum and ε_m is the dielectric constant of a material. Note that in the general case, dielectric polarizability χ is described by a second-rank tensor as dielectric constant $\varepsilon_m = 1+\chi$ (see Chapter 4). As we will see below, however, the polarizability effects in the X-ray domain are very weak, $|\chi| \ll 1$. Consequently, we consider the X-ray polarizability as an isotropic quantity that is reflected in the scalar shape of eq. (15.9).

Furthermore, X-rays are electromagnetic waves having very high frequencies (in the range of ExaHertz, i.e., 10^{18} Hz). At these frequencies, the conductivity of materials is negligible and hence, we can set $\boldsymbol{J} = 0$ in eq. (15.3). We also assume that there are no free electric charges within a medium, i.e., $\rho_f = 0$ in eq. (15.4). After all, we obtain the following system of equations:

$$\nabla \mathrm{x}\boldsymbol{E} = \frac{1}{\varepsilon_0}\nabla \mathrm{x}\left(\frac{\boldsymbol{D}}{1+\chi}\right) \approx \frac{1}{\varepsilon_0}\nabla \mathrm{x}[\boldsymbol{D}(1-\chi)] = -\mu_0\frac{\partial \boldsymbol{H}}{\partial t} \tag{15.10}$$

$$\nabla \cdot \boldsymbol{H} = 0 \tag{15.11}$$

$$\nabla \mathrm{x}\boldsymbol{H} = \frac{\partial \boldsymbol{D}}{\partial t} \tag{15.12}$$

$$\nabla \cdot \boldsymbol{D} = 0 \tag{15.13}$$

Let us apply the rotor (curl) operator (∇x) to both sides of eq. (15.10). Using eq. (15.12) one finds:

$$\frac{1}{c^2}\frac{\partial^2 \boldsymbol{D}}{\partial t^2} = -\nabla \mathrm{x}\nabla \mathrm{x}\boldsymbol{D} + \nabla \mathrm{x}\nabla \mathrm{x}(\chi\boldsymbol{D}) \tag{15.14}$$

where we utilized the definition of the speed of light in vacuum, $c = \frac{1}{\sqrt{\mu_0\varepsilon_0}}$ (see eq. (1.17)). Applying the well-known relationship in vector algebra,

$$\boldsymbol{A}\mathrm{x}\boldsymbol{B}\mathrm{x}\boldsymbol{C} = \boldsymbol{B}(\boldsymbol{A}\cdot\boldsymbol{C}) - (\boldsymbol{A}\cdot\boldsymbol{B})\boldsymbol{C} \tag{15.15}$$

to the first term on the right side of eq. (15.14) yields

$$-\nabla \mathrm{x}\nabla \mathrm{x}\boldsymbol{D} = -\nabla(\nabla\cdot\boldsymbol{D}) + (\nabla\cdot\nabla)\boldsymbol{D} \tag{15.16}$$

Note that the first term on the right-hand side of eq. (15.16) equals zero (see eq. (15.13)). The second term in eq. (15.16) is simply the **Laplace** operator, $(\nabla \cdot \nabla)\boldsymbol{D} = \left(\frac{\partial^2}{\partial x^2} + \frac{\partial^2}{\partial y^2} + \frac{\partial^2}{\partial z^2}\right)\boldsymbol{D} = \nabla^2 \boldsymbol{D}$, already introduced in Chapters 1 and 14 (see e.g., eq. (14.2)). Finally, eq. (15.14) transforms into:

$$\nabla^2 \boldsymbol{D} - \frac{1}{c^2}\frac{\partial^2 \boldsymbol{D}}{\partial t^2} = -\ \nabla \mathrm{x} \nabla \mathrm{x}(\chi \boldsymbol{D}) \tag{15.17}$$

At the beginning, let us find the solution of eq. (15.17) in vacuum, that is, for $\chi = 0$. Substituting a plane wave

$$\boldsymbol{D} = \boldsymbol{D}_0 \exp[i(\boldsymbol{kr} - \omega t)] \tag{15.18}$$

into eq. (15.17), together with $\chi = 0$, yields the well-known dispersion law for electromagnetic waves in vacuum (see eqs. (1.17) and (1.19)):

$$\omega^2 = c^2 \boldsymbol{k}^2 \tag{15.19}$$

Within a material, $\chi \neq 0$, and we must solve eq. (15.17) with a non-zero right-side term. If we are far away from diffraction conditions, we can still use a single-wave solution (15.18), which when substituted into eq. (15.17), yields (in the case of a homogeneous material, χ = constant) the corrected dispersion law for X-ray propagation within a medium:

$$\omega^2 = c^2 \boldsymbol{k}_{\mathrm{m}}^2 (1 - \chi) \tag{15.20}$$

Using eqs. (15.19) and (15.20), we find the refractive index $n = \frac{|\boldsymbol{k}_{\mathrm{m}}|}{|\boldsymbol{k}|} = \frac{k_{\mathrm{m}}}{k}$, which is determined by the ratio between the magnitudes of the respective wave vectors in the medium k_{m} and in vacuum k:

$$n = \frac{k_{\mathrm{m}}}{k} = \frac{1}{\sqrt{1-\chi}} \simeq 1 + \frac{\chi}{2} \tag{15.21}$$

As we know, the refractive index, $n = 1 - \delta$, in the X-ray range is very close to one, with δ being about 10^{-5}–10^{-6}. Correspondingly, the absolute values of the dielectric polarizability χ are on the same order of magnitude, that is, $|\chi| \approx 10^{-5}$–10^{-6}. Nevertheless, this tiny dielectric polarizability is responsible for X-ray diffraction in crystals. The sign of the parameter χ is also of great importance, since for $\chi < 0$, the refractive index $n < 1$ (see eq. (15.21)), and it is possible to arrange the total external reflection of X-rays when they enter a crystal from the vacuum side (see Section 2.3).

Let us find the magnitude and sign of polarizability χ in a simple model that considers the movement of an electron (bounded to an atom) in an external electric field having the amplitude E_0 and angular frequency ω. In this model, already presented in Chapter 6 (see eq. (6.7)), the temporal (t) dependence of one-dimensional electron motion (along the x-coordinate) is described by the force balance equation:

$$\frac{d^2x}{dt^2} - \beta\frac{dx}{dt} + \omega_o^2 x = \frac{e}{m}E_0 \exp(-i\omega t) \tag{15.22}$$

where e and m stand, respectively, for the electron charge and mass, β is the damping constant responsible for the energy dissipation, and ω_0 is the system's resonant frequency. The solution $x(t)$ of eq. (15.22) is an oscillating function:

$$x(t) = x_0 \exp(-i\omega t) \tag{15.23}$$

Substituting the latter expression into eq. (15.22), one finds the amplitude x_0 of the periodic movement of a single electron, as:

$$x_0 = \frac{e}{m}\left(\frac{E_0}{\omega_0^2 - \omega^2 + \beta i\omega}\right) \tag{15.24}$$

The dipole moment of unit volume (i.e., polarization P) is

$$P = e \cdot x \cdot \rho = \frac{e^2}{m}\rho\left(\frac{E_0}{\omega_0^2 - \omega^2 + \beta i\omega}\right)\exp(-i\omega t) \tag{15.25}$$

where ρ is the electron density, that is, the number of electrons per unit volume. Using eq. (15.25), we calculate the amplitude of the electric displacement field D within a material:

$$D = \varepsilon_0\left(E_0 + \frac{P}{\varepsilon_0}\right) = \varepsilon_0 E_0\left[1 + \frac{e^2}{\varepsilon_0 m}\rho\left(\frac{1}{\omega_0^2 - \omega^2 + \beta i\omega}\right)\right] \tag{15.26}$$

Comparing eq. (15.26) with eq. (15.9), we find the dielectric polarizability χ as:

$$\chi = \frac{e^2}{\varepsilon_0 m}\rho\left(\frac{1}{\omega_0^2 - \omega^2 + \beta i\omega}\right) \tag{15.27}$$

Well-above the absorption edge, that is, at $\omega \gg \omega_0$, one obtains

$$\chi = -\frac{e^2}{\varepsilon_0 m}\rho \cdot \frac{1}{\omega^2} = -\frac{r_0\lambda^2\rho}{\pi} \tag{15.28}$$

where $\lambda = 2\pi\frac{c}{\omega}$ is the X-ray wavelength and

$$r_0 = \frac{e^2}{4\pi\varepsilon_0 mc^2} = 2.817 \cdot 10^{-5}\ \text{Å} \tag{15.29}$$

stands for the classical radius of electron r_0 expressed in the SI unit system. To get an impression of the absolute values of dielectric polarizability in the X-ray domain, let us calculate it for a calcite ($CaCO_3$) crystal (having $\rho = 0.815$ electrons per Å^3) and Cu $K\alpha$-radiation ($\lambda = 1.5406$ Å). Substituting these numbers into eq. (15.28) yields $|\chi| = 1.7 \times 10^{-5}$.

Furthermore, it follows from eq. (15.28) that the dielectric polarizability (more exactly, its real part) in the X-ray range (if not very close to an absorption edge) is indeed negative and, correspondingly, the refractive index is less than one (see eq. (15.21)):

$$n = 1 + \frac{\chi}{2} = 1 - \frac{r_0 \lambda^2 \rho}{2\pi} = 1 - \delta \tag{15.30}$$

with

$$\delta = \frac{r_0 \lambda^2 \rho}{2\pi} \tag{15.31}$$

As previously explained in Chapter 2, this fact implies that X-rays experience total external reflection when entering a material at small angles of incidence ($\gamma < \gamma_c$) from the vacuum side. To calculate critical angle γ_c for total external reflection via electron density ρ, let us apply the conservation law for the projections of wave vectors along the entrance surface of a material. Considering that under critical angle conditions, the refracted X-rays propagate (as evanescent waves) along the surface (see Figure 15.1), we find

$$k \cos \gamma_c = k_m = \frac{k}{\sqrt{1-\chi}} \tag{15.32}$$

and

$$\gamma_c = \sqrt{|\chi|} = \lambda \sqrt{\frac{r_0}{\pi} \rho} \tag{15.33}$$

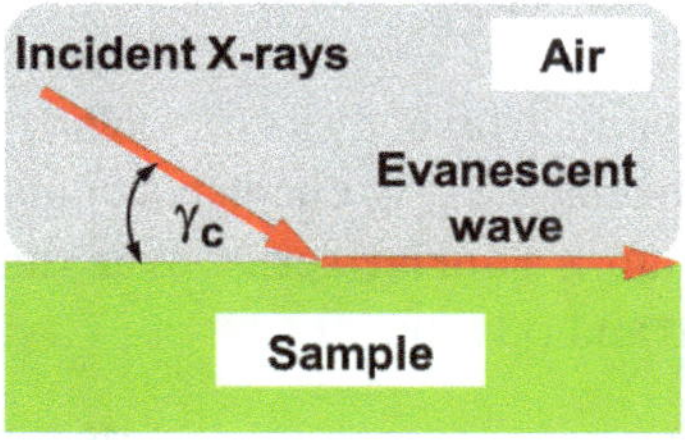

Figure 15.1: Scheme of X-ray propagation under total external reflection.

Therefore, the critical angle γ_c is proportional to the square root of electron density. Again taking a calcite crystal and Cu $K\alpha$-radiation as an example, we calculate $\gamma_c = 0.29°$. Applications of total external reflection for designing the X-ray optics devices are analyzed in Chapter 3. We stress that the critical angle is proportional to the X-ray wavelength, that is, inversely proportional to the X-ray energy. At fixed energy, however, it is determined only by the electron density (see eq. (15.33)). The latter result is widely used to study the near-surface electron density distributions in various materials (not only crystalline, but also liquid and amorphous ones) by X-ray reflectivity (see Appendix 5.B).

Note that the refractive index $n < 1$ means that in a homogeneous medium, the phase velocity of X-rays, $V_p = \frac{\omega}{k_m} = \frac{\omega}{nk} = \frac{c}{n}$, is greater than the velocity of light c in vacuum. This is not forbidden by fundamental physics laws, but the question regarding the group velocity of X-rays, $V_g = \frac{d\omega}{dk_m}$, remains, which according to the theory of relativity should always be smaller than c. Using eqs. (15.19), (15.21), and (15.28) to solve this quandary, we find that

$$k_m = kn = \frac{\omega}{c}\frac{1}{\sqrt{1-\chi}} = \frac{\omega}{c}\frac{1}{\sqrt{1+\frac{p}{\omega^2}}} \tag{15.34}$$

where the frequency-independent parameter $p = 4\pi\rho r_0 c^2 > 0$. Correspondingly, the group velocity is:

$$V_g = \frac{d\omega}{dk_m} = \frac{1}{\frac{dk_m}{d\omega}} = c\frac{\left(1+\frac{p}{\omega^2}\right)^{3/2}}{\left(1+\frac{2p}{\omega^2}\right)} \tag{15.35}$$

Since $(p/\omega^2) = |\chi| \ll 1$, one can expand eq. (15.35) into a **Taylor** series, which yields:

$$V_g = c\left(1-\frac{p}{2\omega^2}\right) = c\left(1+\frac{\chi}{2}\right) < c \tag{15.36}$$

As already mentioned, parameter $p > 0$ ($\chi < 0$) and, therefore, the group velocity is indeed less than c, in agreement with the special theory of relativity.

15.1 Dynamical X-ray diffraction in two-beam approximation

In this section, we again take advantage of the spatial periodicity of dielectric polarizability $\chi(\boldsymbol{r})$, exactly as we did in the case of lattice potential $V(\boldsymbol{r})$, in Chapter 14. Consequently, we expand the dielectric polarizability into a **Fourier** series over the infinite set of scattering vectors $\boldsymbol{Q} = 2\pi\boldsymbol{G}$ (see eqs. (13.11) and (13.17)):

$$\chi(\boldsymbol{r}) = \sum_{\boldsymbol{Q}} \chi(\boldsymbol{Q})\exp(i\boldsymbol{Q}\boldsymbol{r}) \tag{15.37}$$

In turn, **Fourier** coefficients $\chi(\boldsymbol{Q})$ are given by an expression, like eq. (14.7) (integration over the unit cell volume V_c):

$$\chi(\boldsymbol{Q}) = \frac{1}{V_c}\int \chi(\boldsymbol{r})\exp(-i\boldsymbol{Q}\boldsymbol{r})d^3\boldsymbol{r} \tag{15.38}$$

In the two-beam approximation, we only use the three first terms from the series (15.37), that is,

$$\chi(\boldsymbol{r}) = \chi(0) + \chi(\boldsymbol{Q})\exp(i\boldsymbol{Qr}) + \chi(-\boldsymbol{Q})\exp(-i\boldsymbol{Qr}) \tag{15.39}$$

Let us consider two strong waves, the incident one $\boldsymbol{D}_0$, propagating along the direction $\boldsymbol{k}$ close to $\boldsymbol{k}_0$ and the diffracted one $\boldsymbol{D}_Q$, propagating along the direction $(\boldsymbol{k}+\boldsymbol{Q})$, which is close to $(\boldsymbol{k}_0+\boldsymbol{Q})$. We remind the reader that wave vector $\boldsymbol{k}_0$ connects the **Lorentz** point L and the zero node of the reciprocal lattice (see Figure 15.2). Correspondingly, the total wavefield is

$$\boldsymbol{D} = \boldsymbol{D}_0 \exp[i(\boldsymbol{kr} - \omega t)] + \boldsymbol{D}_Q \exp\{i[(\boldsymbol{k}+\boldsymbol{Q})\boldsymbol{r} - \omega t)]\} \tag{15.40}$$

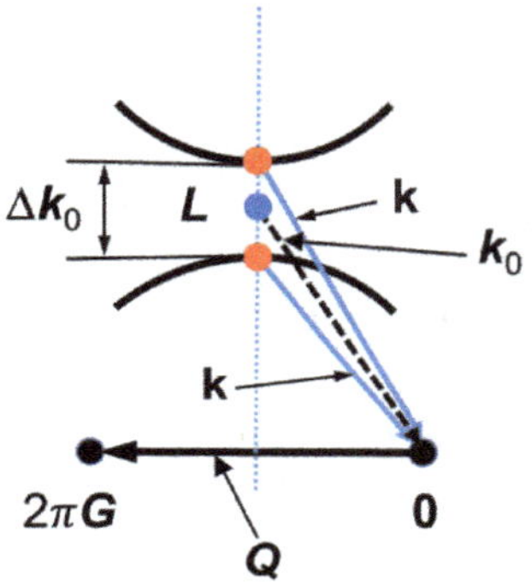

Figure 15.2: Illustration of X-ray wave vectors participating in dynamical diffraction. The **Lorentz** point (blue dot) between the hyperbolic branches of the isoenergetic dispersion surface (solid curves) is marked by the letter L. Wave vector $\boldsymbol{k}_0$ connects the **Lorentz** point L to the **0**-node of the reciprocal lattice. Vectors $\boldsymbol{k}$ connect tie points (red dots) at the isoenergetic surface to the **0**-node of reciprocal lattice.

Substituting eq. (15.39), together with eq. (15.40), into eq. (15.17) yields two equations, which are obtained when separately gathering the terms comprising the periodic functions $\exp[i(\boldsymbol{kr} - \omega t)]$ or $\exp\{i[(\boldsymbol{k}+\boldsymbol{Q})\boldsymbol{r} - \omega t]\}$:

$$\left\{\frac{\omega^2}{c^2} - \boldsymbol{k}^2[1-\chi(0)]\right\}\boldsymbol{D}_0 - \chi(-\boldsymbol{Q})\ \{\boldsymbol{k}\,\mathrm{x}\,[\boldsymbol{k}\,\mathrm{x}\,\boldsymbol{D}_Q]\} = 0 \tag{15.41}$$

$$\left\{\frac{\omega^2}{c^2} - (\boldsymbol{k}+\boldsymbol{Q})^2[1-\chi(0)]\right\}\boldsymbol{D}_Q - \chi(\boldsymbol{Q})\{(\boldsymbol{k}+\boldsymbol{Q})\mathrm{x}\,[(\boldsymbol{k}+\boldsymbol{Q})\mathrm{x}\,\boldsymbol{D}_0]\} = 0 \tag{15.42}$$

We stress that the terms in eqs. (15.41) and (15.42), which contain the vector products, depend on the orientation of vectors $\boldsymbol{D}_0$ and $\boldsymbol{D}_Q$ with respect to wave vectors $(\boldsymbol{k}+\boldsymbol{Q})$ and $\boldsymbol{k}$. In other words, polarization of X-ray waves is now an essential issue. Let us elaborate on this aspect in more detail.

15.1.1 Taking account of X-ray polarization

As we learned in Chapter 5, X-rays are transverse electromagnetic waves, with vector $\boldsymbol{D}$ being perpendicular to the direction of the wave propagation, that is, perpendicular to wave vector $\boldsymbol{k}$. Mathematically, this means that the scalar product

$$(\boldsymbol{k} \cdot \boldsymbol{D}) = 0 \tag{15.43}$$

In other words, X-rays are polarized in the plane perpendicular to the wave vector $\boldsymbol{k}$. Therefore, for each X-ray wave, we must define two polarization projections (i.e., projections of vector $\boldsymbol{D}$) onto the axes of the coordinate system chosen in some way within the above-mentioned plane. For the sake of convenience, this coordinate system is specified as follows. First, we define the scattering plane built on two wave vectors, $\boldsymbol{k}$ and $(\boldsymbol{k} + \boldsymbol{Q})$ (see Figure 15.3).

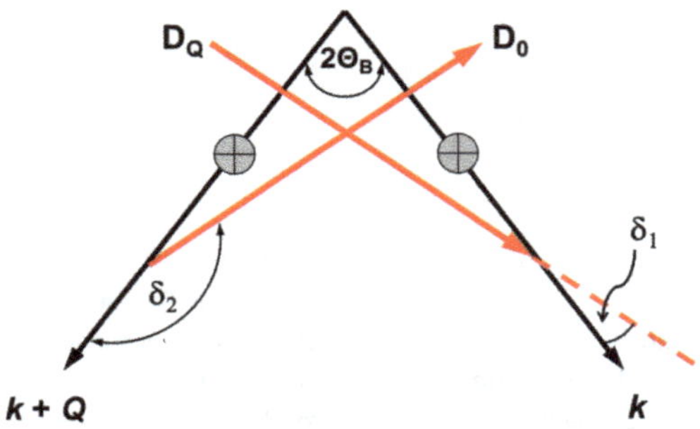

Figure 15.3: Transverse X-ray polarizations situated perpendicular to the scattering plane (symbol ⊗) and in the scattering plane (red arrows).

Next, one axis of our coordinate system is taken to be perpendicular to the scattering plane. Evidently, this choice also fits condition (15.43). The respective projections of vectors $\boldsymbol{D}_0$ and $\boldsymbol{D}_Q$ (so named σ-polarizations already introduced in Section 5.1) are parallel to each other. Using relationship (15.15), we can rewrite eqs. (15.41) and (15.42) for σ-polarized X-rays as

$$\left\{\frac{\omega^2}{c^2} - \boldsymbol{k}^2[1-\chi(0)]\right\}D_0 + [\chi(-\boldsymbol{Q})\boldsymbol{k}^2]D_Q = 0 \tag{15.44}$$

$$\left\{\frac{\omega^2}{c^2} - (\boldsymbol{k}+\boldsymbol{Q})^2[1-\chi(0)]\right\}D_Q + [\chi(\boldsymbol{Q})(\boldsymbol{k}+\boldsymbol{Q})^2]D_0 = 0 \tag{15.45}$$

Note that because the polarization states of the respective X-ray waves, in this case, are both perpendicular to the scattering plane, we can treat the magnitudes of D_0 and D_Q in eqs. (15.44) and (15.45) as regular scalars.

The second axis of our coordinate system is chosen to be situated in the scattering plane, which together with eq. (15.43), leads to the specific arrangement of vectors $\boldsymbol{D}_0$ and $\boldsymbol{D}_Q$ about the respective wave vectors (so named π-polarization also introduced in Section 5.1), shown in Figure 15.3. We stress that vectors $\boldsymbol{D}_Q$ and $\boldsymbol{D}_0$ for π-polarized X-rays are not parallel and, correspondingly, additional care should be taken when further developing eqs. (15.41) and (15.42) in that case. For this purpose, we need to find angles δ_1 and δ_2, respectively, between the pairs of vectors, $\boldsymbol{k}$, $\boldsymbol{D}_Q$, and $(\boldsymbol{k} + \boldsymbol{Q})$, $\boldsymbol{D}_0$ (see Figure 15.3):

$$\begin{aligned} \delta_1 &= 90° - 2\Theta_B \\ \delta_2 &= 90° + 2\Theta_B \end{aligned} \tag{15.46}$$

For π-polarized X-rays, by means of expressions (15.46), eqs. (15.41) and (15.42) are transformed as follows:

$$\left\{\frac{\omega^2}{c^2} - \boldsymbol{k}^2[1-\chi(0)]\right\}D_0 + [\chi(-\boldsymbol{Q})\boldsymbol{k}^2\cos 2\Theta_B]D_Q = 0 \tag{15.47}$$

$$\left\{\frac{\omega^2}{c^2} - (\boldsymbol{k}+\boldsymbol{Q})^2[1-\chi(0)]\right\}D_Q + [\chi(\boldsymbol{Q})(\boldsymbol{k}+\boldsymbol{Q})^2\cos 2\Theta_B]D_0 = 0 \tag{15.48}$$

Also, in this case, after the above-mentioned procedures, one can operate with scalar magnitudes D_0 and D_Q.

We stress that for π-polarized X-rays, there is a trigonometrical factor $\cos 2\Theta_B$ in eqs. (15.47) and (15.48), which provides partial conversion of the incident wave $\boldsymbol{D}_0$ into the diffracted wave $\boldsymbol{D}_Q$, and vice versa. For example, at $\Theta_B = 45°$, $\cos 2\Theta_B = 0$ and, according to eqs. (15.47) and (15.48), there is no diffracted π-polarized wave at all. In other words, at $\Theta_B = 45°$, the diffracted X-rays are purely σ-polarized (as was stated in Section 5.1). Using the optics language, the incident angle $\Theta_B = 45°$ is the **Brewster** angle for X-rays. In fact, partially repeating the relevant analysis in Section 5.1 and recalling the expression for the **Brewster** angle α_B (between the wave vector of incident light and the normal to the surface of a material with refractive index n), we get

$$\tan\alpha_B = n \tag{15.49}$$

Therefore, in the case of X-rays, eq. (15.49) provides $\alpha_B = 45°$ because $n \approx 1$. This means that the angle between the incident wave vector and the crystal surface is also 45°, which coincides with the above obtained value for the **Bragg** angle, Θ_B.

This result is used for producing fully polarized monochromatic X-ray beams by diffraction from the appropriate crystal planes. By applying the **Bragg** law (eq. (13.21)), we find the relationship between the X-ray wavelength and the specific crystal d-spacing, toward complete suppression of the π-polarization (i.e., realizing $\Theta_B = 45°$):

$$d = \frac{\lambda}{\sqrt{2}} \tag{15.50}$$

15.1.2 The four-branch isoenergetic dispersion surface for X-ray quanta

One can unite the systems of eqs. (15.44), (15.45) and (15.47), (15.48) by introducing a new parameter:

$$C = C_1 = 1 \qquad \text{for } \sigma - \text{polarization}$$

$$C = C_2 = \cos 2\Theta_B \quad \text{for } \pi - \text{polarization} \tag{15.51}$$

Using eq. (15.51), dynamical X-ray diffraction is described analytically by two equations

$$\left\{\frac{\omega^2}{c^2} - \boldsymbol{k}^2[1-\chi(0)]\right\}D_0 + [C\chi(-\boldsymbol{Q})\boldsymbol{k}^2]D_Q = 0 \tag{15.52}$$

$$[C\chi(\boldsymbol{Q})(\boldsymbol{k}+\boldsymbol{Q})^2]D_0 + \left\{\frac{\omega^2}{c^2} - (\boldsymbol{k}+\boldsymbol{Q})^2[1-\chi(0)]\right\}D_Q = 0 \tag{15.53}$$

as in the case of dynamical diffraction of scalar wavefields (see eq. (14.17)). Structurally, the basic equations for dynamical diffraction developed right now and in Chapter 14 are very similar, since as we already mentioned, in both cases, we are utilizing the perturbation theory. Therefore, all results obtained in Chapter 14 can be applied to X-rays as well. To proceed, we only need to specify the gap between the branches of the isoenergetic dispersion surface via the corresponding **Fourier** components of dielectric polarizability.

For this purpose, let us define mathematically the already introduced (just after eq. (15.39)) wave vector $\boldsymbol{k}_0$, around which the wave vectors $\boldsymbol{k}$ of the incident X-rays are situated (see Figure 15.2):

$$\boldsymbol{k}_0^2 = (\boldsymbol{k}_0+\boldsymbol{Q})^2 = \frac{\omega^2}{c^2} + \boldsymbol{k}^2\chi(0) \approx \frac{\omega^2}{c^2} + \boldsymbol{k}_0^2\chi(0) \approx \frac{\omega^2}{c^2} + (\boldsymbol{k}_0+\boldsymbol{Q})^2\chi(0) \tag{15.54}$$

Here, we use the fact that $\chi \ll 1$ and vectors $\boldsymbol{k}$ are very close to $\boldsymbol{k}_0$. Now, eqs. (15.52) and (15.53) can be rewritten accordingly:

$$(-\boldsymbol{k}^2 + \boldsymbol{k}_0^2)D_0 + \boldsymbol{k}^2 C\chi(-\boldsymbol{Q})D_Q = 0 \tag{15.55}$$

$$\left[(\boldsymbol{k}_0+\boldsymbol{Q})^2 - (\boldsymbol{k}+\boldsymbol{Q})^2\right]D_Q + \boldsymbol{k}^2 C\chi(\boldsymbol{Q})D_0 = 0 \tag{15.56}$$

Assuming again that $\boldsymbol{k} \approx \boldsymbol{k}_0$, we find

$$\frac{\boldsymbol{k}^2 - \boldsymbol{k}_0^2}{\boldsymbol{k}_0^2}D_0 - C\chi(-\boldsymbol{Q})D_Q = 0 \tag{15.57}$$

$$-C\chi(\boldsymbol{Q})D_0 + \left[\frac{(\boldsymbol{k}+\boldsymbol{Q})^2 - (\boldsymbol{k}_0+\boldsymbol{Q})^2}{\boldsymbol{k}_0^2}\right]D_0 = 0 \tag{15.58}$$

or

$$\frac{\delta \boldsymbol{k} \cdot 2\boldsymbol{k}_0}{\boldsymbol{k}_0^2} D_0 - C\chi(-\boldsymbol{Q})D_Q = 0 \tag{15.59}$$

$$-C\chi(\boldsymbol{Q})D_0 + \left[\frac{\delta \boldsymbol{k} \cdot 2(\boldsymbol{k}_0 + \boldsymbol{Q})}{\boldsymbol{k}_0^2}\right] D_0 = 0 \tag{15.60}$$

where $\delta\boldsymbol{k} = \boldsymbol{k} - \boldsymbol{k}_0$. As in Chapter 14, the system of eqs. (15.59) and (15.60) determines the two-branch isoenergetic surface, which is a function of the wave vector deviation $\delta\boldsymbol{k}$, counted from the **Lorentz** point L (see Figure 15.2). The only difference is the double split of each branch because of the two polarization states, which are characterized by $C_1 = 1$ and $C_2 = cos2\theta_B$ in eqs. (15.59) and (15.60). Consequently, for X-rays, we have a four-branch isoenergetic surface.

To calculate the characteristic gap Δk_0, let us solve eqs. (15.59) and (15.60) for the hyperbola apexes, where vector $\delta\boldsymbol{k}$ is perpendicular to vector $\boldsymbol{Q}$ and $|\delta\boldsymbol{k}| = \Delta k_0/2$. Using Figure 15.2, we find that, for example, in the lower apex

$$\delta\boldsymbol{k} \cdot 2\boldsymbol{k}_0 = \delta\boldsymbol{k} \cdot 2(\boldsymbol{k}_0 + \boldsymbol{Q}) = \Delta k_0 \cdot k_0 \cos\Theta_B \tag{15.61}$$

and, correspondingly

$$\frac{\Delta k_0 \cos\Theta_B}{|\boldsymbol{k}_0|} D_0 - C\chi(-\boldsymbol{Q})D_Q = 0 \tag{15.62}$$

$$-C\chi(\boldsymbol{Q})D_0 + \left[\frac{\Delta k_0 \cos\Theta_B}{|\boldsymbol{k}_0|}\right] D_0 = 0 \tag{15.63}$$

Solving the system of eqs. (15.62) and (15.63) yields:

$$\Delta k_0 = \frac{|\boldsymbol{k}_0| C\sqrt{\chi(\boldsymbol{Q})\chi(-\boldsymbol{Q})}}{\cos\Theta_B} \tag{15.64}$$

In center-symmetric crystals, $\chi(-\boldsymbol{Q}) = \chi^*(\boldsymbol{Q})$, and hence

$$\Delta k_0 = \frac{C|\boldsymbol{k}_0| \cdot |\chi(\boldsymbol{Q})|}{\cos\Theta_B} \tag{15.65}$$

Comparing the latter expression with eq. (14.30), derived for scalar fields, we find complete correspondence if the ratio $\frac{V(Q)}{\mathcal{E}_k}$ between the **Fourier** component of the lattice potential $V(\boldsymbol{Q})$ and the kinetic energy of the particle $\mathcal{E}_k$ is replaced by the **Fourier** component of the dielectric polarizability $\chi(\boldsymbol{Q})$ corrected by polarization factor C, that is,

$$\frac{V(\pm\boldsymbol{Q})}{\mathcal{E}_k} \to C\chi(\pm\boldsymbol{Q}) \tag{15.66}$$

Replacing the ratio $\frac{V(\pm Q)}{\varepsilon_k}$ via expression (15.66) (for $Q\neq 0$) allows us to use all the results obtained in Chapter 14 in the X-ray domain as well. Note that for forward scattering ($Q = 0$), the polarization problem does not exist and the ratio $\frac{V(0)}{\varepsilon_k}$ is replaced by factor $\chi(0)$ for both σ- and π-polarizations.

15.1.3 X-ray extinction length

Finally, let us derive the analytic expression for the extinction length $\tau = 2\pi/\Delta k_0$ in the X-ray domain, which utilizes the X-ray scattering amplitude (structure factor) rather than the dielectric polarizability. Recalling expression (15.28) for dielectric polarizability via electron density $\rho(\boldsymbol{r})$, one finds:

$$\chi(\boldsymbol{Q}) = \frac{1}{V_c}\int \chi(\boldsymbol{r})\exp(i\boldsymbol{Qr})d^3\boldsymbol{r} = -\frac{r_0\lambda^2}{\pi V_c}\int \rho(\boldsymbol{r})\exp(i\boldsymbol{Qr})d^3\boldsymbol{r} \tag{15.67}$$

Integration in (15.67) is performed over the volume V_c of the crystal unit cell. To further proceed, we distinguish in eq. (15.67) between the integration over the electron density within an individual atom (marked below by index j) and the subsequent summation (with the proper phase factors) over all atoms within the unit cell:

$$\chi(\boldsymbol{Q}) = -\frac{r_0\lambda^2}{\pi V_c}\sum_j\left[\int \rho_j(\boldsymbol{r})\exp(i\boldsymbol{Qr})d^3\boldsymbol{r}\right]\exp(i\boldsymbol{Qr}_j) \tag{15.68}$$

where $\rho_j(\boldsymbol{r})$ and $\boldsymbol{r}_j$ stand, respectively, for the electron density distributions and positions of the individual atoms. The first quantity, obtained by integration over the electron density of an individual atom (j), is known as the atomic scattering factor $f_j(Q)$:

$$f_j(Q) = \int \rho_j(\boldsymbol{r})\exp(i\boldsymbol{Qr})d^3\boldsymbol{r} \tag{15.69}$$

Note that the atomic scattering factor $f_j(Q)$ depends on the magnitude of the diffraction vector Q only because the wavefunctions of fully occupied electron shells (major contributors to X-ray diffraction) are spherically symmetric.

In turn, the summation over all atoms within the unit cell provides us with the so-called structure factor $F(\boldsymbol{Q})$ (see also eq. (12.11)):

$$F(\boldsymbol{Q}) = \sum_j f_j\exp(i\boldsymbol{Qr}_j) \tag{15.70}$$

Combining eqs. (15.68)–(15.70) yields:

$$\chi(\boldsymbol{Q}) = -\frac{r_0\lambda^2}{\pi V_c}F(\boldsymbol{Q}) \tag{15.71}$$

Substituting eq. (15.71) into eq. (15.65), one obtains the following expression for the extinction length τ:

$$\tau = \frac{2\pi}{\Delta k_0} = \frac{\pi V_c \cos\Theta_B}{C r_0 \lambda |F(\boldsymbol{Q})|} \tag{15.72}$$

Typically, X-ray extinction lengths are of about a few tens of microns (for CuKα X-rays having energy of 8.048 keV).

Using eq. (14.82), we also calculate the characteristic depth Λ_e for exponential in-depth decay of incident X-ray waves in **Bragg** scattering geometry due to "repumping" into diffraction intensity:

$$\Lambda_e = \frac{\tau}{\pi}\tan\Theta_B = \frac{V_c \sin\Theta_B}{C r_0 \lambda |F(\boldsymbol{Q})|} \tag{15.73}$$

Note that for small-enough **Bragg** angles, $\Theta_B < 45°$, the characteristic depth Λ_e is substantially reduced (as compared to the extinction length τ), being on a micrometer scale. Recalling expression for the scattering amplitude A_p from an individual atomic plane (eq. (12.14)), one can represent eq. (15.73) as $\Lambda_e = \frac{d}{|A_p|}$, that is, in the form already noticed in Section 10.2 (see eq. (10.13)).

Note that the presence of the two-branch (in fact, four-branch) isoenergetic dispersion surface may affect the refractive index of X-rays propagating within a crystal near the **Bragg** diffraction conditions. In that case, the refractive index depends not only on the wave frequency, but also on the direction of the wave vector $\boldsymbol{k}$. Moreover, for some tie points on the dispersion surface, the magnitude of the wave vector $\boldsymbol{k}$ may be larger than that in vacuum, that is, the magnitude of $\boldsymbol{k}_0$ (see Figure 15.2). This implies that the phase velocity V_p of such X-rays, for the same frequency, will be less than the speed of light c and, correspondingly, they can, in principle, be generated by relativistic electrons having the velocity V_e in the range $V_p < V_e < c$ (as for **Cherenkov** radiation discussed in Section 7.6). The generation methods and properties of these X-rays, which are called parametric X-ray radiation, are under investigation.

15.1.4 Isoenergetic dispersion surface for asymmetric reflections

As we already mentioned, all results obtained in Chapter 14 are valid for X-ray diffraction as well, using the replacement (15.66). Up to now, however, analytic expressions were obtained for two scattering geometries –the symmetric **Laue** and **Bragg** diffraction schemes – that is, when the reflecting atomic planes are situated, respectively, perpendicular or parallel to the entrance crystal surface. In the X-ray domain, separating the refraction and diffraction effects in analytical expressions is also commonly accepted. We remind the reader that X-ray refraction leads to the appearance of the terms $\boldsymbol{k}^2\chi(0)$ and $(\boldsymbol{k}+\boldsymbol{Q})^2\chi(0)$ in eqs. (15.52) and (15.53). In the past, this approach also allowed scientists to identify weak refraction effects of X-rays in crystals. In addition, notations in the specialized X-ray diffraction literature are slightly different from those we used here until now. To provide relevant information to readers, in this section, we now derive the shape of the isoenergetic dispersion surface under asymmetric diffraction conditions, that is, when the reflecting atomic planes are inclined with respect to the crystal surface by angles differing from 90° or zero degrees.

We start with the system of eqs. (15.52) and (15.53) and define (to be in line with existing literature) a new entity $\varkappa$ that is the wave vector of X-rays in a vacuum:

$$\varkappa^2 = \frac{\omega^2}{c^2} \tag{15.74}$$

Consequently, the system of basic equations is transformed as follows (considering that $\boldsymbol{k}^2 \approx (\boldsymbol{k}+\boldsymbol{Q})^2 \approx \boldsymbol{\varkappa}^2$):

$$\frac{\boldsymbol{k}^2 - \boldsymbol{\varkappa}^2}{\boldsymbol{\varkappa}^2} D_0 = \chi(0) D_0 + C\chi(-\boldsymbol{Q}) D_Q \tag{15.75}$$

$$\frac{(\boldsymbol{k}+\boldsymbol{Q})^2 - \boldsymbol{\varkappa}^2}{\boldsymbol{\varkappa}^2} D_Q = \chi(0) D_Q + C\chi(\boldsymbol{Q}) D_0 \tag{15.76}$$

Within a crystal, wave vectors $\boldsymbol{k} = \boldsymbol{\varkappa} + \boldsymbol{q}_c$ differ from vector $\boldsymbol{\varkappa}$ in vacuum by some small quantities, $\boldsymbol{q}_c$. In the case of an infinite interface between vacuum and a crystal, the "force" acting on the X-rays when they enter the crystal from the vacuum side is along the normal $\boldsymbol{n}_s$ to the crystal surface. This implies that:

$$\boldsymbol{k} = \boldsymbol{\varkappa} + \boldsymbol{q}_c = \boldsymbol{\varkappa} + \zeta|\boldsymbol{\varkappa}|\boldsymbol{n}_s \tag{15.77}$$

where ζ is some small parameter specifying the change in the X-ray wave vectors. Correspondingly,

$$\frac{\boldsymbol{k}^2 - \boldsymbol{\varkappa}^2}{\boldsymbol{\varkappa}^2} = \frac{(\boldsymbol{k}-\boldsymbol{\varkappa})(\boldsymbol{k}+\boldsymbol{\varkappa})}{\boldsymbol{\varkappa}^2} = \frac{2\zeta|\boldsymbol{\varkappa}|(\boldsymbol{n}_s \cdot \boldsymbol{\varkappa})}{\boldsymbol{\varkappa}^2} = 2\zeta\cos\gamma_0 \tag{15.78}$$

$$\frac{(\boldsymbol{k}+\boldsymbol{Q})^2 - \boldsymbol{\varkappa}^2}{\boldsymbol{\varkappa}^2} = \frac{(\boldsymbol{k}+\boldsymbol{Q}-\boldsymbol{\varkappa})(\boldsymbol{k}+\boldsymbol{Q}+\boldsymbol{\varkappa})}{\boldsymbol{\varkappa}^2} = \frac{(\boldsymbol{Q}+\zeta|\boldsymbol{\varkappa}|\boldsymbol{n}_s)(2\boldsymbol{\varkappa}+\boldsymbol{Q}+\zeta|\boldsymbol{\varkappa}|\boldsymbol{n}_s)}{\boldsymbol{\varkappa}^2} = \frac{\boldsymbol{Q}^2 + 2\boldsymbol{Q}\boldsymbol{\varkappa}}{\boldsymbol{\varkappa}^2} + 2\zeta\cos\gamma_Q \tag{15.79}$$

Here, γ_0 and γ_Q are the angles between the normal $\boldsymbol{n}_s$ to the entrance crystal surface and the wave vectors, $\boldsymbol{\varkappa}$ and $(\boldsymbol{Q}+\boldsymbol{\varkappa})$, respectively (see Figure 15.4).

As we show later, the parameter ζ is linked to the excitation (tie) points on the dispersion surface. Note that when deriving eqs. (15.78) and (15.79), we kept only linear terms with respect to the small parameter ζ. This approximation is valid if the angles γ_0 and γ_Q are not very close to 90°. Otherwise, as for grazing incidence diffraction ((GID), we must also keep the next terms in the expansion in series, that is, those of the order of ζ^2.

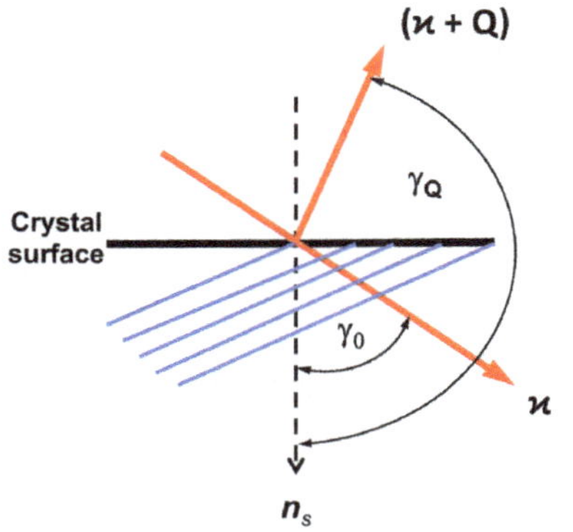

Figure 15.4: Scheme of asymmetric X-ray diffraction.

Further consideration should be given to the first term on the right-hand side of eq. (15.79), which depends on the angular deviation $\Delta\theta = \theta - \theta_B$ of the incident X-ray beam from the exact **Bragg** angle θ_B. Using Figure 15.5 we find:

$$\frac{\boldsymbol{Q}^2 + 2\boldsymbol{Q}\boldsymbol{\varkappa}}{\boldsymbol{\varkappa}^2} = \frac{Q^2 + 2Q\varkappa\cos(90^\circ + \Theta_B + \Delta\theta)}{\varkappa^2} = \frac{Q^2 - 2Q\varkappa\sin\Theta_B}{\varkappa^2} - \frac{2Q\varkappa\cos\Theta_B}{\varkappa^2}\Delta\theta \tag{15.80}$$

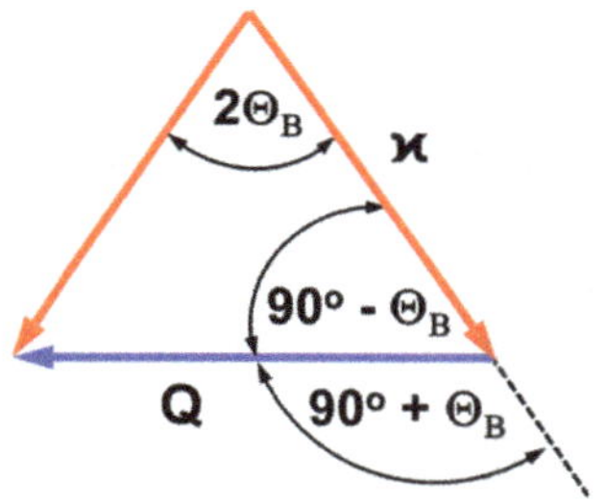

Figure 15.5: Illustration of deriving eq. (15.80). Diffraction vector ***Q*** is indicated by blue arrow.

Applying the **Bragg** law in the form of eq. (13.19) finally yields:

$$\frac{\boldsymbol{Q}^2 + 2\boldsymbol{Q}\boldsymbol{\varkappa}}{\boldsymbol{\varkappa}^2} = -2\Delta\theta\sin(2\Theta_B) = \alpha_d \tag{15.81}$$

With the aid of expressions (15.78), (15.79), and (15.81), the basic system of eqs. (15.75)–(15.76) transforms as follows:

$$2\zeta\cos\gamma_0 D_0 = \chi(0)D_0 + C\chi(-\boldsymbol{Q})D_Q \tag{15.82}$$

$$(\alpha_d + 2\zeta\cos\gamma_Q)D_Q = \chi(0)D_Q + C\chi(\boldsymbol{Q})D_0 \tag{15.83}$$

For any angular deviation α_d, the solution of this system of equations (i.e., solution of the quadratic secular equation) determines four tie points at the hyperbolic dispersion surface (two for each polarization state being indicated by C_1 or C_2, see eq. (15.51)). These tie points are characterized by parameters ζ_1 and ζ_2:

$$\zeta_{1,2} = \frac{[\chi(0)(1+\beta_d) - \alpha_d] \pm \sqrt{[\chi(0)(1-\beta_d) - \alpha_d]^2 + 4\beta_d C^2 \chi(\mathbf{Q})\chi(-\mathbf{Q})}}{4\cos\gamma_Q} \tag{15.84}$$

with

$$\beta_d = \frac{\cos\gamma_Q}{\cos\gamma_0} \tag{15.85}$$

Let us check that eq. (15.84) provides the correct results for the previously analyzed symmetric **Laue** and **Bragg** scattering geometries (Figure 15.6).

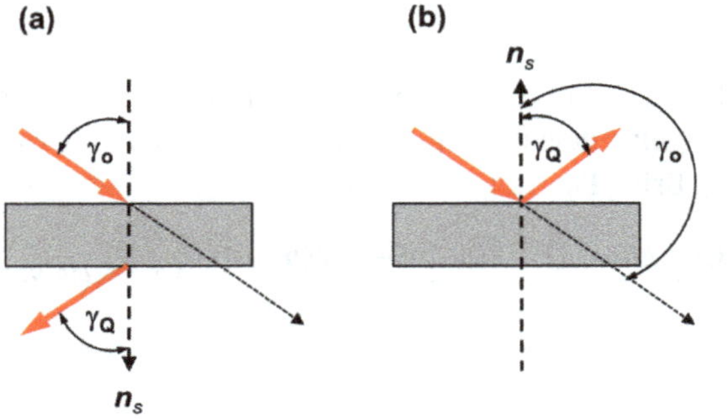

Figure 15.6: Illustration of angular parameters, γ_0 and γ_Q, which enter eqs. (15.82)–(15.85) in the case of symmetric **Laue** (a) and symmetric **Bragg** (b) scattering geometries.

In the symmetric **Laue** case, the parameter $\beta_d = 1$ and $\cos\gamma_Q = \cos\Theta_B$ (see Figure 15.6a). Therefore,

$$\zeta_{1,2} = \frac{[2\chi(0) - \alpha_d] \pm \sqrt{\alpha_d^2 + 4C^2\chi(\mathbf{Q})\chi(-\mathbf{Q})}}{4\cos\Theta_B} \tag{15.86}$$

The gap Δk_0 between the two apexes of the hyperbola ($\alpha_d = 0$) equals:

$$\Delta k_0 = \varkappa(\zeta_1 - \zeta_2) \approx |\mathbf{k}_0|(\zeta_1 - \zeta_2) = \frac{C|\mathbf{k}_0|\sqrt{\chi(\mathbf{Q})\chi(-\mathbf{Q})}}{\cos\Theta_B} \tag{15.87}$$

which agrees with eq. (15.64).

In the symmetric **Bragg** case, $\beta_d = -1$ and $\cos\gamma_Q = \sin\Theta_B$ (see Figure 15.6b). Hence, eq. (15.84) transforms into:

$$\zeta_{1,2} = \frac{-\alpha_d \pm \sqrt{[2\chi(0) - \alpha_d]^2 - 4C^2\chi(\mathbf{Q})\chi(-\mathbf{Q})}}{4\sin\Theta_B} \tag{15.88}$$

Total reflection of X-rays from a crystal means that the ζ-value (15.88) contains an imaginary part, which exists in the angular range:

$$[2\chi(0) - \alpha_d]^2 < 4C^2\chi(\boldsymbol{Q})\chi(-\boldsymbol{Q}) \tag{15.89}$$

or

$$|2\chi(0) - \alpha_d| < 2C\sqrt{\chi(\boldsymbol{Q})\chi(-\boldsymbol{Q})} \tag{15.90}$$

Recalling the definition (15.81), one can convert the α_d-interval into the real angular range $\Delta\theta$, in which the X-ray diffraction measurements are performed. On this angular scale, the full width of the total reflection, that is, the full width of the **Darwin** plateau Γ_D equals (using also eq. (15.64)):

$$\Gamma_D = 2\frac{C\sqrt{\chi(\boldsymbol{Q})\chi(-\boldsymbol{Q})}}{2\sin\Theta_B\cos\Theta_B} = 2\frac{C|\boldsymbol{k}_0|\sqrt{\chi(\boldsymbol{Q})\chi(-\boldsymbol{Q})}}{2|\boldsymbol{k}_0|\sin\Theta_B\cos\Theta_B} = 2\frac{\Delta k_0}{Q} = 2\frac{d}{\tau} \tag{15.91}$$

which coincides with eq. (14.78). Taking for rough estimations $\tau = 50$ µm and $d = 2$ Å yields $\Gamma_D = 8\cdot10^{-6}$ rad or 1.7 s of arc. By choosing weaker reflections with smaller d-spacings and larger extinction lengths, the profile width can be reduced below one second of arc.

Solving eqs. (15.82) and (15.83), together with boundary conditions for a particular scattering geometry and crystal shape, allows us, in principle, to calculate the X-ray diffraction profiles, that is, the distribution of the diffraction intensity as a function of angular deviation $\Delta\theta = \theta - \theta_B$.

15.2 The phase-shift plates for producing circularly polarized X-rays

As we learned in Section 5.4, establishing the methods for producing circularly polarized X-rays is a crucial step toward studying magnetic circular dichroism and the related quantum-mechanical effects. We also know that in visible light optics, circular polarization is routinely produced using birefringent crystals. The latter deliver extraordinary and ordinary rays having orthogonal polarizations (see Section 5.4). In the case of equal amplitudes of these polarization states, circular polarization is generated by introducing the proper phase shift ($\pi/2$), which is accumulated when both waves are travelling across a certain crystal thickness (see eq. (5.38)). As we know, such a phase-shifting device is called a quarter-wave plate, whose thickness is determined by the light wavelength and the difference between the extraordinary and ordinary refractive indices. We also know that severe problems of X-ray optics are caused by the smallness of the refractive index in the X-ray domain, the index being very close to one. It turns out, however, that despite the smallness of the refractive index itself, in the framework of dynamical diffraction, it is possible to organize sufficient difference between the refractive indices for the orthogonal (σ and π) polarization states, and make effective quarter-wave phase-shifting plates.

One (perhaps the simplest) option was developed by the staff of AT&T Bell Laboratories and implemented at the Cornell High-Energy Synchrotron Source (CHESS). In this method, linearly polarized synchrotron X-rays enter a Si crystal at the exact **Bragg** position, with the polarization vector oriented at 45° with respect to the scattering plane, thus providing equal amplitudes of the out-of-plane (σ) and in-plane (π) polarization states. In the symmetric **Laue** scattering geometry, at the exact **Bragg** angular position, the excitation (tie) points of the dispersion surface are at hyperbolic apexes (see Figure 15.7).

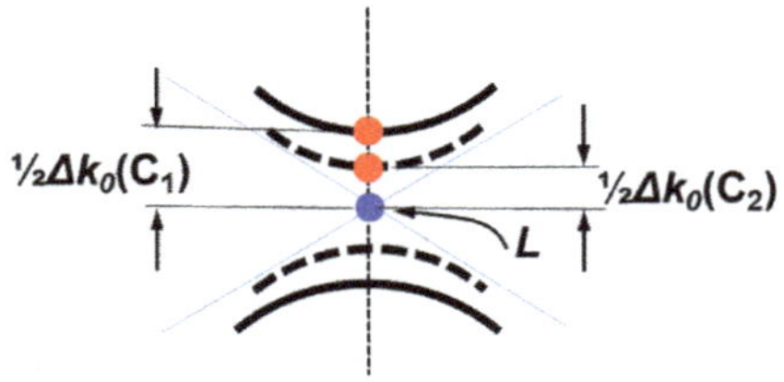

Figure 15.7: Illustration showing how circular X-ray polarization is produced by mixing the σ- and π-polarized X-ray waves. The branches of the isoenergetic surfaces for out-of-plane (σ) and in-plane (π) polarizations are marked by solid and dashed curves, respectively. The essential tie points are colored in red, while the **Lorentz** point L is in blue.

Considering the two upper branches of the dispersion surface, one for the out-of-plane polarization (solid line, $C_1 = 1$) and the second one for the in-plane polarization (dashed line, $C_2 = \cos 2\theta_B$), one finds that the phase difference $\Delta\varphi$ is determined by the difference $1/2[\Delta k_0(C_1) - \Delta k_0(C_2)]$ of the respective half gaps (counted from the **Lorentz** point L) as well as by the plate thickness T, that is,

$$\Delta\varphi = \frac{1}{2}[\Delta k_0(C_1) - \Delta k_0(C_2)]\ T \tag{15.92}$$

Using eq. (15.87) yields

$$\begin{aligned}\Delta\varphi &= \frac{|\boldsymbol{k}_0|\sqrt{\chi(\boldsymbol{Q})\chi(-\boldsymbol{Q})}}{2\cos\Theta_B}(C_1 - C_2)T = \frac{|\boldsymbol{k}_0|\sqrt{\chi(\boldsymbol{Q})\chi(-\boldsymbol{Q})}}{2\cos\Theta_B}[1 - \cos(2\theta_B)]T \\ &= 2\pi T\frac{\sqrt{\chi(\boldsymbol{Q})\chi(-\boldsymbol{Q})}}{\lambda\cos\Theta_B}\sin^2\theta_B\end{aligned} \tag{15.93}$$

Note again that for center-symmetric crystals, $\sqrt{\chi(\boldsymbol{Q})\chi(-\boldsymbol{Q})} = |\chi(\boldsymbol{Q})|$. Expressing X-ray polarizability $\chi(\boldsymbol{Q})$ via structure factor $F(\boldsymbol{Q})$ (see eq. (15.71)), one obtains the phase shift

$$\Delta\varphi = 2\pi T \frac{|F(\boldsymbol{Q})| r_o \lambda}{\pi V_c \cos \Theta_B} \sin^2\theta_B \tag{15.94}$$

or

$$\Delta\varphi = 2\pi \frac{T}{\tau} \sin^2\theta_B \tag{15.95}$$

via extinction length τ, given by eq. (15.72). For a quarter-wave plate, $\Delta\varphi = \frac{\pi}{2}$ and, hence, its thickness $T_{1/4}$ equals

$$T_{1/4} = \frac{\tau}{4 sin^2 \theta_B} \tag{15.96}$$

We remind the reader that in the symmetric **Laue** case, the X-ray energy flow is along the diffractive planes (see Figure 14.8 in Section 14.2). Therefore, if even $T_{1/4}$ prevails over the X-ray absorption length, the wave propagation proceeds via anomalous transmission (i.e., in the **Borrmann** regime).

An alternative approach was developed at the European Synchrotron Radiation Facility (ESRF) and is based on fine-tuning the respective phase shift by angular deviation from the exact **Bragg** position. To calculate the phase shift $\Delta\varphi$ in the **Laue** scattering geometry, we must now use eq. (15.86) for non-zero angular deviation ($\alpha_d \neq 0$) that yields

$$\begin{aligned}\Delta\varphi &= \frac{\Delta k_0 (C_1 - C_2)}{2} T = \frac{|\boldsymbol{k}_0|}{2} (\zeta_1 - \zeta_2) T \\ &= \pi \frac{T}{\lambda} \left\{ \frac{[2\chi(0) - \alpha_d] \pm \sqrt{\alpha_d{}^2 + 4 C_1{}^2 \chi(\boldsymbol{Q}) \chi(-\boldsymbol{Q})}}{4 \cos \Theta_B} - \frac{[2\chi(0) - \alpha_d] \pm \sqrt{\alpha_d{}^2 + 4 C_2{}^2 \chi(\boldsymbol{Q}) \chi(-\boldsymbol{Q})}}{4 \cos \Theta_B} \right\}\end{aligned} \tag{15.97}$$

Substituting $C_1 = 1$ and $C_2 = \cos 2\theta_B$ into eq. (15.97), one finds for the two upper branches of the isoenergetic dispersion surface,

$$\Delta\varphi = \pi \frac{T}{4\lambda \cos \Theta_B} \left[\sqrt{\alpha_d{}^2 + 4\chi(\boldsymbol{Q})\chi(-\boldsymbol{Q})} - \sqrt{\alpha_d{}^2 + 4\cos^2(2\theta_B)\chi(\boldsymbol{Q})\chi(-\boldsymbol{Q})} \right] \tag{15.98}$$

Multiplying both the denominator and the numerator by $\left[\sqrt{\alpha_d{}^2 + 4\chi(\boldsymbol{Q})\chi(-\boldsymbol{Q})} + \sqrt{\alpha_d{}^2 + 4\cos^2(2\theta_B)\chi(\boldsymbol{Q})\chi(-\boldsymbol{Q})} \right]$ gives

$$\Delta\varphi = \pi \frac{T}{4\lambda \cos \Theta_B} \frac{4|\chi(\boldsymbol{Q})|^2 \sin^2(2\theta_B)}{\left[\sqrt{\alpha_d{}^2 + 4|\chi(\boldsymbol{Q})|^2} + \sqrt{\alpha_d{}^2 + 4\cos^2(2\theta_B)|\chi(\boldsymbol{Q})|^2} \right]} \tag{15.99}$$

Here and below, we again use the simplification, $\chi(\boldsymbol{Q})\chi(-\boldsymbol{Q}) = |\chi(\boldsymbol{Q})|^2$, as in center-symmetric crystals. At rather large angular deviations, $\alpha_d > 2|\chi(\boldsymbol{Q})|$

$$\Delta\varphi \simeq \pi \frac{T}{2\lambda \alpha_d \cos \Theta_B} |\chi(\boldsymbol{Q})|^2 \sin^2 2\theta_B \tag{15.100}$$

With the aid of relationships (15.71) and (15.81), eq. (15.100) can be rewritten as

$$\Delta\varphi = T \frac{r_0^2 \lambda^3 |F(\boldsymbol{Q})|^2}{2\pi V_c^2 \Delta\theta} \sin \theta_B \tag{15.101}$$

Correspondingly, the thickness $T_{1/4}$ of a quarter-wave plate ($\Delta\varphi = \frac{\pi}{2}$) equals

$$T_{1/4} = \frac{\pi^2 V_c^2}{r_0^2 \lambda^3 |F(\boldsymbol{Q})|^2 \sin \theta_B} \Delta\theta \tag{15.102}$$

We stress that it varies, being linearly proportional to the angular deviation, $\Delta\theta$.

15.3 X-ray beam compression using highly asymmetric reflections

Here, we describe the diffraction-based approach utilizing highly asymmetric reflections, which allows us to significantly change the widths of the X-ray beams. In asymmetric scattering geometry, the sizes of the incident (S_i) and the diffracted (S_d) beams are dissimilar, depending on the entrance (ω_i) and exit (ω_d) beam angles with respect to the crystal surface. With the aid of Figure 15.8, we find that

$$\omega_i + \omega_d = 2\theta_B \tag{15.103}$$

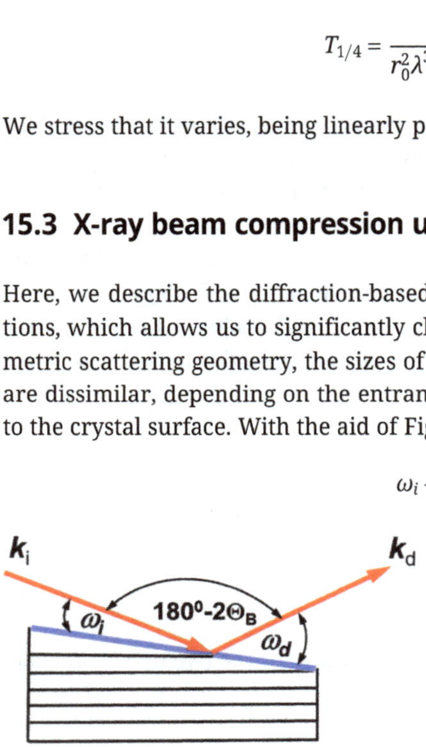

Figure 15.8: Angular interrelations for asymmetric reflections. Wave vectors of the incident and the diffracted X-rays are indicated as $\boldsymbol{k}_i$ and $\boldsymbol{k}_d$, respectively.

If the entrance angle ω_i is smaller than the exit angle ω_d (see Figure 15.9), then

$$S_d = \frac{\sin(\omega_d)}{\sin(\omega_i)} S_i = \frac{\sin(2\theta_B - \omega_i)}{\sin(\omega_i)} S_i \tag{15.104}$$

In the case of $\omega_i \ll 1$ (highly asymmetric reflection) and ordinary **Bragg** angles of about $\theta_B \approx 45°$,

$$S_d \approx \frac{S_i}{\omega_i} \tag{15.105}$$

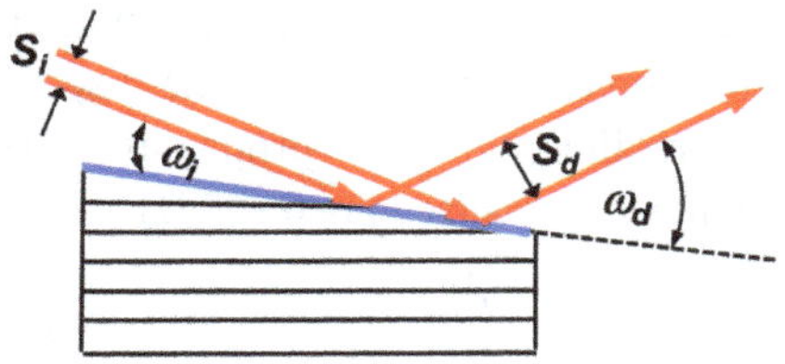

Figure 15.9: Expanding the X-ray beam ($S_d > S_i$) with the aid of asymmetric reflection.

and we obtain a substantial increase in the diffracted beam size, $S_d \gg S_i$, compared to the size of the incident beam (S_i).

In the opposite situation, that is, when $\omega_d < \omega_i$ (see Figure 15.10),

$$S_d = \frac{\sin(\omega_d)}{\sin(\omega_i)} S_i = \frac{\sin(\omega_d)}{\sin(2\theta_B - \omega_d)} S_i \tag{15.106}$$

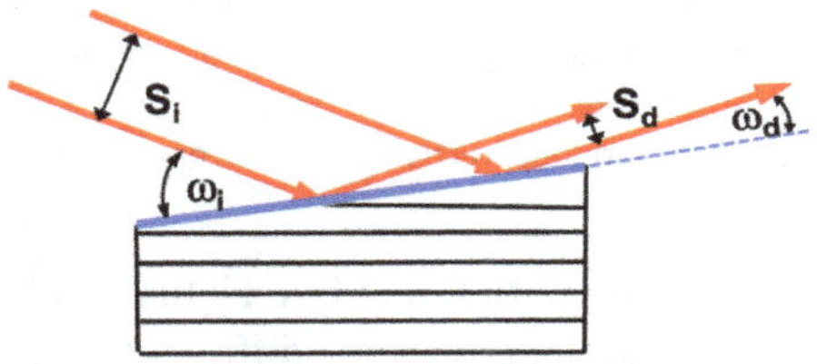

Figure 15.10: Compressing the X-ray beam ($S_d < S_i$) with the aid of asymmetric reflection.

If $\omega_d \ll 1$ and again $\theta_B \approx 45°$, then

$$S_d \approx \omega_d S_i \tag{15.107}$$

that is, there is a substantial compression of the diffracted beam ($S_d \ll S_i$) compared to the size of the incident beam (S_i). At $\omega_d \simeq 1^{\circ} \simeq \frac{1}{57.3}$ rad, the compression effect is nearly a factor of 50. This means that if S_i is initially restricted by slits to 500 microns, at the exit, we will get $S_d \simeq 10$ microns.

15.4 The use of multicrystal diffractometers for phase-contrast X-ray imaging

As mentioned in Section 9.4, some methods for phase-contrast X-ray imaging utilize superb angular resolution of multicrystal diffractometers. Let us start with the double-crystal diffractometer, a unique instrument, which, for example, played a key role in the discovery of the **Compton** effect (the 1927 Nobel Prize in Physics). Detailed analysis of the working principles of this apparatus was presented in 1937 by **Jesse DuMond**.

Note that the diffractive crystal alone operates as the dispersive optical element (like a glass prism for visible light) in the case when there are different wavelengths in the spectrum of incident radiation, for example, $K\alpha_1$ and $K\alpha_2$ emission lines coming from a sealed X-ray tube. Certainly, there is no correlation between the direction of the X-ray propagation and the wavelength λ before the diffraction event. The dispersion is governed by the **Bragg** law, $2d \sin\theta = \lambda$, which links the X-ray propagation direction (via diffraction angle 2θ) to the radiation wavelength λ (see Figure 15.11)

$$\theta = \arcsin\left(\frac{\lambda}{2d}\right) \tag{15.108}$$

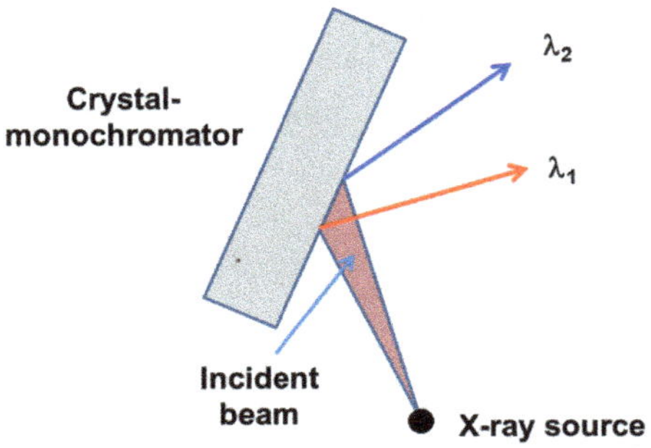

Figure 15.11: A single crystal placed at the diffractive **Bragg** position serves as a dispersive optical element for X-rays, linking the X-ray wavelength to the direction of wave propagation.

Placing the second crystal at some distance, parallel to the first one (see Figure 15.12), does not change the situation, in the sense that, all wavelengths being diffracted from the first crystal will be reflected automatically from the second crystal. It is often said that dispersions introduced by the individual crystals cancel out each other.

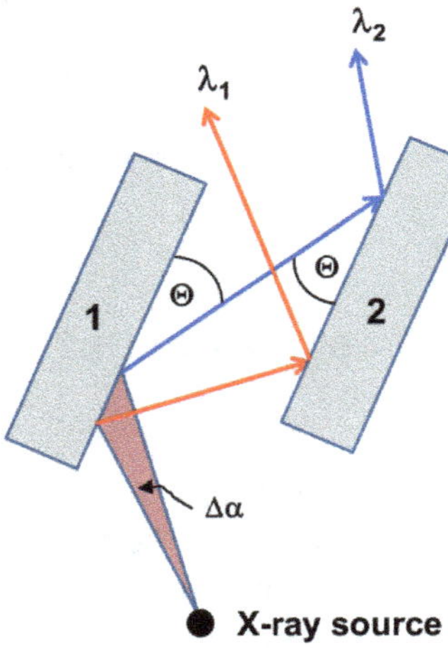

Figure 15.12: Principal scheme of the double-crystal monochromator, built of identical single crystals (1 and 2) in a parallel setting.

In contrast, placing the second crystal in the so-called non-parallel (or antiparallel) setting (see Figure 15.13) leads to a stronger dispersion because of the summing of individual contributions.

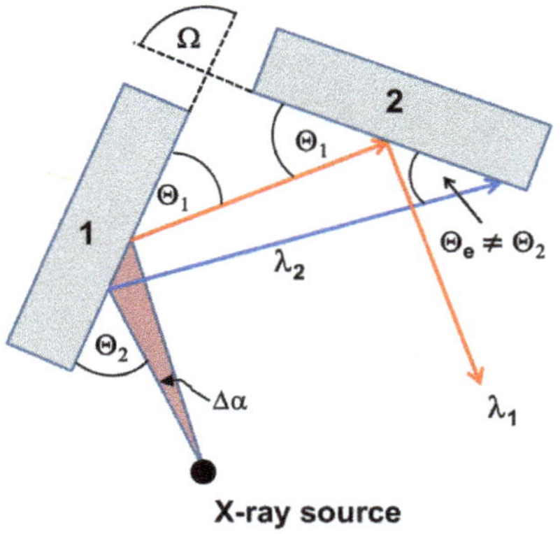

Figure 15.13: Double-crystal setup using a dispersive (nonparallel) setting of individual crystals (1 and 2).

As a result, only one wavelength (λ_1) will go through this setup – namely, the one dictated by the angle Ω between the individual crystals

$$\Omega = 180^{\circ} - 2\theta_1 = 180^{\circ} - 2\arcsin\frac{\lambda_1}{2d} \tag{15.109}$$

where θ_1 is the **Bragg** angle for λ_1. As is seen in Figure 15.13, X-rays with other wavelengths will meet the second crystal at the wrong entrance angles θ_e, since

$$\theta_e = 180^{\circ} - \Omega - \theta = 2\theta_1 - \theta \neq \theta \tag{15.110}$$

where θ is the Bragg angle for wavelength $\lambda \neq \lambda_1$. Correspondingly, the transmission of all these wavelengths (except λ_1) will be eliminated.

We stress that the non-dispersive parallel setting is the most popular because of one unique feature: all wavelengths pass through the setup if the crystals are strictly parallel, but no one wavelength passes when the crystals are slightly misaligned. In other words, a gentle rotation (ω_r-rocking) of the second crystal about an axis perpendicular to the plane of the drawing in Figure 15.14 practically stops the X-ray transmission at rocking angles exceeding some very small values.

In other words, the incoming wavelength spectrum (see Figure 15.15a) is convoluted into a very narrow rocking curve profile $I(\omega_r)$ (Figure 15.15b). Its angular width is mostly determined by the **Darwin** plateau (eq. (15.91)), which typically is a few seconds of arc, but can be reduced to a few tenths of a second of arc or even smaller.

Adding a third crystal (crystal-analyzer) converts the double-crystal diffractometer into a triple-crystal (or triple-axis) diffractometer (see Figure 15.16) with improved angular resolution, mostly by suppressing the rocking curve tails and cutting the inelastic scattering intensity.

Multicrystal diffractometers are successfully used to measure subtle distortions of diffraction profiles induced by static and dynamic deformation fields of different origin. As a working example, we demonstrate here the phonon-modulated diffraction profile due to X-ray interaction with surface acoustic waves (SAWs) generated in

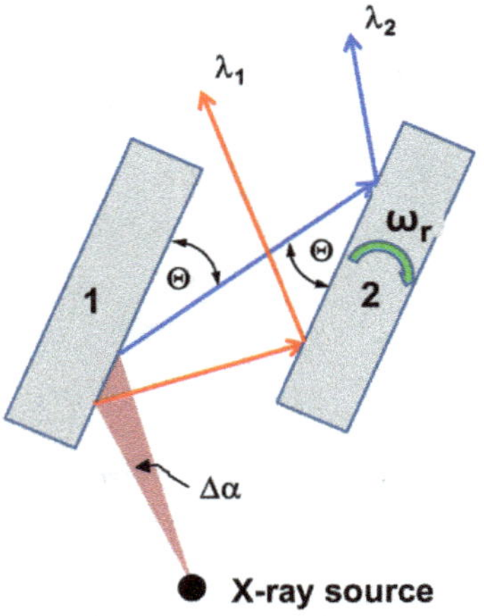

Figure 15.14: Non-dispersive (parallel) setting of individual crystals (1 and 2) within a double-crystal setup showing rotation direction ω_r (gentle rocking around the axis perpendicular to the drawing).

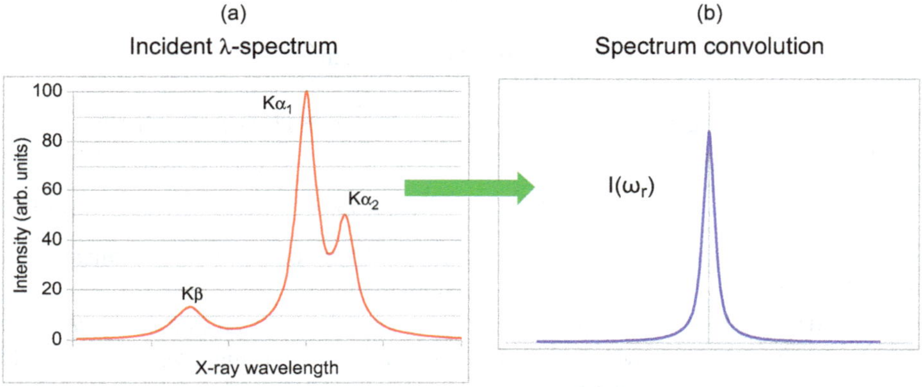

Figure 15.15: Illustration of how all spectral components (a) of the incoming radiation to the double-crystal diffractometer are convoluted (b) into a narrow single-peak rocking curve $I(\omega_r)$.

$LiNbO_3$ single crystals (see Figure 15.17). SAWs were produced with the aid of interdigital electrodes (IDE) deposited on top of $LiNbO_3$ crystalline plates (see bottom panel in Figure 15.17). The X-ray/phonon interaction changes the quasi-wave vector conservation law from that described by eq. (13.18) to

$$\boldsymbol{k}_d - \boldsymbol{k}_i = 2\pi\boldsymbol{G} \pm m\boldsymbol{q}_{\mathrm{ph}} \tag{15.111}$$

where $m = 1, 2, 3, \ldots$ is the number of phonons created or absorbed during X-ray/phonon interaction. In turn, the phonon wave vector $\boldsymbol{q}_{\mathrm{ph}}$ is expressed via the SAW wavelength Λ_{ph} as

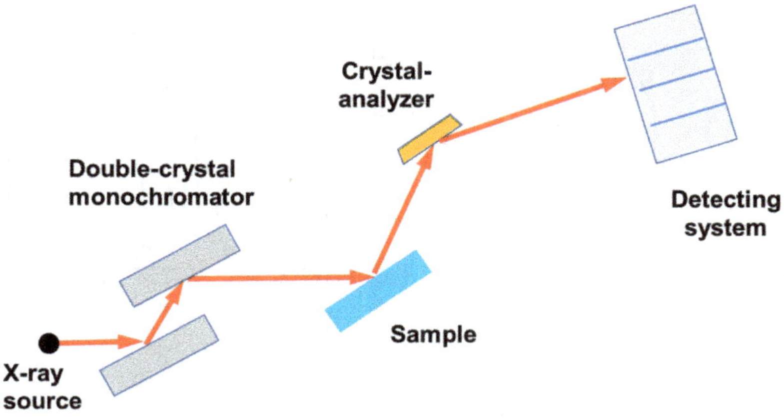

Figure 15.16: Principle scheme of a triple-crystal diffractometer.

$$|\boldsymbol{q}_{\mathrm{ph}}| = \frac{2\pi}{\Lambda_{\mathrm{ph}}} \tag{15.112}$$

Correspondingly, the angular distance between adjacent satellites in the measurement geometry used ($\boldsymbol{q}_{\mathrm{ph}} \perp \boldsymbol{G}$) is

$$\Delta\theta_s = \frac{|\boldsymbol{q}_{\mathrm{ph}}|}{|\boldsymbol{G}|} = \frac{d}{\Lambda_{\mathrm{ph}}} \tag{15.113}$$

which, in our case of $\lambda = 0.59$ Å, $d = 0.733$ Å, $\Lambda_{\mathrm{ph}} = 6$ μm, yields $\Delta\theta_s \simeq 2.56$ s of arc.

The usage of a double-crystal monochromator, composed of (333)-cut Si crystals in a parallel setting, and perfect $LiNbO_3$ crystal-analyzer allowed the Israeli-American research team at Advanced Photon Source (APS) of Argonne National Laboratory to obtain very narrow rocking curves with FWHM equal to 1.4 s of arc (Figure 15.17a). With such a well-collimated and monochromatized X-ray beam, reflected from the surface of the SAW-excited $LiNbO_3$ crystal, phonon-induced angular satellites are clearly resolved (Figure 15.17b). Note that the measured angular separation between the adjacent satellites agrees well with that predicted by eq. (15.113).

As previously stated in Section 9.4, superior angular resolution of multicrystal diffractometers is employed for phase-contrast X-ray imaging. In its simple variant, a transparent object is placed between individual crystals of the double-crystal diffractometer (see Figure 9.7). If the crystals are slightly misaligned, the directly transmitted beam will be heavily suppressed, in proportion to the rocking angle ω_r. The latter is determined by the choice of the appropriate working point on the rocking curve profile shown in Figure 15.15b. Correspondingly, the image will be formed by the slightly refracted X-rays

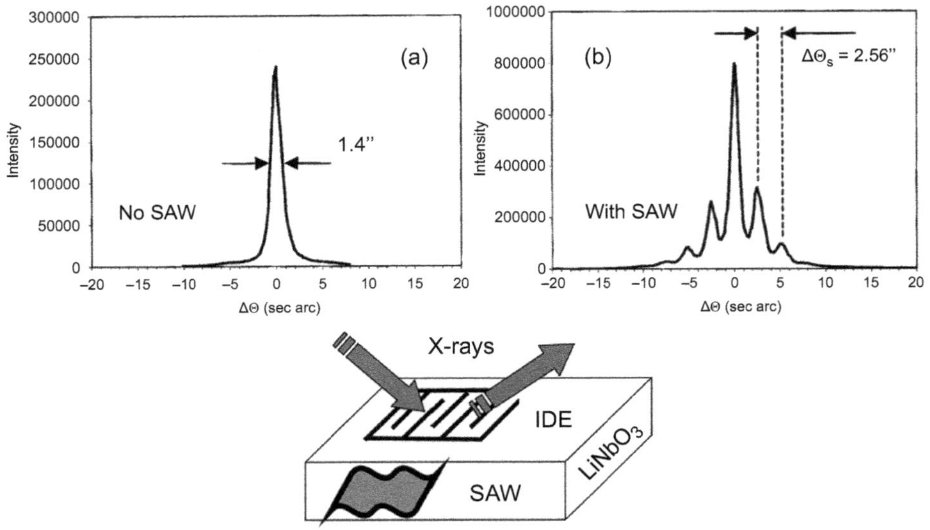

Figure 15.17: Results of the X-ray diffraction measurements with single $LiNbO_3$ crystals revealing the X-ray -phonon interaction. The (060)-$LiNbO_3$ rocking curves: (a) with no SAW; (b) under 0.58 GHz SAW excitation. SAWs were generated using interdigital electrodes (IDE) deposited on top of $LiNbO_3$ plates (see bottom panel). Figures (a) and (b) are reproduced from *Rev. Sci. Instr.* 73(3), 1643 (2002) with the permission of AIP Publishing.

and the interference between the deviated and non-deviated components. In this way, the image offers information on the spatial distribution of the refractive index, rather than on the X-ray absorption. Taking images at different working points, the optimal imaging conditions can be found. For weakly absorbed samples, the obtained phase contrast is much better (sharper) than in conventional absorption-based X-ray images.

Chapter 16 Optical phenomena in photonic structures

Currently, “photonics” is a very wide field with a variety of applications, described in thousands of papers and tens of books. Here, we provide only a brief account of the main ideas, focusing on the extraordinary optical phenomena. The term “photonic materials” or “photonic structures” unites artificial material structures in which photons of certain wavelengths are not able to propagate. In other words, in these structures, there are forbidden energy (wavelength) gaps for photons, as they exist for electrons in semiconductors and insulators. In principle, this situation (actually for wave vector gaps at constant energy, rather than for energy gaps) has been well-known in the field of X-ray diffraction for more than a hundred years. Between 1914 and 1916, **Charles Galton Darwin** and **Paul Ewald** developed the dynamical theory of X-ray diffraction and showed that under **Bragg** reflection conditions, there are forbidden states (wave vector gaps on the isoenergetic dispersion surface) for X-rays in a crystal (see Chapter 15). The mathematical treatment of the problem is very similar to the analytical approach, which was developed later in solid state physics for electron wave states (see eq. (14.22) and Figure 14.2 in Chapter 14). These wave vector gaps arise because of strong interaction of the transmitted and diffracted X-rays via periodically modulated crystal polarizability; the latter due to the translational symmetry in crystals. The modulation periods are determined by the unit cell sizes. The arising gaps in the reciprocal space cause total reflection of X-rays from perfect crystals in proximity to the **Bragg** angles. Note that the interaction of X-rays with matter is rather weak and, hence, the arising gaps are very narrow. Optical photons interact much more strongly with materials, so similar effects in an optical range are expected to be much more pronounced. The problem is how to arrange the energy/wavelength gaps for photons within a material, rather than the wave vector gaps on the isoenergetic dispersion surface. To create energy/wavelength gaps, first, one needs to organize the artificial material's periodicity on the photon wavelength scale. By producing such periodicity, completely new horizons of light manipulation open up. In 1987, two seminal papers, which paved the way to an exciting novel branch of solid-state optics called photonics, were simultaneously (and independently) published.

16.1 Key ideas

One of these papers, authored by **Eli Yablonovitch**, suggested a spatially modulated dielectric structure, with the forbidden energy gap for electromagnetic radiation overlapping with the electron bandgap. If the electromagnetic bandgap extends beyond the electron band edge by more than the thermal energy at a given temperature, then the electron–hole radiative recombination might be severely inhibited. In this way, undesirable spontaneous emission, which limits the performance of semiconductor

https://doi.org/10.1515/9783111140100-017

lasers, would be prohibited. To realize this idea, **Yablonovitch** proposed the three-dimensional modulation of a dielectric structure with periodicity of $\lambda/2$, bearing in mind the destructive interference of electromagnetic waves of wavelength λ, propagating within the structure. The periodicity of the dielectric constant ε_{m} and, hence, the refractive index $n = \sqrt{\varepsilon_{\mathrm{m}}}$ were achieved by utilizing etch pits, which form the desired periodic pattern (see Figure 16.1).

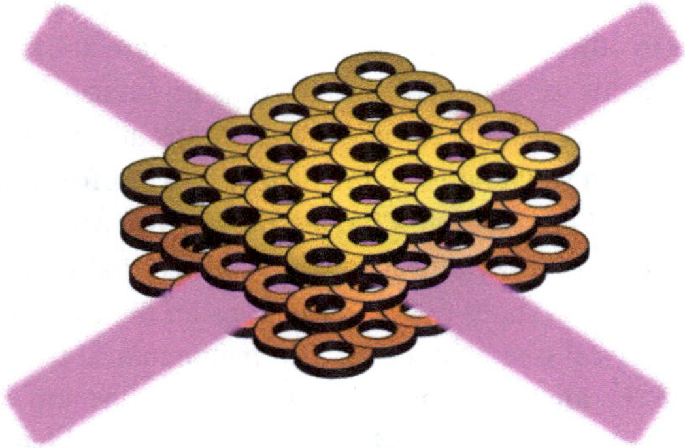

Figure 16.1: Schematics of a photonic structure with three-dimensional periodic modulation of a dielectric constant and, hence, the refractive index.

The second paper was written by **Sajeev John**, who investigated the possible light localization (see also Section 16.2) in disordered dielectric superlattices, such as the **Anderson** localization of electrons in disordered systems. The latter is revealed as a disorder-induced phase transition in the electron transport from the classical diffusion (drift) regime to a localized state, converting material to an insulator. We stress that **Anderson** localization originates in the interference of electron waves that encounter multiple scattering by defects.

Based on these ideas, **John** showed that **Bragg** reflections in photonic structures give rise to the forbidden frequency bands where optical propagation cannot occur, and the intensity of an electromagnetic wave decays exponentially with the distance into the medium. He suggested that this phenomenon can be observed in carefully prepared three-dimensional photonic superlattices with real and positive dielectric constants. He calculated that if the adjacent layers in a superlattice have substantially different dielectric constants, $\varepsilon_{\mathrm{m}a}$ and $\varepsilon_{\mathrm{m}b}$, then the photon density of states (DOS) will reveal a gap around the angular frequency,

$$\omega_{\mathrm{g}} = 2\pi c \frac{\sqrt{\varepsilon_{\mathrm{m}a} - \varepsilon_{\mathrm{m}b}}}{\Lambda_p \varepsilon_{\mathrm{m}a}} \tag{16.1}$$

where Λ_p is the superlattice period and c is the speed of light (see Figure 16.2). Photons with this frequency are unable to propagate through the system.

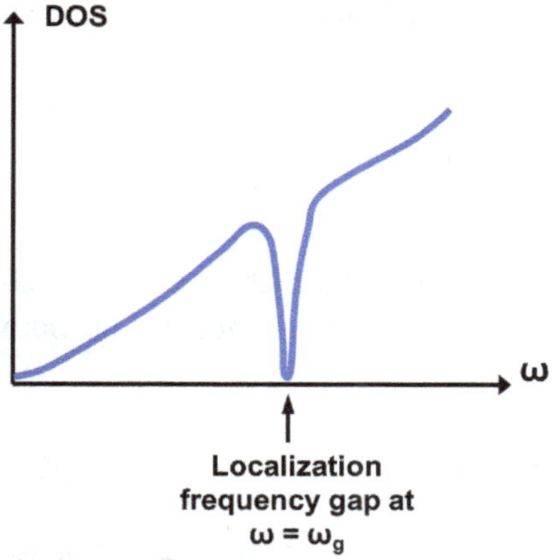

Figure 16.2: Illustration of the gap in the photon density of states (DOS) at $\omega = \omega_g$ (eq. (16.1)), arising due to strong photon localization in a disordered superlattice.

In the last two decades, several splendid proposals for making materials in which electromagnetic waves cannot propagate at certain wavelengths have been realized. Among them are photonic crystal (PC)-based optical devices: lasers, amplifiers, waveguides, dielectric mirrors for antenna, sensors, and so on. One interesting project considers the possibility of modifying the wavelength distribution of **Planck** blackbody radiation by photonic bandgap structures, aimed at suppressing its infrared part and redirecting the thermal energy toward the visible spectral region. This can be crucial for improving the efficiency of solar cells and light emitters. A few types of photonic devices will be discussed in more detail in Section 16.4.

16.2 Light localization in photonic structures

The most fundamental and intriguing phenomenon in the field of photonics is, undoubtedly, possible light localization. As we already said, the prototype of light localization is **Anderson** localization of electron waves in crystals possessing some degree of disorder. This phenomenon is named after **Philip Anderson,** who in 1958 published the seminal paper "Absence of diffusion in certain random lattices." **Anderson** showed that electron localization, which stops electron conductivity and converts a conducting material to an insulator, is a quantum-mechanical effect originating in the interference between different electron trajectories arising from multiple coherent scattering of electrons by lattice defects. For this fundamental result, **Philip Anderson** was awarded the 1977 Nobel Prize in Physics.

The next critical step was made at the end of the 1980s by the Dutch team led by **Ad Lagendijk**. They theoretically studied light propagation in a semi-infinite medium with transverse disorder, i.e., that existing in the (x, y)-plane normal to the light propagation direction z. By solving the so-called paraxial equation (see the next Section 16.2.1), they showed that such a system indeed exhibits strong two-dimensional (x, y) light localization after propagating substantial distance in the z-direction.

16.2.1 Helmholtz and paraxial wave equations

In Section 1.1, we introduced the wave equation (see eq. (1.25)), which interrelates spatial ($\boldsymbol{r}$ (x, y, z)) and temporal (t) characteristics of the respective wavefields for both electrons and photons

$$\frac{\partial^2 \Psi(\boldsymbol{r}, t)}{\partial t^2} = \left[V_\text{p}(x, y, z)\right]^2 \nabla^2 \Psi(\boldsymbol{r}, t) \tag{16.2}$$

In the case of light propagation in a nonmagnetic, nonconducting medium, the phase velocity $V_\text{p}(x, y, z)$ (eq. (1.27)) equals

$$V_\text{p}(x, y, z) = \frac{c}{\sqrt{\varepsilon_\text{m}(x, y, z)}} = \frac{c}{n(x, y, z)} \tag{16.3}$$

with the refractive index

$$n(x, y, z) = \sqrt{\varepsilon_\text{m}(x, y, z)} = \frac{c}{V_\text{p}(x, y, z)} \tag{16.4}$$

where c is the light speed in vacuum.

Previously, in Section 1.1, we showed that one possible solution of the wave equation is the plane wave, $\Psi = \Psi_0 \exp[i(\boldsymbol{kr} - \omega t)]$, which provides the linear dispersion law, that is, the proportionality between light frequency ω and the magnitude of the wave vector $\boldsymbol{k}$. Bearing in mind the possible light or electron localization, however, the plane wave cannot be the right solution. Therefore, one needs to separate the temporal ($\boldsymbol{r}$) and spatial (t) parts of the wave function $\Psi(\boldsymbol{r}, t)$.

The conventional way to do this is to represent the latter as the product of the respective spatial $U(\boldsymbol{r})$ and temporal $W(t)$ components,

$$\Psi(\boldsymbol{r}, t) = U(\boldsymbol{r}) \cdot W(t) \tag{16.5}$$

Substituting expression (16.5) into wave equation (16.2) yields

$$W(t) \cdot \nabla^2 U(\boldsymbol{r}) - \frac{n^2}{c^2} \frac{\partial^2 W(t)}{\partial t^2} U(\boldsymbol{r}) = 0 \tag{16.6}$$

For monochromatic waves (constant frequency ω), we set

$$W(t) = \exp(-i\omega t) \tag{16.7}$$

Correspondingly, eq. (16.5) is converted into

$$\Psi(\boldsymbol{r},t)=U(\boldsymbol{r})\exp(-i\omega t) \tag{16.8}$$

Note that presentation (16.8) is somehow similar to that one (1.32) used for deriving the eikonal equation in Section 1.2. The only difference is that in the latter, the spatial part of the wave function is additionally separated into its amplitude and phase factor components. With the aid of definition (16.8), eq. (16.6) is transformed into the well-known **Helmholtz** equation for eigenvalues of the **Laplace** operator (∇^2):

$$\nabla^2 U(\boldsymbol{r})+k^2 U(\boldsymbol{r})=0 \tag{16.9}$$

with

$$k^2=\omega^2\frac{n^2}{c^2}=n^2k_0^2 \tag{16.10}$$

Further simplification is achieved using paraxial waves

$$U(\boldsymbol{r})=B_{\mathrm{p}}(\boldsymbol{r})\exp(ikz) \tag{16.11}$$

Substituting expression (16.11) into eq. (16.9), one obtains

$$\frac{\partial^2 B_{\mathrm{p}}}{\partial x^2}+\frac{\partial^2 B_{\mathrm{p}}}{\partial y^2}+\frac{\partial^2 B_{\mathrm{p}}}{\partial z^2}+2ik\frac{\partial B_{\mathrm{p}}}{\partial z}=0 \tag{16.12}$$

If function B_{p} changes slowly with distance z, that is $\frac{\partial^2 B_{\mathrm{p}}}{\partial z^2}\ll 2k\frac{\partial B_{\mathrm{p}}}{\partial z}$, then we can neglect the term containing the second derivative over z, which finally yields the paraxial equation in its classical form,

$$\frac{\partial^2 B_{\mathrm{p}}}{\partial x^2}+\frac{\partial^2 B_{\mathrm{p}}}{\partial y^2}+2ik\frac{\partial B_{\mathrm{p}}}{\partial z}=0 \tag{16.13}$$

or

$$i\frac{\partial B_{\mathrm{p}}}{\partial z}=-\frac{1}{2k}\left(\frac{\partial^2 B_{\mathrm{p}}}{\partial x^2}+\frac{\partial^2 B_{\mathrm{p}}}{\partial y^2}\right) \tag{16.14}$$

If, however, function B_{p} changes more rapidly with distance z and $\frac{\partial^2 B_{\mathrm{p}}}{\partial z^2}\simeq k^2 B_{\mathrm{p}}$, then eq. (16.12) transforms into

$$\frac{\partial^2 B_{\mathrm{p}}}{\partial x^2}+\frac{\partial^2 B_{\mathrm{p}}}{\partial y^2}+k^2 B_{\mathrm{p}}+2ik\frac{\partial B_{\mathrm{p}}}{\partial z}\simeq 0 \tag{16.15}$$

In the case of transverse disorder in the (x,y)-plane, we can separate in the refractive index n, the randomly fluctuating part $\Delta n(x,y)$ and the term n_{av}, averaged over disorder-induced fluctuations:

$$n = n_{av} + \Delta n(x,y) \tag{16.16}$$

Consequently, using eqs. (16.10) and (16.16), one obtains

$$k^2 B_p = n^2 k_0^2 \approx k_0^2 n_{av}^2 \left[1 + 2\frac{\Delta n(x,y)}{n_{av}}\right] \approx k^2 \left[1 + 2\frac{\Delta n(x,y)}{n_{av}}\right] \tag{16.17}$$

Furthermore, with the aid of eqs. (16.10) and (16.17), the key eq. (16.15) can be rewritten as

$$i\frac{\partial B_p}{\partial z} = -\frac{1}{2k}\left(\frac{\partial^2 B_p}{\partial x^2} + \frac{\partial^2 B_p}{\partial y^2}\right) - k\left[\frac{1}{2} + \frac{\Delta n(x,y)}{n_{av}}\right] B_p \tag{16.18}$$

The latter structurally resembles the time-dependent (two-dimensional) **Schrödinger** equation (see eq. (14.1)), in which time t is replaced by the coordinate z. The role of the potential $V(x,y)$ is played by the fluctuating part of the refractive index ($-\Delta n(x,y)$), induced by the transverse disorder.

As we know from quantum mechanics, the solutions of the two-dimensional **Schrödinger** equation describe the localized states only. Direct numerical simulations using a **Schrödinger**-like eq. (16.18) with a **Gaussian**-shaped entrance beam and statistical distribution of $\frac{\Delta n(x,y)}{n_{av}}$ demonstrate strong two-dimensional localization of light after propagating a significant distance (thousands of wavelengths) in the z-direction.

Experimental observations of **Anderson** localization of light have been reported by several research groups. Here, we mention the results of the Israeli group headed by **Mordechai (Moti) Segev**, who observed transverse light localization caused by random fluctuations of refractive index in the triangular two-dimensional photonic lattice. The latter was created optically within a photorefractive SBN:60 ($Sr_{0.6}Ba_{0.4}Nb_2O_6$) crystal with the aid of the intentionally induced local changes in the refractive index. We stress that in this experiment, it was possible to precisely control "lattice" disorder by utilizing a complementary speckled beam, which was generated by a conventional laser beam passed through a diffuser. Adding the speckled beam to the primary interference pattern, which creates a photonic lattice, allowed the **Segev** group to control the disorder level by changing the intensity of the speckled beam.

16.3 Slowing light

The term "slowing light" means decreasing its group velocity,

$$V_g = \frac{d\omega}{dk} \tag{16.19}$$

in some frequency intervals. It is a well-known phenomenon in solid state physics related to the resonant properties of a material's polarizability χ, dielectric permittivity ε, and, correspondingly, the refractive index n near the absorption edges characterized by a set of frequencies ω_0. Let us recall the expression for materials polarizability from Chapter 15 (eq. (15.27)) near a certain edge having resonant frequency ω_0:

$$\chi = \frac{e^2}{\varepsilon_0 m}\rho \cdot \frac{1}{\omega_0^2 - \omega^2 + \beta i\omega} = \frac{A}{\omega_0^2 - \omega^2 + \beta i\omega} \tag{16.20}$$

where frequency-independent parameter $A = \frac{e^2}{\varepsilon_0 m}\rho$. Polarizability χ has the real ($\mathrm{Re}\,\chi = \chi_\mathrm{r}$) and the imaginary ($\mathrm{Im}\,\chi = \chi_\mathrm{im}$) parts,

$$\chi_\mathrm{r} = A\frac{\omega_0^2 - \omega^2}{(\omega_0^2 - \omega^2)^2 + (\beta\omega)^2}; \quad \chi_\mathrm{im} = -iA\frac{\beta\omega}{(\omega_0^2 - \omega^2)^2 + (\beta\omega)^2} \tag{16.21}$$

both revealing resonant behavior close to $\omega = \omega_0$ (see Figures 16.3 and 16.4). Since the refractive index in dielectrics is expressed directly via polarizability χ (eqs. (1.30) and (6.6))

$$n = \sqrt{\varepsilon_m} = \sqrt{1+\chi} \tag{16.22}$$

then refractive index n also has real and imaginary parts that was already mentioned in Chapter 6 (see eq. (6.16)). Correspondingly, the refractive index behaves resonantly near the edge frequency ω_0, which can be used to manipulate the group velocity, for example, toward realizing the well-controlled delay lines for optical communication systems.

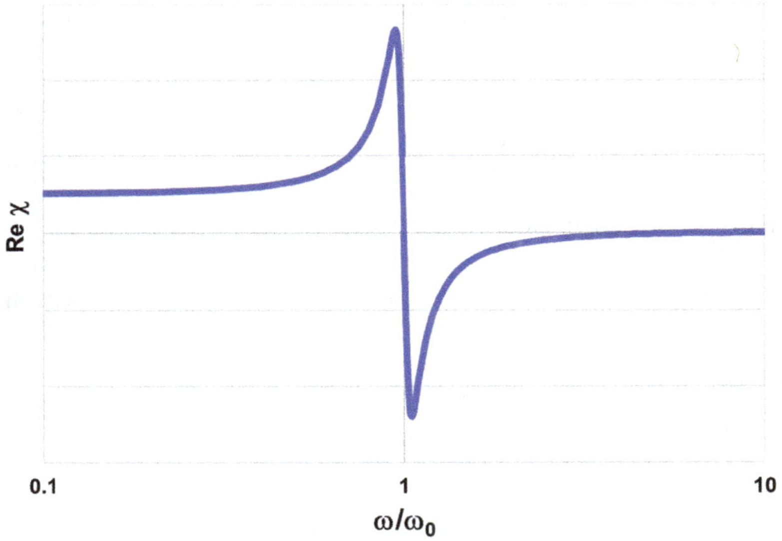

Figure 16.3: Resonant behavior of the real part of polarizability ($\mathrm{Re}\chi = \chi_r$) near $\omega = \omega_0$ (eq. 16.21)).

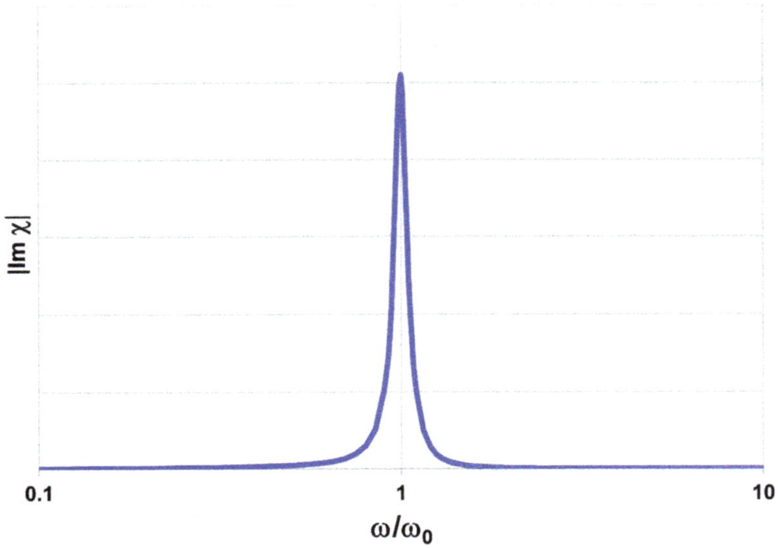

Figure 16.4: Resonant behavior of the absolute value of the imaginary part of polarizability ($|\mathrm{Im}\chi| = |\chi_{im}|$) near $\omega = \omega_0$ (eq. 16.21)).

Actually, expressing the wave vector k in a material via a wave vector in vacuum, $k_0 = \frac{\omega}{c}$ and refractive index n, as

$$k = nk_0 = n\frac{\omega}{c} = \frac{\omega}{c}\sqrt{1+\chi} \tag{16.23}$$

one can find the group velocity

$$V_g = \frac{d\omega}{dk} = \frac{1}{dk/d\omega} \tag{16.24}$$

using eqs. (16.19), (16.21)–(16.23). In fact,

$$\frac{dk}{d\omega} = \frac{1}{c}\frac{d(\omega\sqrt{1+\chi}\,)}{d\omega} = \frac{1}{c}\left[\sqrt{1+\chi} + \omega\frac{1}{2\sqrt{1+\chi}}\frac{d\chi}{d\omega}\right] = \frac{1}{c\sqrt{1+\chi}}\left[1+\chi+\frac{\omega}{2}\frac{d\chi}{d\omega}\right] \tag{16.25}$$

Recalling eq. (16.20), we obtain

$$\frac{d\chi}{d\omega} = -A\frac{(i\beta - 2\omega)}{(\omega_0^2 - \omega^2 + \beta i\omega)^2} \tag{16.26}$$

Substituting eq. (16.26) into (16.25) yields

$$\frac{dk}{d\omega}=\frac{1}{c\sqrt{1+\chi}}\left[1+\chi-\frac{A}{2}\frac{(i\beta\omega-2\omega^2)}{(\omega_0^2-\omega^2+\beta i\omega)^2}\right]=\frac{1}{c\sqrt{1+\chi}}\left[1+\chi-\frac{\chi}{2}\frac{(i\beta\omega-2\omega^2)}{(\omega_0^2-\omega^2+\beta i\omega)}\right] \tag{16.27}$$

For low frequencies ($\omega \ll \omega_0$), for example, in the case of visible light far from the absorption edges,

$$\frac{dk}{d\omega}=\frac{1}{c\sqrt{1+\chi}}(1+\chi)=\frac{\sqrt{1+\chi}}{c}=\frac{n}{c} \tag{16.28}$$

Correspondingly, the group velocity

$$V_g=\left(\frac{dk}{d\omega}\right)^{-1}=\frac{c}{n} \tag{16.29}$$

is equal to the phase velocity in a material.

At the opposite limit (($\omega \gg \omega_0$), eq. (16.27) transforms into

$$\frac{dk}{d\omega}=\frac{1}{c\sqrt{1+\chi}} \tag{16.30}$$

providing the group velocity,

$$V_g=\left(\frac{dk}{d\omega}\right)^{-1}=c\sqrt{1+\chi} \tag{16.31}$$

which coincides with our previous result obtained for X-rays in Chapter 15 (see eq. (15.36) derived for $\chi \ll 1$).

In the general case, it is worth operating with the group refractive index (or simply group index) n_g, which determines the group velocity as

$$V_g=\left(\frac{dk}{d\omega}\right)^{-1}=\frac{c}{n_g} \tag{16.32}$$

Using eqs. (16.22)–(16.25), and (16.32), we find that

$$n_g=n+\omega\frac{dn}{d\omega} \tag{16.33}$$

Now it is clear that slowing light ($n_g > n$) is achieved when

$$\frac{dn}{d\omega}>0 \tag{16.34}$$

Correspondingly, V_g becomes smaller than the phase velocity $V_p = \frac{c}{n}$. Using eqs. (16.22), (16.23), (16.26), and (16.33), one can express the difference $(n_g - n)$ as

$$n_g - n = \omega \frac{dn}{d\omega} = \frac{\omega}{2(1+\chi)^{\frac{1}{2}}} \frac{d\chi}{d\omega} = -\frac{\omega}{2n} A \frac{(i\beta - 2\omega)}{\left(\omega_0^2 - \omega^2 + \beta i\omega\right)^2} \tag{16.35}$$

We stress again that, generally, refractive indices are complex variables, in which real and imaginary parts are interconnected via the **Kramers–Kronig** relationships,

$$n_r = \frac{1}{\pi} P_v \int_{-\infty}^{\infty} \frac{n_{im}(\omega')}{\omega' - \omega} d\omega' \tag{16.36}$$

$$n_{im} = -\frac{1}{\pi} P_v \int_{-\infty}^{\infty} \frac{n_r(\omega')}{\omega' - \omega} d\omega'$$

where improper integrals are characterized by their **Cauchy** principal values (P_v). To calculate the group velocity, we must use the real part of the group index n_g in eq. (16.35). Note that the **Kramers–Kronig** relationships are the consequence of the causality principle in the frequency domain, which, in turn, follows from analytic properties of the function $\varepsilon(\omega)$. Because of this, the real part of the refractive index typically has a rapid jump in the vicinity of an absorption edge, which is also anticipated from eq. (16.35). The real part of the difference $(n_g - n)$, calculated with the aid of eq. (16.35) near the resonant frequency ω_0, is plotted in Figure 16.5. This plot indeed has a maximum at $\omega = \omega_0$, and reveals the regions of slow and fast light in which $n_g - n > 0$ and $n_g - n < 0$, respectively.

A great benefit of photonic structures and metamaterials is an ability to fine-tune the resonance frequencies to the desirable values by playing with the samples' geometry and the sizes of their structural features, as described in Section 6.4.

16.4 Photonic devices

Photonic devices are designed to manipulate photons toward gaining, storing, and transmitting information, as is done with electrons in conventional electronics. Certainly, the use of lasers that provide a huge photon flux has a great impact on photonics development. We remind the reader that the lasing principle was described in Section 8.2.1. The crucial role of optical fibers (see Section 2.2) and the doped-fiber amplifiers for the long-distance information transmission, which is vital for operating telecommunication systems, including the internet, should also not be ignored. The principle of optical amplification was invented by **Gordon Gould** in 1957. He filed his patent in 1959, entitled "Light amplifiers employing collisions to produce population inversions," which was finally issued in 1988.

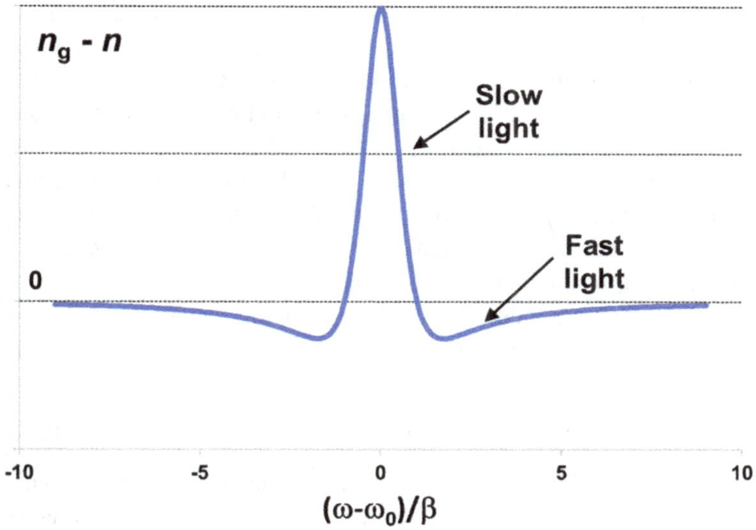

Figure 16.5: Resonant behavior of the difference ($n_g - n$) near $\omega = \omega_0$, calculated with the aid of eq. (16.35) and revealing the regions of slow and fast light.

For long-distance transmission, a doped-fiber amplifier magnifies an optical signal directly, without converting it into an electrical signal. In some sense, it works as a laser, but with no resonance cavity. Signal amplification is achieved through stimulated emission of an additional light by dopants (e.g., erbium ions) placed in the fiber core (see Figure 2.10 in Chapter 2). Specifically, a relatively high-power laser beam is mixed with the input signal (Figure 16.6). The input signal and the excitation (pump) light must have significantly different wavelengths. The laser beam excites the dopants to their higher-energy state. When the input signal photons, differing in wavelength from the

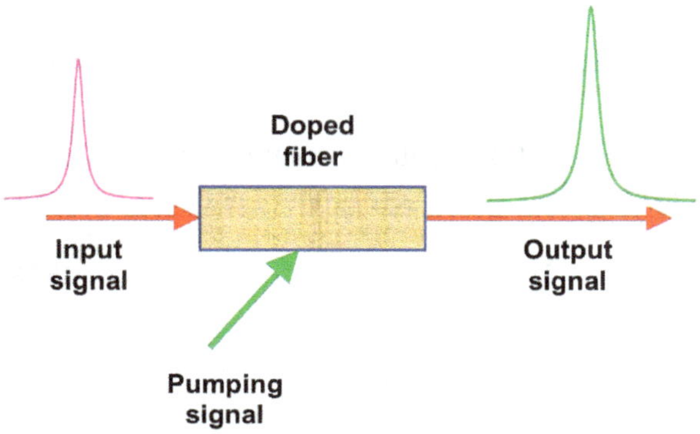

Figure 16.6: Acting of a doped fiber amplifier.

pumping light, meet up with the excited dopants, the latter transfer some of their energy to the incoming signal and return to their lower-energy state. We stress that the dopants generate coherent photons being in phase with the input signal. Therefore, the signal is amplified only along the direction of propagation, and extra power is guided in the same fiber mode as the incoming signal.

Another wide-scale application of photonics is in the field of sensing. Most sensing devices detect the change in the resonant frequency of the sensor element during measurement or the phase shift throughout the electromagnetic wave propagation. In medicine, for example, ring resonators are very popular for measuring biomedical markers in blood using refractive index sensing. The resonant wavelengths λ_{m} of a ring resonator are determined by the product of the refractive index $n(\lambda_{\mathrm{m}}, p)$ and the ring circumference $l(p)$,

$$m\lambda_{\mathrm{m}}(p) = n(\lambda_{\mathrm{m}}, p)l(p) \tag{16.37}$$

where m is an integer number. We see that eq. (16.37) is simply the condition of constructive interference across the ring, which is sensitive to the change of the parameter of interest p. The wavelength change can be induced by the variation of the refractive index n and also the length l, the latter, for example, due to temperature variations. To transfer the incoming and outgoing signals, the waveguides are on-chip-incorporated together with the ring resonators (see Figure 16.7). We will discuss the waveguide and on-chip technology later in this section.

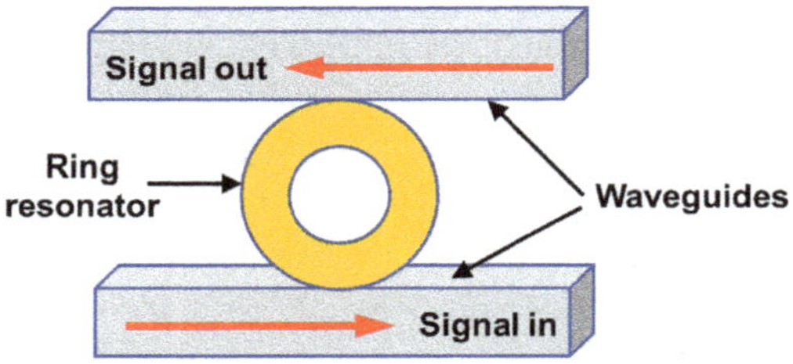

Figure 16.7: Schematic illustration of a ring resonator coupled with waveguides for light transmission.

One can further extend the analysis of eq. (16.37) by differentiating it over the physical parameter, p. This operation yields

$$m\frac{\partial \lambda_{\mathrm{m}}}{\partial p} = \frac{\partial n}{\partial p}l + \frac{\partial \lambda_{\mathrm{m}}}{\partial p}\frac{\partial n}{\partial \lambda_{\mathrm{m}}}l + n\frac{\partial l}{\partial p} \tag{16.38}$$

or

$$\frac{\partial \lambda_{\mathrm{m}}}{\partial p} = \frac{\frac{\partial n}{\partial p} l + n \frac{\partial l}{\partial p}}{m - \frac{\partial n}{\partial \lambda_{\mathrm{m}}} l} \tag{16.39}$$

Introducing a new parameter $\lambda_i = \lambda_{\mathrm{m}}$ for $p = 0$ and using eq. (16.37), we finally find

$$\frac{\partial \lambda_{\mathrm{m}}}{\partial p} = \left(\frac{n}{n_{\mathrm{g}}}\right)\left(\frac{\lambda_i}{n}\frac{\partial n}{\partial p} + \frac{\lambda_i}{l}\frac{\partial l}{\partial p}\right) \tag{16.40}$$

where the group index

$$n_{\mathrm{g}} = n - \lambda_i \frac{\partial n}{\partial \lambda_{\mathrm{m}}} \tag{16.41}$$

is quite similar to that introduced earlier (see eq. (16.33)). Note that eq. (16.40) provides the linearized feedback of the sensor near its unaffected state ($p = 0$). If the circumference l does not change, the feedback signal becomes linearly proportional to the first derivative of the refractive index, $\frac{\partial n}{\partial p}$:

$$\frac{\partial \lambda_{\mathrm{m}}}{\partial p} = \left(\frac{1}{n_{\mathrm{g}}}\right)\left(\lambda_i \frac{\partial n}{\partial p}\right) \tag{16.42}$$

When tiny index changes must be detected, **Mach–Zehnder** (M-Z) interferometers (also on-chip integrated) are used. The M-Z interferometer converts the phase change in one of its arms to the amplitude variation (see Section 8.3). As an important advantage, we indicate that the outcome signal is insensitive to temperature variations, which equally affect optical passes in both arms. Note that the combination of a silicon-based M-Z interferometer with the thin film of nonlinear $LiNbO_3$ material deposited on top of it allows the fabrication of an efficient electro-optical modulator, which can handle high optical power of about 100 mW, and offers a high-speed modulation bandwidth of 110 GHz at 1.55 μm.

As already mentioned, the essential photonic components are different types of waveguides. Conventional light waveguides utilize total internal reflection to prevent light propagation (and related energy losses) in undesirable directions. As we learned in Section 2.2, the critical angle α_c for total reflection depends on the ratio between the refractive indices of the core, that is, an optically denser medium (n_{d}) and a less dense medium (n_{l}), which is called cladding:

$$\sin \alpha_{\mathrm{c}} = \frac{n_{\mathrm{l}}}{n_{\mathrm{d}}} \tag{16.43}$$

Here, angle α_c is between the incident wave vector and the normal to the interface between these two media. In many cases, the complementary angle $\gamma_c = 90^0 - \alpha_c$ counted from the interface itself is more informative. If so, eq. (16.43) transforms into

$$\cos \gamma_c = \frac{n_l}{n_d} \tag{16.44}$$

Optimal waveguide operation, including miniaturization of its possible size, depends strongly on the refractive index "contrast":

$$\xi_c = \frac{\Delta n}{n_d} = \frac{n_d - n_l}{n_d} = 1 - \frac{n_l}{n_d} \tag{16.45}$$

Correspondingly,

$$\cos \gamma_c = \frac{n_l}{n_d} = 1 - \xi_c \tag{16.46}$$

In earlier technologies, the index contrast was produced by ion implantation or metal diffusion into the subsurface layers of nonlinear crystals such as lithium niobate ($LiNbO_3$). The contrast values, achievable in this way, are, however, very modest; at best, about $\xi_c \approx 0.01$. Substituting the latter into eq. (16.46) and expanding the cosine-function on the left-hand side of eq. (16.46) into the **Taylor** series, $\cos \gamma_c \approx 1 - \frac{(\gamma_c)^2}{2}$, yields respectively small critical angle,

$$\gamma_c \approx \sqrt{2\xi_c} \approx 8.1^\circ \tag{16.47}$$

Modern on-chip integrated circuits utilize the high index contrast $\xi_c = 1 - \frac{n_l}{n_d} \approx 0.57$ between the bulk silicon ($n_d \approx 3.5$) and silicon oxide (SiO_2) cladding layer ($n_l \approx 1.5$). These numbers are valid for the wavelength of 1.55 μm, for which the losses in optical fibers are the smallest. Correspondingly, the critical angle, extracted with the aid of eq. (16.46),

$$\gamma_c = \arccos(1 - \xi_c) \approx 64.6^\circ \tag{16.48}$$

is high enough for most practical applications. Specifically, this means that all waves, having wave vectors inclined by angles $\gamma < \gamma_c$ with respect to the oxide/silicon interface, will be enclosed within a waveguide and propagate with almost no energy losses. A photonic-integrated circuit utilizes waveguides to confine and guide light, similar to what optical fibers do in optical communication systems, but in the planar chip geometry.

Fabrication of on-chip integrated circuits was the greatest achievement of nanotechnology in the second half of the twentieth century, which completely changed the design rules for modern microelectronics and optoelectronics. A crucial step was the invention of the metal-oxide-semiconductor field-effect transistor (MOSFET) by **Mohamed Atalla and Dawon Kahng** at Bell Laboratories in 1959. A MOSFET has a source, gate, and drain terminals; the gate being built of a metallic or polysilicon layer atop an SiO_2 insulator (see Figure 16.8). The gate region works as a metal-oxide-semiconductor (MOS) capacitor, whose voltage regulates the current in the conducting channel between the source and drain. For this reason, a MOSFET is sometimes called a MOS transistor. An important twist, the complementary metal-oxide-semiconductor (CMOS) technology, was realized by **Frank Wanlass** at Fairchild Semiconductor in the 1960s. CMOS technology is a type of MOSFET fabrication process that employs complementary and symmetrical pairs of *p*-type and *n*-type MOSFETs for logic functions. Later, mass-production planar technology was introduced into transistor fabrication, which included Si oxidation, photolithography, diffusion of suitable dopants, and surface metallization, as the stages in a unified technological process of integrated circuit production. Applying planar technology to large-size Si wafers allowed the simultaneous production of a huge number of identical transistors that dramatically increased the yield of reliable elements for microelectronic devices. The practical realization of this concept was made primarily by two engineers, **Robert Noyce** and **Jack Kilby**. For this achievement, **Jack Kilby** was awarded the 2000 Nobel Prize in Physics. Already in 2011, 99% of integrated circuits were fabricated using CMOS technology.

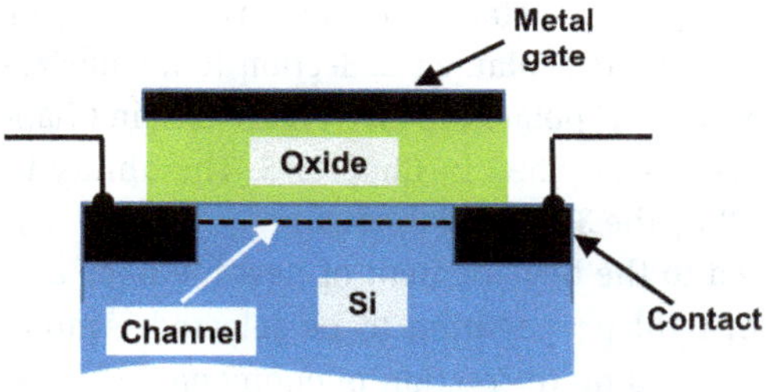

Figure 16.8: Scheme of a MOSFET.

Incorporation of these ideas into advanced optoelectronics started in the middle of the 1980s, while the silicon-on-insulator waveguides were fabricated at the end of the 1980s. For years, photonics has been challenged by the telecommunication industry, and starting from the 1990s, many research groups intensively worked on combining electronics and photonics elements. A major advantage of this idea is that photonic integrated circuits can be fabricated with the aid of the well-established CMOS infrastructure of the semiconductor industry, enabling reduced cost and reliable mass-production. As was already mentioned, the high refractive index contrast between silicon and silicon oxide facilitates the fabrication of very small photonic devices.

Silicon, however, is not an optically active material since it is an indirect semiconductor, which cannot be used to generate light at 1.55 μm. To overcome this difficulty, germanium and silicon–germanium alloys that are also compatible with CMOS fabrication technology are used to provide this function. Otherwise, direct semiconductors of the III–V type can be integrated into silicon photonic-integrated circuits.

Correspondingly, besides silicon photonics, other platforms for integrated photonics exist, each with its specific advantages, such as that using indium phosphide for fabricating lasers and optical detectors, as well as some passive elements. It should be mentioned that certain discrete optical components are employed with integrated circuits, for example, diode lasers based on III(In, Al, Ga)–V(As, P) materials and the aforementioned electro-optical modulators using lithium niobate ($LiNbO_3$).

Just as for conventional microelectronics, a miniaturization trend triggers a rapid decrease in the characteristic sizes of photonic devices below a photon wavelength scale λ. At the same time, this means that the rules of geometrical optics break down because they can be applied to light propagation up to the feature sizes of $S_f \geq 10$ μm, that is, when one can still say that $\lambda \ll S_f$. To manipulate light at the nanoscale ($S_f \leq 100$ nm), we must use novel principles and devices, which essentially comprise the field of nanophotonics.

In this context, let us remember that photons have three basic characteristics: frequency/energy, polarization, and propagation direction that is defined by wave vector $\boldsymbol{k}$. For example, frequency selectivity, the ability to filter light by its colors (as in flat display technology) can be obtained by using dyes, surface plasmonic resonances, or frequency/energy bandgaps in photonic crystals, as explained in Section 16.1. Polarization selectivity is achieved by means of "wire grid" polarizers (see Figure 5.22 in Chapter 5) or by exploiting birefringent materials described in Chapter 5. The ability to filter light by its polarization is widely used by the 3D cinema industry.

The progress of nano-photonics has led to the development of new methods and devices that enable specific directions of light propagation to be selected. Among them are narrow-band selection methods, based on diffraction phenomena and plasmonic resonances. Broad-band angular selectivity is achieved using birefringent materials, metallic gratings, and photonic crystals, in which the refractive index varies periodically on the wavelength scale. Another approach utilizes metamaterials (mentioned in Section 6.4); the metal and dielectric building blocks of the latter being arranged to create a subwavelength periodicity.

Certainly, here, we touched upon only the tip of the iceberg called "photonics." The remainder, hidden under the "sea" of a great number of scientific results, patents, and research proposals, waits to be described in another book.

List of scientists

Abbe, Ernst (1840–1905): German physicist (Chs. 3, 10, 11).

Aizenberg, Joanna (b. 1960): Russian-born American physical chemist and biologist (Ch. 3).

Anderson, Philip Warren (1923–2020): American physicist, 1977 Nobel Prize in Physics "for fundamental theoretical investigations of the electronic structure of magnetic and disordered systems" (Ch. 16).

Arago, François (1786–1853): French mathematician, physicist, and astronomer (Chs. 5, 12).

Archard, John Frederick (1918–1989): British engineer (Ch. 5).

Archimedes (c. 287 BC–c. 212 BC): Ancient Greek mathematician, physicist, engineer, astronomer, and inventor (Ch. 11).

Aspect, Alain (b. 1947): French physicist, 2022 Nobel Prize in Physics "for experiments with entangled photons, establishing the violation of **Bell**'s inequalities and pioneering quantum information science" (Ch. 8).

Atalla, Mohamed (1924–2009): Egypt-born American engineer (Ch. 16).

Auger, Pierre Victor (1899–1993): French physicist (Ch. 14).

Avogadro, Amedeo (1776–1856): Italian scientist (Chs. 5, 8).

Baez, Albert (1912–2007): Mexican-American physicist (Chs. 3, 11, 12).

Barkla, Charles Glover (1877–1944): British physicist, 1917 Nobel Prize in Physics "for his discovery of the characteristic **Röntgen** radiation of the elements" (Preface).

Bartholinus, Erasmus (1625–1698): Danish physician (Ch. 5).

Baruchel, Jose (b. 1946): French physicist (Ch. 9).

Basov, Nicolay (1922–2001): Soviet physicist, 1964 Nobel Prize in Physics "for fundamental work in the field of quantum electronics, which has led to the construction of oscillators and amplifiers based on the maser-laser principle" (Ch. 8).

Becquerel, Edmond (1820–1891): French physicist (Ch. 7).

Beliaevskaya, Elena (unidentified): Russian scientist (Ch. 9).

Bell, John Stewart (1928–1992): Northern Irish physicist (Ch. 8).

Bessel, Friedrich Wilhelm (1784–1846): German mathematician and astronomer (Ch. 10).

Betzig, Robert Eric (b. 1960): American physicist, 2014 Nobel Prize in Chemistry for "the development of super-resolved fluorescence microscopy" (Ch. 11).

https://doi.org/10.1515/9783111140100-018

Bloembergen, Nicolaas (1920–2017): Dutch-American physicist, 1981 Nobel Prize in Physics (shared with **Arthur Schawlow**) "for their contribution to the development of laser spectroscopy" (Ch. 8).

Bohr, Niels Henrik David (1885–1962): Danish physicist, 1922 Nobel Prize in Physics "for his services in the investigation of the structure of atoms and of the radiation emanating from them" (Preface).

Boltzmann, Ludwig Eduard (1844–1906): Austrian physicist (Ch. 8).

Bonse, Ulrich (1928–2022): German physicist. (Chs. 8, 9, 12).

Borrmann, Gerhard (1908–2006): German physicist (Introduction, Chs. 14, 15).

Bragg, William Henry (1862–1942): British physicist, 1915 Nobel Prize in Physics (shared with his son **William Lawrence Bragg** "for their services in the analysis of crystal structure by means of X-rays" (Introduction, Chs. 5, 7, 8, 10, 12, 13, 14, 15, 16).

Bragg, William Lawrence (1880–1971), son of **William Henry Bragg**: Australian-born British physicist, 1915 Nobel Prize in Physics "for their services in the analysis of crystal structure by means of X-rays" (Introduction, Chs. 5, 7, 8, 10, 12, 13, 14, 15, 16).

Brewster, David (1781–1868): Scottish scientist and inventor (Chs. 5, 15).

Brillouin, Léon Nicolas (1889–1969): French physicist (Chs. 13, 14).

Cauchy, Augustin-Louis (1789–1857): French mathematician, engineer, and physicist (Chs. 14, 16).

Chalfie, Martin Lee (b. 1947): American scientist, 2008 Nobel Prize in Chemistry "for the discovery and development of the green fluorescent protein, GFP" (Ch. 11).

Cherenkov, Pavel (1904–1990): Soviet physicist, 1958 Nobel Prize in Physics "for the discovery and the interpretation of the **Cherenkov** effect" (Chs. 7, 15).

Clauser, John Francis (b. 1942): American physicist, 2022 Nobel Prize in Physics "for experiments with entangled photons, establishing the violation of **Bell**'s inequalities and pioneering quantum information science" (Ch. 8).

Clausius, Rudolf (1822–1888): German scientist (Ch. 6).

Compton, Arthur Holly (1892–1962): American physicist, 1927 Nobel Prize in Physics "for his discovery of the effect named after him" (Ch. 15).

Cormack, Allan MacLeod (1924–1998): South African-American physicist, 1979 Nobel Prize in Physiology or Medicine "for the development of computer-assisted tomography" (Ch. 12).

Daguerre, Louis-Jacques-Mandé (1787–1851): French artist and photographer (Ch. 7).

Darwin, Charles Galton (1887–1962): English physicist (Chs. 14, 15, 16).

David, Christian (b. 1965): Dutch-Swiss physicist (Ch. 9).

de Broglie, Louis Victor Pierre Raymond (1892–1987): French physicist, 1929 Nobel Prize in Physics "for his discovery of the wave nature of electrons" (Preface, Chs. 3, 10).

Descartes, René (Renatus Cartesius) (1596–1650): French philosopher and scientist (Chs. 4, 15).

Doppler, Christian (1803–1853): Austrian physicist (Ch. 6).

Drude, Paul Karl Ludwig (1863–1906): German physicist (Ch. 5).

Dumke, William (b. 1930): American physicist (Ch. 8).

DuMond, Jesse William Monroe (1892–1976): American physicist (Ch.15).

Egger, David (b. 1936): American neurobiologist (Ch. 11).

Einstein, Albert (1879–1955): The greatest German-born physicist, 1921 Nobel Prize in Physics "for his services to theoretical physics, and especially for his discovery of the law of the photoelectric effect" (Preface, Chs. 8, 11).

Ewald, Paul Peter (1888–1985): German crystallographer and physicist (Chs. 13, 14, 16).

Fabry, Maurice Paul Auguste Charles (1867–1945): French physicist (Introduction, Ch. 8).

Faraday, Michael (1791–1867): Great English scientist (Ch. 5).

Fermat, Pierre de (1607–1665): French mathematician (Introduction, Chs. 2, 3).

Fermi, Enrico (1901–1954): Great Italian physicist, 1938 Nobel Prize in Physics "for his demonstrations of the existence of new radioactive elements produced by neutron irradiation, and for his related discovery of nuclear reactions brought about by slow neutrons" (Ch. 8).

Foucault, Jean Bernard Léon (1819–1868): French physicist (Ch. 5).

Fourier, Joseph Jean-Baptiste (1768–1830): Great French mathematician and physicist (Chs. 10, 11, 12, 14, 15).

Frank, Ilya (1908–1990): Soviet physicist, 1958 Nobel Prize in Physics "for the discovery and the interpretation of the **Cherenkov** effect" (Ch. 7).

Fraunhofer, Joseph Ritter von (1787–1826): German physicist and inventor (Introduction, Chs. 10, 12).

Fresnel, Augustin-Jean (1788–1827): French physicist (Preface, Introduction, Chs. 5, 10, 12).

Friedrich, Walter (1883–1968): German physicist (Preface, Ch. 10).

Gabor, Dennis (1900–1979): Hungarian-British electrical engineer and physicist, 1971 Nobel Prize in Physics "for his invention and development of the holographic method" (Introduction, Ch. 7).

Galilei, di Vincenzo Bonaiuti de' Galilei (1564–1643): Italian astronomer, physicist, and engineer (Preface).

Gauss, Johann Carl Friedrich (1777–1855): Great German mathematician and physicist (Ch. 16).

Glan, Paul (1846–1898): German scientist (Ch. 5).

Göbel, Herbert (unidentified): German engineer and inventor (Chs. 3, 10).

Gould, Gordon (1920–2005): American physicist and inventor (Ch. 16).

Gregory, James (1637–1675): Scottish mathematician and astronomer (Ch. 7).

Hall, Robert Noel (1919-2016): American engineer (Ch. 8).

Hamilton, William Rowan (1805–1865): Irish mathematician and physicist (Ch. 2).

Hart, Michael (b. 1938): English physicist (Chs. 8, 9, 12).

Heisenberg, Werner Karl (1901–1976): German physicist, 1932 Nobel Prize in Physics "for the creation of quantum mechanics" (Preface, Ch. 3).

Hell, Stefan Walter (b. 1962): Romanian-German physicist, 2014 Nobel Prize in Chemistry "for the development of super-resolved fluorescence microscopy" (Ch. 11).

Helmholtz, Hermann Ludwig Ferdinand (1821–1894): German physician and physicist (Ch. 16).

Herschel, John Frederick William (1792–1871): English scientist (Ch. 5).

Hoffman, Robert (unidentified): American inventor (Ch. 9).

Hooke, Robert (1635–1703): English scientist (Preface, Ch. 7).

Hounsfield, Godfrey Newbold (1919–2004): English electrical engineer, 1979 Nobel Prize in Physiology or Medicine "for the development of computer-assisted tomography" (Ch. 12).

Huygens, Christiaan (1629–1695): Great Dutch scientist (Ch. 10).

Ingal, Victor (unidentified): Russian scientist (Ch. 9).

Isaacson, Michael (b. 1944): American physicist (Ch. 11).

John, Sajeev (b. 1957): American physicist (Ch. 16).

Kahng, Dawon (1924–2009): Korean-American engineer (Ch. 16).

Kao, Charles (1933–2018): China-born English-American electrical engineer and physicist, 2009 Nobel Prize in Physics "for groundbreaking achievements concerning the transmission of light in fibers for optical communication" (Ch. 2).

Kerr, John (1824–1907): Scottish physicist (Ch. 5).

Kilby, Jack St. Clair (1923–2005): American electrical engineer, 2000 Nobel Prize in Physics "for his part in the invention of the integrated circuit" (Ch. 16).

Kirkpatrick, Paul (1894–1992): American physicist and inventor (Chs. 3, 11, 12).

Knipping, Paul (1883–1935): German physicist (Preface, Ch. 10).

Köhler, August Karl Johann Valentin (1866–1948): German scientist (Ch. 3).

Kramers, Hendrik Anthony "Hans" (1894–1952): Dutch physicist (Ch. 16).

Kronecker, Leopold (1823–1891): German mathematician (Ch. 13).

Kronig, Ralph (1904–1995): German physicist (Ch. 16).

Lagendijk, Ad (b. 1947): Dutch physicist (Ch. 16).

Lamb, Horace (1849–1934): British mathematician and physicist (Ch. 6).

Land, Edwin Herbert (1909–1991): American scientist and inventor (Ch. 5).

Laplace, Pierre-Simon (1749–1827): One of the greatest French scientists (Chs. 1, 14, 15, 16).

Larson, Bennett (Ben) (b. 1941): American physicist (Ch. 12).

Laue, Max Theodor Felix von (1879–1960): German physicist, 1914 Nobel Prize in Physics "for his discovery of the diffraction of X-rays by crystals" (Preface, Chs. 10, 14, 15).

Leeuwenhoek, Anton Philips van (1632–1723): Dutch microbiologist and microscopist (Preface).

Leith, Emmett Norman (1927–2005): American electrical engineer (Ch. 7).

Lippmann, Jonas Ferdinand Gabriel (1845–1921): Franco-Luxembourgish physicist and inventor, 1908 Nobel Prize in Physics "for his method of reproducing colors photographically based on the phenomenon of interference" (Introduction, Ch. 7).

Lorentz, Hendrik Antoon (1853–1928): Dutch physicist, 1902 Nobel Prize in Physics (shared with **Pieter Zeeman**) "in recognition of the extraordinary service they rendered by their researches into the influence of magnetism upon radiation phenomena" (Chs. 8, 13, 14, 15).

Mach, Ludwig (1868–1951): Austrian physicist (Introduction, Chs. 8, 16).

Malus, Étienne-Louis (1775–1812): French scientist and engineer (Ch. 5).

Mandelstam, Leonid (1879–1944): Soviet physicist (Ch. 6).

Maupertuis, Pierre Louis Moreau de (1698–1759): French mathematician and philosopher (Ch. 2).

Maxwell, James Clerk (1831–1879): Great English physicist (Preface, Introduction, Chs. 1, 5, 6, 7, 14, 15).

Michelson, Albert Abraham (1852–1931): American physicist, 1907 Nobel Prize in Physics “for his optical precision instruments and the spectroscopic and metrological investigations carried out with their aid” (Ch. 8).

Miller, William Hallowes (1801–1880): Welsh mineralogist (Ch. 13).

Minsky, Marvin (1927–2016): American scientist and inventor (Ch. 11).

Moerner, William Esco (b. 1953): American physical chemist, 2014 Nobel Prize in Chemistry “for the development of super-resolved fluorescence microscopy” (Ch. 11).

Momose, Atsushi (b. 1962): Japanese physicist (Chs. 9, 12).

Mössbauer, Rudolf Ludwig (1929–2011): German physicist, 1961 Nobel Prize in Physics “for his researches concerning the resonance absorption of gamma radiation and his discovery in this connection of the effect which bears his name” (Ch. 7).

Mossotti, Ottaviano-Fabrizio (1791–1863): Italian physicist (Ch. 6).

Neumann, Franz Ernst (1798–1895): German scientist (Ch. 4).

Newton, Isaac (1643–1727): The greatest British scientist (Preface, Introduction, Chs. 2, 3, 6, 7, 8, 10).

Nicol, William (1768–1851): Scottish geologist and physicist (Ch. 5).

Nièpce, Joseph Nicéphore (1765–1833): French inventor (Ch. 7).

Nipkow, Paul Julius Gottlieb (1860–1940): German inventor (Ch. 11).

Noether, Emmy (1882–1935): German mathematician (Ch.13).

Nomarski, Georges (1919–1997): Polish-French physicist (Chs. 5, 9).

Noyce, Robert Norton (1927–1990): American physicist (Ch. 16).

Ohm, George Simon (1789–1854): German physicist and mathematician (Ch. 1).

Pendry, John Brian (b. 1943): English physicist (Chs. 6, 11).

Pêrot, Jean-Baptiste Alfred (1863–1925): French physicist (Introduction, Ch. 8).

Petran, Mojmir (1923–2022): Czech inventor (Ch. 11).

Pfeiffer, Franz (b. 1972): German physicist (Ch. 12).

Planck, Max Karl Ernst Ludwig (1858–1947): German physicist, 1918 Nobel Prize in Physics "in recognition of the services he rendered to the advancement of physics by his discovery of energy quanta" (Preface, Chs. 3, 8, 10, 13, 14, 16).

Pockels, Friedrich Carl Alwin (1865–1913): German physicist (Ch. 5).

Pohl, Dieter (b. 1938): Swiss physicist (Ch. 11).

Poisson, Siméon Denis (1781–1840): French mathematician, physicist, and engineer (Ch. 12).

Porod, Günther (1919–1984): Austrian physicist (Ch. 5).

Poulsen, Henning Friis (b. 1961): Danish physicist (Ch. 12).

Poynting, John Henry (1852–1914): English physicist (Ch. 6).

Prokhorov, Alexander (1916–2002): Australian-born Soviet physicist, 1964 Nobel Prize in Physics "for fundamental work in the field of quantum electronics, which has led to the construction of oscillators and amplifiers based on the maser-laser principle" (Ch. 8).

Prokudin-Gorsky, Sergey (1863–1944): Russian chemist and photographer (Ch. 7).

Raman, Chandrasekhara Venkata (1888–1970): Indian scientist, 1930 Nobel Prize in Physics "for his work on the scattering of light and for the discovery of the effect named after him" (Ch. 11).

Rayleigh (Strutt, John William (Lord Rayleigh)) (1842–1919): British physicist, 1904 Nobel Prize in Physics "for his investigations of the densities of the most important gases and for his discovery of argon in connection with these studies" (Chs. 10, 12).

Rochon, Alexis-Marie de (Abbé Rochon) (1741–1817): French astronomer and physicist (Ch. 5).

Rodenburg, John Marius (b. 1960): English physicist (Ch. 12).

Röntgen, Wilhelm Konrad (1845–1923): German physicist, 1901 Nobel Prize in Physics "in recognition of the extraordinary services he has rendered by the discovery of the remarkable rays subsequently named after him" (Preface, Chs. 10, 12).

Rothen, Alexandre (1900–1987): Swiss-American physical chemist (Ch. 5).

Ruska, Ernst August Friedrich (1906–1988): German physicist, 1986 Nobel Prize in Physics "for his fundamental work in electron optics, and for the design of the first electron microscope" (Ch. 10).

Schawlow, Arthur Leonard (1921–1999): American physicist, 1981 Nobel Prize in Physics (shared with **Nicolaas Bloembergen**) "for their contribution to the development of laser spectroscopy" (Ch. 8).

Schrödinger, Erwin Rudolf Josef Alexander (1887–1961): Austrian physicist, 1933 Nobel Prize in Physics "for the discovery of new productive forms of atomic theory" (Preface, Chs. 14, 16).

Segev, Mordechai (Moti) (b. 1958): Israeli physicist (Ch. 16).

Shimomura, Osamu (1928–2018): Japanese organic chemist and marine biologist, 2008 Nobel Prize in Chemistry "for the discovery and development of the green fluorescent protein, GFP" (Ch. 11).

Snell (Snellius), Willebrord (1580–1626): Dutch astronomer and mathematician (Preface, Chs. 2, 5, 6, 7).

Snigirev, Anatoly (b. 1957): Russian physicist (Ch. 9).

Synge, Edward Hutchinson (1890–1957): Irish physicist (Ch. 11).

Talbot, William Henry Fox (1800–1877): English scientist and inventor (Introduction, Chs. 9, 12).

Tamm, Igor (1895–1971): Soviet physicist, 1958 Nobel Prize in Physics "for the discovery and the interpretation of the **Cherenkov** effect" (Ch. 7).

Taylor, A. M. (unidentified): British physicist (Ch. 5)

Taylor, Brook (1685–1731): English mathematician (Chs. 12, 15, 16).

Tesla, Nikola (1856–1943): Serbian-American inventor (Ch. 6).

Thibault, Pierre (b. 1978): Canadian-born Swiss-French physicist (Ch. 12).

Townes, Charles Hard (1915–2015): American physicist, 1964 Nobel Prize in Physics "for fundamental work in the field of quantum electronics, which has led to the construction of oscillators and amplifiers based on the maser-laser principle" (Ch. 8).

Tsien, Roger Yonchien (1952–2016): American biochemist, 2008 Nobel Prize in Chemistry "for the discovery and development of the green fluorescent protein, GFP" (Ch. 11).

Upatnieks, Juris (b. 1936): Latvian-born American physicist and inventor (Ch. 7).

Verdet, Marcel Émile (1824–1866): French physicist (Ch. 5).

Veselago, Victor (1929–2018): Soviet physicist (Ch. 6).

Wanlass, Frank Marion (1933–2010): American electrical engineer (Ch. 16).

Wollaston, William Hyde (1766–1828): English chemist and physicist (Ch. 5).

Yablonovitch, Eli (b. 1946): Austrian-born American physicist and engineer (Ch. 16).

Young, Thomas (1773–1829): British scientist (Preface, Ch. 10).

Zehnder, Ludwig Louis Albert (1854–1949): Swiss physicist (Introduction, Chs. 8, 16).

Zeilinger, Anton (b. 1945): Austrian physicist, 2022 Nobel Prize in Physics "for experiments with entangled photons, establishing the violation of **Bell**'s inequalities and pioneering quantum information science" (Ch. 8).

Zeiss, Carl (1816–1888): German businessman and optical instruments maker (Ch. 3).

Zeldovich, Yaakov (1914–1987): Soviet physicist (Ch. 14).

Zernike, Frits (1888–1966): Dutch physicist, 1953 Nobel Prize in Physics "for his demonstration of the phase contrast method, especially for his invention of the phase contrast microscope" (Introduction, Chs. 7, 9).

Index

absorption 117, 120–121, 123, 125, 131
absorption edge 83
active medium 120, 123
Advanced Photon Source (APS) 253
air 21, 26–27, 30
amorphous materials 67
amorphous silica 187–188
analyzer 70–73, 79
Anderson localization 256–257, 260
angular deviation 243, 245, 247–248
angular divergence 98
angular magnification 35
angular resolution 139
angular selectivity 270
anisotropic media 42
antiphase 75
aperture 40–41
apex 213–214, 216–217
astigmatism 23
asymmetric reflection 241, 248–249
atomic network 202–203
Avogadro number 131

bandgap 80
beam splitter 109
Bessel function 153
best accommodation distance 38
biaxial crystal 54–55
biological sample 138
bird feather 104
birefringence 49, 52–56, 66–67, 73
Boltzmann statistics 121
Borrmann effect 213, 217–218
boundary 50, 52, 58–59, 61, 74–76, 78, 80
boundary condition 74–76, 78
Bragg angle 103
Bragg scattering geometry 223–228
Bragg–Fresnel lens 181–182
branch 212, 224–225
Brewster angle 49–51, 56, 79
bright field 135–137
brilliance 186, 189
Brillouin zone boundary 204
butterfly wing 104

calcite 26
calcite microlenses 26
camouflage 105
Canada balsam 56–57, 59, 61
capillary 20
capillary optics 31
causality principle 264
caustics 24
cavity 117, 119, 125
center-symmetric crystal 209, 212–213
characteristic depth 241
characteristic gap 210, 212
characteristic length 86
charge carrier 80
charge-coupled device (CCD) 162
charged particle 112
Cherenkov radiation 112–113
chitin 104
chromatic aberration 23, 27, 30
circular polarization 63–65, 67–70
circular X-ray polarization 70
circularly polarized X-rays 245
cladding 267–268
classical radius of electron 27
Clausius–Mossotti relationship 84
clockwise 64, 67–68
coherence 39
coherent beam 114, 125
coherent illumination 39, 41
colored photograph 106
compact lenses 27
complementary metal-oxide-semiconductor
(CMOS) technology 269
Compton effect 249
computer-assisted tomography 183
concave 32
condenser 38–39
confocal microscopy 159, 162, 164–165
conservation law 12, 14
constructive interference 98, 100, 102–103, 108, 112
converging lens 26, 31
convex 32
coordinate axis 45
coordinate system 43, 45–46

https://doi.org/10.1515/9783111140100-019

copper 87, 90
copper sulfate 143
core 265, 267
Cornell High-Energy Synchrotron Source (CHESS) 246
counterclockwise 64, 67–68
critical angle 16, 18–20
critical thickness 146
cross section 46–47
crystal symmetry 42, 45
crystallographic direction 47
cubic 47–48
cubic crystal 49, 53
current density 3
curvature 22–23, 25, 28, 31
cylindrical surface 23

damping coefficient 84
Darwin plateau 226–227
de Broglie relationship 158
degeneracy of states 204
degeneracy point 204
degeneration point 209–211
delay line 261
density of states (DOS) 256–257
Descartes coordinate system 229
destructive interference 99, 102–103
dichroism 56, 67, 69
dielectric materials 42
dielectric permittivity 3
dielectric polarizability 83–84
differential interference contrast 135–136
diffracted beam 51
diffraction 29, 35, 37, 39
diffraction angle 103
diffraction grating 39
diffraction physics 205
diffraction profile 218, 222, 225–227
diffraction vector 147
dipole moment 83–85, 90
dispersion law 4–5
dispersive optical element 249–250
displacement field 3
distortion 34
divergence 3
doped-fiber amplifier 264–265
Doppler effect 92
double split 239
double-crystal diffractometer 139
double-crystal monochromator 150
d-spacing 103
dynamical diffraction 130
dynamical X-ray diffraction 144

eikonal equation 7
elastic scattering 7
electrical conductivity 3
electron density 19
electron gas 90
electron microscope 90
electron wave 205, 207, 210
electron wavelength 158
electro-optical modulator 267, 270
electro-optics 126
ellipse 29–30
ellipse equation 30
ellipsoid of rotation 47
elliptic polarization 63, 77
elliptic surface 29–30
Emmy Nöther's theorem 196
energy flow 92–93, 215–216
energy gap 209–210, 212
energy losses 85–86, 96
entrance aperture 159
entrance crystal surface 214, 218, 223–224
entrance pupil 155–157
E-ray 57, 59–62, 70–71
erbium 265
etalon 115–116
European Synchrotron Radiation Facility (ESRF) 140, 163, 184, 247
evanescent wave 17
Ewald sphere 201–204
exit aperture 159–160
extinction length 176
extraordinary ray 56–57, 59
extraordinary refractive index 52–53, 56
eye 38

Faraday effect 69
far-field 141
Fermi energy 124
f-factor 157
filament 162
fluorescence 162–166, 168
fluorescence microscopy 164–166
fluorescent radiation 163
focal distance 24–27, 34, 38

focal length 12, 25–28, 31–32, 37, 40
focal plane 39–40
focal point 21–22, 24, 29–30, 32–37
focal spot 23, 27, 30–31, 34
focusing element 21–22
fourfold rotation axis 43
Fourier series 166
Fourier-coefficient 206
Fourier-component 151
Fraunhofer diffraction 141, 151–152
free electron laser 189
Fresnel diffraction 172–173, 179–180, 190–191
Fresnel equations 74–76, 82
Fresnel zone 172, 174–183
Fresnel zone plate 155, 181, 182
full-field 185
fundamental constant 84

Gaussian 260
geometrical optics 3, 6
geometrical progression 144
germanium 269
Glan–Foucault prism 61
Glan–Taylor prism 61
Glan–Thompson prism 60–61
glass 14, 16, 19–20
Göbel mirror 29
gradient 8
grazing incidence diffraction 207
green fluorescent protein 164
ground state 121, 123
group index 263–264, 267
group velocity 92–93, 234
guanine 105

half a wavelength 100
half-wave plate 65–66
hard X-rays 177, 181
Heisenberg principle 35
Helmholtz equation 259
hexagonal 46–48
hole 26–27
holography 109, 111–112
homogeneity of space 196
homogeneous medium 37
hyperbolic 212–213, 215–216

illumination 38–39
image 25, 32–35, 37–41
image plane 37, 40–41
incident beam 51, 60–61
index contrast 268–269
indium phosphide 270
induced birefringence 67
in-plane polarization 50, 76
integrated diffraction intensity 222–223
interface 50–51, 57–58, 60, 74, 78–79
interference 97–101, 104, 106, 110, 112
interference pattern 99–100, 110
interferometer 114–118, 125–126, 129, 131
International System of Units (SI) 116
interplanar spacing 103
inverse occupation 123
inversion 43–44
invisible cloak 95
iridescence 101
iso-energetic dispersion surface 210, 218, 224–225
isotropic 42, 47–48

jellyfish 164

K-B mirror 30
Kerr cell 67
Kerr constant 67
Kerr effect 67
kinematic approximation 144, 146
kinematic diffraction 146
kinetic energy 158
Kirkpatrick–Baez mirror 29
Kramers–Kronig relationships 264

Laplace operator 4
lattice node 197
lattice parameter 131
lattice potential 205–206, 208–210, 212
Laue geometry 214
least action principle 10
left-handed material 92
lens 99–100
life science 159, 164
light dispersion 23
light ellipsometry 77
light emission 120, 123–125
light energy 17
light extinction 70
light focusing 21, 27, 36
light frequency 4
light localization 256–258, 260

light modulator 126
light ray 22, 29, 37
light velocity 4
$LiNbO_3$ 252–254
linear electro-optic effect 53
linear focus 23
linear polarization 49, 62–65, 68
locus 24
longitudinal coherence 97
longitudinal magnification 34
Lorentz point 204
Lorentzian 118–119

magnetic circular dichroism 69
magnetic constant 3
magnetic coupling 96
magnetic induction 3
magnetic permeability 3
magnification 33–35, 37–38
magnifying glass 38
main optic axis 52–53, 56, 59, 62, 65, 68
Malus law 72
matrix 43–46, 48
Maxwell's equations 3
medium 21, 26
mercury 106, 108, 111
mesoporous silica 162
metal 26
metallic nanoparticle 105
metal-oxide-semiconductor field-effect transistor (MOSFET) 268
metal-oxide-semiconductor (MOS) capacitor 269
microscope 35, 37–39, 41
Miller indices 199, 201
mirage 6, 8–9
mirror plane 44–45
modulation-contrast microscopy 136, 138
momentum 8
momentum conservation law 196
monochromatic light 99
monoclinic 44–46
MOS transistor 269
multicrystal diffractometer 249, 253
multilayer 103–104

nano-photonics 270
nanotechnology 159, 164, 169
nano-tomography 187–188
near-field microscopy 159, 165–166, 168
near-field scanning optical microscopy (NSOM) 168
negative magnetic constant 95
negative material 93, 95–96
negative refractive index 93
Newton formula 33–34
Nicol prism 56–58
Nipkow disc 160–161
Nomarski microscope 135–136
Nomarski prism 61–62
non-homogeneous medium 196–197
normal incidence 86–87, 99, 102
numerical aperture 37

object 32–39, 41
object plane 39, 41
objective 31, 37–38
objective lens 159, 163
ocular 37–38
on-chip integrated circuit 268
on-chip technology 266
opaque medium 90
optical activity 68
optical axis 22–23, 29, 32, 34, 39
optical coatings 102
optical communication system 125–126
optical fiber 17
optical instrument 37
optical path 10–11, 13
optical power 25, 27, 32
optical system 22–23, 25, 32, 34–35, 40–41
O-ray 57, 59, 61–62, 70
ordinary ray 56–58
ordinary refractive index 52–53, 56
orthogonality conditions 199–200
orthorhombic 44–46
out-of-plane polarization 50

parabola 22, 28
parabolic equation 22, 28
parabolic mirror 28–29
parallel atomic planes 144, 147
parallel beam 23, 29, 39, 41
parametric X-ray radiation 241
paraxial equation 257, 259
paraxial wave 258–259
Pendellösung effect 215
phase contrast 134, 138
phase difference 62–63, 65–67, 71–72

phase gradient 137
phase object 133, 135, 137–138, 140
phase shift 133–135, 137–138, 140
phase velocity 6
phase-contrast imaging 193–195
phase-contrast microscope 133
phase-contrast microscopy 109
phonon 80
photoelastic effect 53
photographic emulsion 109–110
photographic plate 106–108, 110–111
photolithography 181, 183
photon 49
photonic crystal 257, 270
photonic devices 257, 269–270
photonic integrated circuit 268–270
photonic lattice 260
photonic materials 255
photonic structure 104
photonic structures 255–257, 264
photonics 255, 257, 264, 266, 269–270
pinhole 159, 163
planar technology 269
Planck blackbody radiation 257
plane polarization 50, 62
plane wave 4–7
plasma frequency 89–91
plasmon 80
plasmonic nanostructure 96, 105
Pockels effect 66–67
point focus 23, 29
point group 44–45
polarizability 18
polarization 3
polarization microscopy 62
polarization rotation 65, 69
polarization vector 49, 65–66, 68, 71
polarizer 70–73, 79
polaroid 56
Porod law 81
Poynting vector 92
precession 69
prism 57, 59, 61, 70
ptychography 188–190

quadratic electro-optic effect 53, 67
quantum confinement 169
quantum dot 164, 169
quantum entanglement 126–127, 129
quarter-wave plate 65, 70–71
quartz 68
quasi-momentum conservation law 199–200

rainbow 14–15
Raman spectrometer 162, 168
Rayleigh criterion 154–155
reciprocal lattice 196, 199–204
reciprocal space 199, 204
reduced Planck constant 196
reference wave 109–111
reflection 6
reflection coefficient 75–76, 78–79
reflective grating 142
reflectivity 144, 148
reflectivity coefficient 89
refraction 5–6
refractive index 3, 5–6, 8
resolving power 35, 37
resonant frequency 84, 96
rhombohedral 46–48
right-handed material 93
ring resonator 266
Rochon prism 59–60, 62
rocking curve 251–254
rotation 43–47
rotation of the polarization plane 62, 65, 68–69
rutile 53

scanning electron microscope (SEM) 158
scanning transmission X-ray microscope 185
scattered wave 206
scattering amplitude 141, 144, 146, 151–152
scattering plane 16
Schrödinger equation 205–206
sealed X-ray tube 250
secular equation 209, 213, 218
silica 18
silicon oxide (SiO_2) 268
silicon–germanium alloy 269
silicon-on-insulator waveguide 269
silver 90
sixfold rotation axis 43–44, 46
skin effect 86
slowing light 260, 263
soap bubbles 101
soft X-rays 177, 182, 185–186

solid state physics 209
source 21–22, 25, 29, 38, 40
spatial resolution 132
spectral interval 97, 104
specular reflection 12, 19
speed of light 88
spherical aberration 22–23, 27
spherical lens 24–26, 31
spherical mirror 12
split-ring resonator 96
spontaneous emission 121–122
$Sr_{0.6}Ba_{0.4}Nb_2O_6$ 260
stealth 99
stimulated emission 120, 122–123, 125
stimulated emission depletion (STED) 165
structural color 104–105
structure factor 175
sugar 69
superlattice period 256
surface acoustic wave 184–185
surface plasmon 105
symmetric Bragg case 244
symmetric Laue case 244, 247
symmetry plane 43, 45
symmetry system 42, 44–48
synchrotron 27, 29–30

Talbot carpet 192
Talbot distance 192
Talbot effect 140
Talbot interferometer 194
Taylor series 173
tensor 42, 44–48
tetragonal 46–48
thickness fringes 218, 221–222
thin film 77, 80
thin lens 31–32
three-dimensional X-ray diffraction microscopy 187
threefold rotation axis 43–44
tie point 213, 218, 225
topaz 55
total external reflection 18–20
total internal reflection 16, 18
total reflection 147
tourmaline 56
transformation law 42, 44–46
translational symmetry 143
transmission electron microscope 158
transmission electron microscopy (TEM) 217
transmission X-ray microscope 185
transmissive grating 141–142
transmitted intensity 72
transparent medium 90
transverse coherence 98
transverse disorder 257, 259–260
transverse electromagnetic wave 235
triaxial ellipsoid 46
triclinic 44–46
two-beam approximation 204
twofold rotation axis 43

uniaxial crystal 51–53, 55, 65–67, 71
unit cell 199, 203

vacuum 86, 88–89, 93
vector of reciprocal lattice 200, 202
Verdet constant 69
visible light 142, 148, 154, 158

water 14, 16
wave equation 4–6
wave function 7
wave surface 7
wavefield 39–40
waveguide 17
wavelength 6
wave vector 4–5, 7–8
wavevector gap 255
wire grid 70
Wollaston prism 59–61

X-ray beam compression 248
X-ray energy 19
X-ray focusing 19
X-ray optics 26, 155
X-ray reflectivity 77, 80–81
X-ray standing wave 217–218
X-ray tomography 131
X-ray/phonon interaction 252

π-polarization 50, 53, 74, 76, 79
σ-polarization 50, 53, 79

www.ingramcontent.com/pod-product-compliance
Lightning Source LLC
Chambersburg PA
CBHW061342210326
41598CB00035B/5864
* 9 7 8 3 1 1 1 1 3 9 6 9 2 *